Gene Transcription

The Practical Approach Series

SERIES EDITORS

D. RICKWOOD
Department of Biology, University of Essex
Wivenhoe Park, Colchester, Essex CO4 3SQ, UK

B. D. HAMES
Department of Biochemistry and Molecular Biology,
University of Leeds, Leeds LS2 9JT, UK

Affinity Chromatography
Anaerobic Microbiology
Animal Cell Culture (2nd Edition)
Animal Virus Pathogenesis
Antibodies I and II
Behavioural Neuroscience
Biochemical Toxicology
Biological Data Analysis
Biological Membranes
Biomechanics—Materials
Biomechanics—Structures and Systems
Biosensors
Carbohydrate Analysis
Cell–Cell Interactions
Cell Growth and Division
Cellular Calcium
Cellular Neurobiology
Centrifugation (2nd Edition)
Clinical Immunology
Computers in Microbiology
Crystallization of Nucleic Acids and Proteins
Cytokines
The Cytoskeleton
Diagnostic Molecular Pathology I and II
Directed Mutagenesis
DNA Cloning I, II, and III
Drosophila
Electron Microscopy in Biology
Electron Microscopy in Molecular Biology
Electrophysiology
Enzyme Assays
Essential Molecular Biology I and II
Experimental Neuroanatomy
Fermentation
Flow Cytometry
Gel Electrophoresis of Nucleic Acids (2nd Edition)
Gel Electrophoresis of Proteins (2nd Edition)
Gene Targeting
Gene Transcription
Genome Analysis
Glycobiology
Growth Factors
Haemopoiesis
Histocompatibility Testing
HPLC of Macromolecules
HPLC of Small Molecules
Human Cytogenetics I and II (2nd Edition)
Human Genetic Disease Analysis

Immobilised Cells and Enzymes
Immunocytochemistry
In Situ Hybridization
Iodinated Density Gradient Media
Light Microscopy in Biology
Lipid Analysis
Lipid Modification of Proteins
Lipoprotein Analysis
Liposomes
Lymphocytes
Mammalian Cell Biotechnology
Mammalian Development
Medical Bacteriology
Medical Mycology
Microcomputers in Biochemistry
Microcomputers in Biology
Microcomputers in Physiology
Mitochondria
Molecular Genetic Analysis of Populations
Molecular Imaging in Neuroscience
Molecular Neurobiology
Molecular Plant Pathology I and II
Molecular Virology
Monitoring Neuronal Activity
Mutagenicity Testing
Neural Transplantation
Neurochemistry
Neuronal Cell Lines
Nucleic Acid and Protein Sequence Analysis
Nucleic Acid Hybridisation
Nucleic Acids Sequencing
Oligonucleotides and Analogues
Oligonucleotide Synthesis
PCR
Peptide Hormone Action
Peptide Hormone Secretion
Photosynthesis: Energy Transduction
Plant Cell Culture
Plant Molecular Biology
Plasmids
Pollination Ecology
Postimplantation Mammalian Embryos
Preparative Centrifugation
Prostaglandins and Related Substances
Protein Architecture
Protein Engineering
Protein Function
Protein Phosphorylation
Protein Purification Applications
Protein Purification Methods
Protein Sequencing
Protein Structure
Protein Targeting
Proteolytic Enzymes
Radioisotopes in Biology
Receptor Biochemistry
Receptor–Effector Coupling
Receptor–Ligand Interactions
Ribosomes and Protein Synthesis
Signal Transduction
Solid Phase Peptide Synthesis
Spectrophotometry and Spectrofluorimetry
Steroid Hormones
Teratocarcinomas and Embryonic Stem Cells
Transcription Factors
Transcription and Translation
Tumour Immunobiology
Virology
Yeast

Gene Transcription

A Practical Approach

Edited by

B. DAVID HAMES

and

STEPHEN J. HIGGINS

Department of Biochemistry and Molecular Biology,
University of Leeds, UK

Oxford University Press, Walton Street, Oxford OX2 6DP
Oxford New York Toronto
Delhi Bombay Calcutta Madras Karachi
Kuala Lumpur Singapore Hong Kong Tokyo
Nairobi Dar es Salaam Cape Town
Melbourne Auckland Madrid
and associated companies in
Berlin Ibadan

Oxford is a trade mark of Oxford University Press

A Practical Approach is a registered trade mark
of the Chancellor, Masters, and Scholars of the University of Oxford
trading as Oxford University Press

Published in the United States
by Oxford University Press Inc., New York

A catalogue record for this book is available from the British Library

Library of Congress Cataloging in Publication Data
ISBN 0–19–963292–8 (hbk.)
ISBN 0–19–963291–X (pbk.)

Typeset by Footnote Graphics, Warminster, Wilts
Printed in Great Britain by Information Press Ltd, Eynsham, Oxon

Preface

The precise scope of molecular biology is often debated, but most researchers would surely agree with Keith Yamamoto in his Introduction to this volume that the essential core of the subject, and its undoubted successes, lie in its powerful methodology. Nowhere is this more true than in studies of the regulation of gene expression. In 1984, we edited a book for the Practical Approach Series which was designed to bring together all the key techniques associated with transcription and translation. The popularity of *Transcription and translation: a practical approach* was testimony to the skill of the contributors and their shared dedication with us in achieving this aim. Since that time, however, the field has enlarged enormously and the techniques and experimental approaches employed have undergone radical expansion and revision—so much so that a simple update of this single volume would have been without merit. Rather, we decided that the only reasonable way forward was to replace it with separate books each dedicated to one of the three major phases of gene expression; transcription, RNA processing, and translation. This volume is the first of the three; a second book, on RNA processing, is nearing completion, and a third volume, on translation, is planned.

Gene transcription: a practical approach concentrates on RNA polymerase II transcription of eukaryotic protein coding genes and covers all the major current procedures and approaches in this field, including not only the analysis of transcription *in vitro* and *in vivo* but also the identification, purification, and characterization of transcription factors and their interaction with specific DNA target sites. Also notable is the inclusion of an appendix which lists all known transcriptional control *cis*-elements and *trans*-factors for polymerase II genes, complete with a full bibliography.

We wish to express our warm thanks to all the authors who made this book possible, in the confident belief that their efforts will be rewarded by the popularity we fully expect this volume, like its predecessor, will enjoy. It is a book that will undoubtedly be an essential guide for the newcomer to this area and a valuable addition to the laboratories and office bookshelves of more experienced researchers.

Leeds
June 1992

DAVID HAMES
STEVE HIGGINS

Contents

Contributors

DANIEL M. BECKER
Stanford Law School, Stanford, CA94365, USA.

ANDREW R. CLARK
Department of Medicine, University of Birmingham, Queen Elizabeth Hospital, Birmingham B15 2TH, UK.

NIALL DILLON
NIMR, The Ridgeway, Mill Hill, London NW7 1AA, UK.

KEVIN DOCHERTY
Department of Medicine, University of Birmingham, Queen Elizabeth Hospital, Birmingham B15 2TH, UK.

TARIQ ENVER
Leukaemia Research Fund Centre, The Institute of Cancer Research, Chester Beatty Laboratories, Fulham Road, London SW3 6BJ, UK.

JOHN D. FIKES
Cytel Corporation, 3525 John Hopkins Court, San Diego, CA 92121, USA.

MICHAEL J. GARABEDIAN
Department of Biochemistry and Biophysics, University of California Medical Center, San Francisco CA94143, USA.

FRANK GROSVELD
NIMR, The Ridgeway, Mill Hill, London NW7 1AA, UK.

B. DAVID HAMES
Department of Biochemistry and Molecular Biology, University of Leeds, Leeds LS2 9JT, UK.

STEPHEN J. HIGGINS
Department of Biochemistry and Molecular Biology, University of Leeds, Leeds LS2 9JT, UK.

STEPHEN P. JACKSON
Wellcome/CRC Institute, Tennis Court Road, Cambridge CB2 1QR, UK.

JOSHUA LaBAER
Department of Biochemistry and Biophysics, University of California Medical Center, San Francisco CA94143, USA.

WEI-HONG LIU
Department of Biochemistry and Biophysics, University of California Medical Center, San Francisco CA94143, USA.

JOSEPH LOCKER
University of Pittsburgh, School of Medicine, Department of Pathology, Scaife Hall, Room 777A, Pittsburgh, PA 15261, USA.

PHILIP J. MASON
Department of Haematology, RPMS Hammersmith Hospital, Ducane Road, London W12 0NN, UK.

UELI SCHIBLER
Department of Molecular Biology, University of Geneva, Sciences II, Quai Ernest-Ansermet 30, CH-1211 Geneva 4, Switzerland.

FELIPE SIERRA
Nestle Company, SA 1350, Orbe, Switzerland.

JAY R. THOMAS
Department of Biochemistry and Biophysics, University of California Medical Center, San Francisco CA94143, USA.

JIAN-MIN TIAN
Department of Molecular Biology, University of Geneva, Sciences II, Quai Ernest-Ansermet 30, CH-1211 Geneva 4, Switzerland.

DAVID WILKINSON
Laboratory of Eukaryotic Molecular Genetics, NIMR, The Ridgeway, Mill Hill, London NW7 1AA, UK.

JEFFREY G. WILLIAMS
ICRF Fund, Clare Hall Laboratory, Blanche Lane, South Mimms, Potters Bar EN6 3LD, UK.

KEITH R. YAMAMOTO
Department of Biochemistry and Biophysics, University of California Medical Center, San Francisco, CA94143, USA.

Abbreviations

AdML	adenovirus major late promoter
APH	aminoglycoside 3′-phosphotransferase
BrdU	bromodeoxyuridine
BSA	bovine serum albumin
CAD	carbanoyl-phosphate synthase–aspartate transcarbamylase-dihydroorotase
CAT	chloramphenicol acetyl transferase
cDNA	complementary DNA
CIP	calf intestine alkaline phosphatase
CPRG	chlorophenol red-β-D-galactopyranoside
dATP	deoxyadenosine triphosphate
dCTP	deoxycytidine triphosphate
DEAE	diethyl amino ethyl
DEPC	diethyl pyrocarbonate
DFMO	D-difluoromethylornithine
dGTP	deoxyguanosine triphosphate
DHFR	dihydrofolate reductase
dIMP	deoxyinosine monophosphate
DMEM	Dulbecco's modified Eagle's medium
DMS	dimethyl sulphate
DMSO	dimethyl sulphoxide
DNase	deoxyribonuclease
dNTP	deoxyribonucleoside triphosphate
DTT	dithiothreitol
dTTP	deoxythymidine triphosphate
E1A	Early region 1A (gene product of adenovirus)
EDTA	ethylene diamine tetraacetic acid
ES	embryonic stem (cell)
FOA	5-fluoroorotic acid
Gal4	transcriptional activator protein of the yeast *Gal1-Gal10* divergent promoter
GlcNAc	*N*-acetylglucosamine
HAT	hypoxanthine–aminopterin–thymidine medium
HBS	Hepes-buffered saline
HCG	human chorionic gonadotrophin
HPLC	high-performance liquid chromatography
HSV	herpes simplex virus
IPTG	isopropyl-β-D-thiogalactoside
kb	kilobase

LCR	locus control region
LTR	long terminal repeat sequence
MCR	multiple cloning region
M-MLV	Moloney murine leukaemia virus
mRNA	messenger RNA
nt	nucleotide
ONPG	*O*-nitrophenyl-β-D-galactopyranoside
ori	origin of replication
PALA	*N*-phosphoacetyl-L-aspartate
PBS	phosphate-buffered saline
PCR	polymerase chain reaction
PCV	packed cell volume
PMS	pregnant mare's serum
PMSF	phenylmethylsulphonyl fluoride
PP_i	inorganic pyrophosphate
rATP	adenosine triphosphate
rCTP	cytidine triphosphate
rGTP	guanosine triphosphate
RNase	ribonuclease
rNTP	ribonucleoside triphosphate
RT	reverse transcriptase
RT–PCR	reverse transcription–polymerase chain reaction
rUTP	uridine triphosphate
SCS	specific chromatin structure
SDS	sodium dodecyl sulphate
SDS-PAGE	SDS-polyacrylamide gel electrophoresis
SSC	standard saline citrate
SV	simian virus
TAE	Tris–acetate–EDTA buffer
TBE	Tris–borate–EDTA buffer
TCA	trichloroacetic acid
TE	Tris–EDTA buffer
TEMED	*N,N,N′,N′*-tetramethylenediamine
TK	thymidine kinase
tRNA	transfer RNA
UV	ultraviolet
WGA	wheat germ agglutinin
X-gal	5-bromo-4-chloro-3-indolyl-β-D-galactoside
XGPRT	xanthine–guanine phosphoribosyl transferase
YAC	yeast artificial chromosome

Introduction

Eukaryotic gene transcription: big leaps, no bounds

KEITH R. YAMAMOTO

Molecular biology, beneath its trendiness and jargon, is basically a set of tools. That is, it is not an intellectual niche in the body of biological knowledge, but instead a collection of methodologies that defines and supports reductionist strategies—approaches that seek simplified experimental systems and minimized experimental variables. Be that as it may, this set of tools has absolutely revolutionized biological research. And perhaps nowhere is the influence of this technology more evident than in studies of eukaryotic transcription and the regulation of its initiation.

Historically, studies of eukaryotic transcription were considered weak relatives of the elegant investigations carried out in prokaryotes. Thus, powerful molecular genetic strategies devised to study *Escherichia coli* and its viruses produced concepts and paradigms that formed and dominated our notions of transcription in complex systems. Lacking such genetic armament, studies in eukaryotes were constrained to biochemistry, and out of the cold room emerged, for example, the discovery of three discrete RNA polymerases, each functionally distinct. Arduous fractionations and painstaking reconstitutions yielded, grudgingly, a glimpse of the complexity of the polymerases, crude fractions containing activities essential for initiation, and a few putative regulatory proteins.

Then came cloning and sequencing, mutagenesis and transfection, domain mapping and chimeric proteins. The trickle of regulatory and initiation factors, polymerase subunits, and DNA sequence elements at which they act, grew rapidly to a torrent. With ready access to rare proteins and sequences, and to reverse genetic schemes for manipulating them, the nature of questions and quality of answers accessible in eukaryotic gene expression changed dramatically. The long-range regulatory effects of enhancer elements were discovered first in mammalian cells and subsequently in bacteria, one of the earliest examples of a concept canonized in eukaryotes and uncovered only later in prokaryotes. It became apparent that promoters and enhancers were occupied by dynamic, multiprotein structures assembled along pathways defined by the developmental and physiological status of the particular cell.

Although much remains to be done along this front, these were profound leaps, both practical and conceptual.

The methodological advances that have opened doors to the machinery and mechanisms of transcriptional regulation have also established an appreciation of the extent to which that machinery is directly involved in specifying cellular processes. I mean this in two ways. First, students of transcription *per se*, such as devotees of polymerase enzymology, or sequence-specific DNA recognition, or the mechanics of intiation, have been joined by investigators from virology, cancer biology, endocrinology, neurobiology, developmental genetics, and a host of other subdisciplines who have discovered, sometimes to their chagrin, that their favourite genes and factors turn out to be transcription components. Second, the now vast catalogue of cloning, sequencing, and transfection efforts has revealed that core polymerase machinery has been conserved from bacteria to mammals, that sets of related but functionally distinct regulatory factors are encoded in large multigene families, and, perhaps most startling, that regulators are commonly competent to function in cells from species that diverged as long as a billion years ago. Hence, species boundaries seem not to apply—steroid receptors from rats, for example, regulate appropriately the transcriptional machinery of *Drosophila*, yeast, and plants. Clearly, transcriptional regulation is an ancient process, the components of which serve as primary determinants of phenotypic differences between cells.

The same technologies that have brought eukaryotic transcriptional regulation to this exciting new level of sophistication and understanding have also illuminated daunting problems that are not addressed (or may in fact be exacerbated) by present approaches. How will we detemine, for example, exactly which factors reside *in vivo* in a multiprotein initiation complex or enhancer complex under a given set of conditions? And even when each factor can be identified unequivocally, how will we determine which of the components relieves a rate-limiting step in initiation, which of them acts indirectly or passively, and which, while present in the complex, may play no functional role at all? What is the meaning of 'promoter bash' experiments in which mutations of multiple protein-binding sites typically seem to imply that *each* element is responsible for one-half to three-quarters of the initiation activity from the promoter? If absolute and relative levels of regulator expression are important determinants of the mode and direction of regulation, as present findings imply, do transient transfection assays generate artifacts that mask bona fide regulatory mechanisms? How will we assess the potential roles of chromatin components in primary regulatory mechanisms?

We are able to recognize these complexities only through knowledge gained from present experimental approaches. That is, problems newly perceived do not indict current methodologies; they mean only that we must remain mindful of the limitations of present procedures and seek ways to overcome them. Clearly, much is still to be done and learned with today's approaches; indeed, it will undoubtedly be the utilization and modification of these techniques that will spawn the next generation of breakthroughs.

This, then, is a book for today—a concise summary of procedures for the analysis, preparation, and manipulation of the transcription machinery and its regulatory components. While much effort has been expended to bring together precise and rigorous descriptions of useful protocols, practitioners should come to consider them as guidelines, not rigid rules. It is commonly through the evolution of novel methods that we gain new insight; thus, tomorrow's crucial technological advances could be hidden within these pages.

1

Assay of gene transcription *in vivo*

PHILIP J. MASON, TARIQ ENVER, DAVID WILKINSON, and JEFFREY G. WILLIAMS

1. Introduction

All methods of analysing gene transcription *in vitro* rely on the ability of nucleic acid molecules to hybridize, one to another. Nucleic acid hybridization can be used to determine the primary structure and concentration of an individual RNA species by a variety of procedures.

(a) *Northern transfer* can be used to give an estimate of the length of an RNA transcript.

(b) *Polymerase chain reaction (PCR)* can be used to amplify selected segments of an RNA for gel electrophoretic analysis or DNA sequence determination.

(c) *Nuclease S1 mapping* or *RNase protection* can be used to identify the positions of the 5′ and 3′ termini of a gene and of any introns within it.

(d) *Primer extension* can be used to determine the position of the 5′ terminus of an RNA.

(e) *In situ hybridization* can be used to visualize the pattern of gene expression of individual cells within a tissue and tissues within an organism.

The first four techniques, together with *RNA dot blots* and *RNA slot blots* can also be used quantitatively to determine the relative concentration of a specific RNA sequence in different RNA populations.

We will not describe methods for isolating and handling RNA (see ref. 1 for a discussion of these methods). However, one cannot overemphasize that the success of subsequent analysis depends critically upon avoiding non-specific degradation of the RNA under study. RNA molecules are *very* susceptible to degradation. The major danger is of cleavage by ribonucleases during extraction and purification. Therefore, procedures designed to minimize this problem must always be followed. At high temperature, and especially at neutral or alkaline pH, RNA is degraded *very* rapidly. Thus hybridization should be performed for the minimum possible time, in a buffer of slightly acidic pH, and at the lowest temperatures allowing efficient hybridization.

Formamide destabilizes duplex molecules, so that a lower temperature can be used for hybridization. Formamide-containing buffers are therefore widely used for RNA analysis.

2. Northern transfer

2.1 Procedures

Northern transfer (2, 3) is used to analyse specific RNA molecules in a complex RNA population. The RNA is separated by electrophoresis in an agarose gel under denaturing conditions, transferred to a nitrocellulose or nylon filter and specific RNA species are then detected by hybridization with a radioactively labelled probe. The technique is very sensitive, detecting as little as 1 pg RNA per band. The intensity of the autoradiographic signal is a measure of the concentration of the specific RNA and the distance migrated by the RNA species can be used to determine its molecular weight. Total cellular, or total cytoplasmic, or purified polyadenylated RNA can be used.

Gel electrophoresis is performed using conditions which disrupt RNA secondary structure. This greatly improves resolution and allows an accurate estimation of the length of the RNA molecule. A number of methods of denaturation are in common use. These include treatment of the RNA with formaldehyde (4, 5), glyoxal (6), or methyl mercuric hydroxide (7), but only the formaldehyde procedure is presented here (*Protocol 1*) because we find it to be the most simple and effective.

The percentage of agarose in the gel will depend on the size of the RNA species being studied. A 1.4% gel is appropriate for RNAs of 0.5–2.0 kb, while a 1% gel gives better resolution of larger species. The amount of RNA loaded per lane will depend on the abundance of the RNA species being studied. Abundant mRNA species (>0.1% of the mRNA population) can be readily detected in 5–20 μg of total cellular RNA. For detection of rare mRNAs it is advisable to first purify poly$(A)^+$ RNA and then load 1–10 μg per lane. The RNA is denatured by heating at 65°C in a solution containing formamide and formaldehyde prior to loading on the gel and is maintained in a denatured condition during electrophoresis by incorporating formaldehyde within the gel.

A variety of nitrocellulose or nylon filters can be used for the transfer. We have found Genescreen Plus (NEN) to perform as well as any, paricularly for several re-probings of the same filter with different labelled probes. After transfer, the filter is washed briefly in low salt buffer, dried, and baked at 80°C to fix the RNA to the filter.

Finally, the RNAs of interest are detected on the filter by hybridization with a ^{32}P-labelled complementary DNA probe. Several suitable procedures exist for radiolabelling DNA probes but a rapid method which routinely generates high specific activity probes is oligo-labelling (*Protocol 2*). Here the

DNA template is hybridized with random hexanucleotides which then act as primers for the synthesis of a radiolabelled second strand by *E. coli* DNA polymerase I (Klenow fragment). After the removal of unincorporated radioactive precursor (see *Protocol 3*), the radiolabelled DNA probe is added to the filter containing the immobilized RNA. This hybridization step is performed in the presence of formamide using a modification of the method of Wahl *et al.* (8); see *Protocol 4*.

An example of a Northern blot produced using this methodology is shown in *Figure 2*.

Protocol 1. Electrophoresis of RNA in denaturing gels and Northern transfer

Reagents

All aqueous solutions should be treated with diethyl pyrocarbonate (DEPC) to destroy RNase. This is done by making solutions 0.2% in DEPC and then autoclaving (or incubating for 1 h at 65°C) to break down residual DEPC.

- Agarose (low electro-endosmosis grade; from BRL)
- 10 × Mops buffer stock [0.2 M Mops (Sigma), 50 mM NaOAc, 0.01 M EDTA]. Adjust to pH 7.0 with acetic acid. This solution turns yellow after autoclaving or exposure to light but it still works
- Formaldehyde; 40% (v/v) analytical reagent grade
- Genescreen Plus transfer membrane (NEN)
- RNA size markers (e.g. from BRL)
- Deionized formamide. Some analytical grade or ultra-pure products can be used directly. If in doubt, deionize the formamide by stirring 5 g of mixed bed ion-exchange resin with 100 ml formamide for 1–2 h at room temperature. Filter and store at −20°C
- RNA loading buffer (50% glycerol, 1 mM EDTA, 0.01% bromophenol blue)
- Ethidium bromide stock; 10 mg/ml
- 10 × SSC stock (1.5 M NaCl, 0.15 M trisodium citrate).

Method

Always wear disposable gloves and use sterile glass and plastic ware.

1. Prepare a horizontal 1.0% or 1.4% agarose gel with the following composition per 100 ml:

	1.0% gel	**1.4% gel**
agarose	1.0 g	1.4 g
10 × Mops stock	10 ml	10 ml
formaldehyde (40% v/v)	17 ml	17 ml
H_2O	76 ml	76 ml.

To prepare the gel, dissolve the agarose in the Mops/water mixture by boiling, cool to 50°C, then add the formaldehyde and pour the gel immediately (in a fume hood because formaldehyde vapours can be toxic). The gel should be 4 mm in depth with sample slots 0.8 cm wide.

Protocol 1. *Continued*

2. Once set, place the gel in the electrophoresis apparatus located in a fume hood or in a covered tank using 1 × Mops buffer as the electrophoresis buffer.
3. To the RNA sample (≤20 μg) in 12 μl water in a plastic microcentrifuge tube add:

deionized formamide	25 μl
10 × Mops	5 μl
formaldehyde	8 μl.

   Mix by vortexing briefly and incubate at 65°C for 5 min.
4. Chill on ice, add 5 μl RNA loading buffer and mix.
5. Also prepare one sample of size markers such as the RNA ladder provided by BRL. Use about 3 μg of these and treat them exactly as for the RNA samples.
6. Load the RNA samples into appropriate wells. Load the size markers at one end of the gel.
7. Start the electrophoresis. For a gel of 20 cm in length, perform the electrophoresis at 100 V for 3–4 h, by which time the marker dye should have migrated about 15 cm.
8. Carefully remove the gel from the tank and rinse briefly (5 min) in distilled water to remove excess formaldehyde.
9. Cut off the end of the gel containing the size markers and stain by immersion in 5 μg/ml ethidium bromide for 30 min. (*NB* Wear gloves during this step as ethidium bromide is a potential carcinogen.)
10. Photograph the visualized RNA bands under UV transillumination (9). (*NB* Protect your eyes with suitable goggles or a visor.)
11. Set up the Northern transfer using the rest of the gel in a conventional gel-blotting apparatus (see *Figure 1*) using 10 × SSC as the transfer medium. Cut a piece of Genescreen Plus (NEN) to the size of the gel, and mark side B with pencil (side B is the concave side because of the natural curvature of the material). Wet the filter in distilled water (side B will now have become convex) and then soak in 10 × SSC for 15 min. Place side B in contact with the gel, place filter paper and paper towels on top of the gel (*Figure 1*) and leave to blot for 16–24 h.
12. Dismantle the apparatus, mark the positions of the wells on the filter with permanent ink, rinse the filter in 2 × SSC for 15 min, blot dry, and bake at 80°C for 2 h.

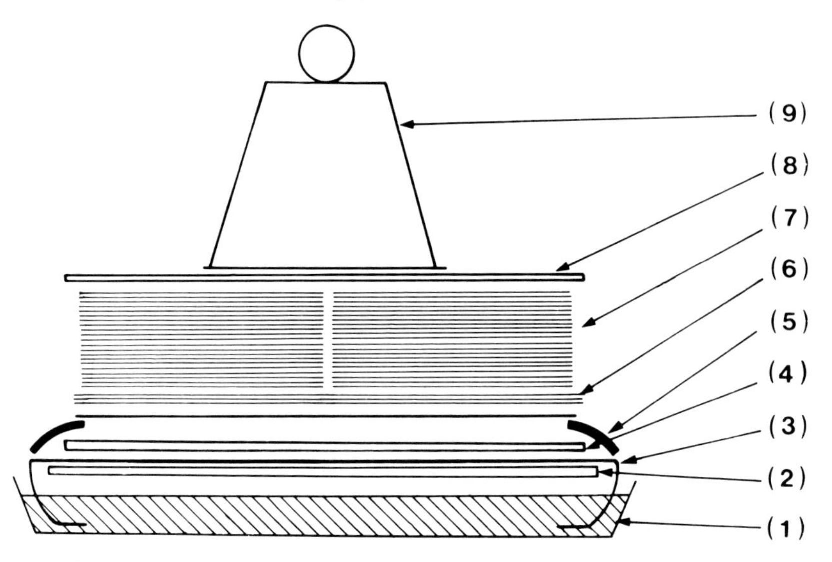

Figure 1. Cross-section of a Northern transfer apparatus. (1) Tray filled with 20 × SSC, (2) glass plate supported by two sides of the tray, (3) wick of three sheets of Whatman 3MM paper, (4) gel, (5) Parafilm round all sides of the gel, (6) three sheets of Whatman 3MM paper, (7) paper towels, (8) glass plate, (9) weight. From ref. 34.

Protocol 2. Oligo-labelling of the DNA probe[a]

Reagents

- DNA probe (100 ng at ≥20 μg/ml in water)
- 10 × labelling buffer (0.5 M Tris–HCl pH 7.8, 50 mM $MgCl_2$, 0.1 M 2-mercapto-ethanol)
- TE buffer (10 mM Tris–HCl pH 8.0; 1 mM EDTA)
- Random hexanucleotide primers stock in TE buffer (90 A_{260} units/ml). Prepare this by suspending 50 A_{260} units of the primers [p(dN)6 sodium salt from PL-Biochemicals] in 0.555 ml TE buffer
- 1.0 M Hepes pH 6.6
- 200 μM dNTP solutions of dATP, dGTP, and TTP
- 10 mg/ml bovine serum albumin (BSA)
- [α-^{32}P]dCTP (10 mCi/ml; 5000 Ci/mM)
- DNA polymerase I (Klenow fragment) 2.5 units/μl
- Sephadex G50 spun columns.[b]

Method

1. Heat the DNA probe (100 ng in 5 μl water) in a boiling water bath for 3 min, then chill immediately in ice/water.
2. To the DNA add the following:

1 M Hepes pH 6.6	5.0 μl
10 × labelling buffer	2.5 μl

Protocol 2. *Continued*

200 μM dGTP	2.5 μl
200 μM dATP	2.5 μl
200 μM dTTP	2.5 μl
1:50 dilution of random hexanucleotide primers	7.0 μl
10 mg/ml BSA	1.0 μl.

3. Add 25 μl [α-^{32}P]dCTP and 1.0 μl (2.5 units) of DNA polymerase I (Klenow fragment).
4. Mix gently and incubate for 2.5 h at room temperature.
5. Separate unincorporated [α-^{32}P]dCTP from the radiolabelled probe by Sephadex G50 spun column chromatography (see *Protocol 3*).

[a] From ref. 10
[b] See *Protocol 3.*

Protocol 3. Sephadex G50 spun column chromatography

Preparation of columns

1. Take a 1 ml plastic syringe without the plunger and plug the bottom with a little polymer wool (Interpet).
2. Fill the syringe to the top with a slurry of Sephadex G50 (Sigma G50-150) that has been swollen in TE buffer.
3. Place the filled syringe inside a 10 ml plastic tube and centrifuge at room temperature for 4 min at 1500*g*. The syringe should now contain about 0.7 ml dry Sephadex G50.
4. Discard the eluate and fill the syringe to the top with TE buffer and repeat the centrifugation step.
5. Add TE buffer and centrifuge once more. At this stage columns can be stored at 4°C for several days.

Use of columns

1. Remove the cap from a 0.5 ml microcentrifuge tube and place it in the 10 ml tube beneath the G50 column to collect the eluate.
2. Load the labelling reaction on to the column and centrifuge at room temperature for 4 min at 1500*g*. The 0.5 ml microcentrifuge tube should contain the labelled DNA in a volume equal to the volume loaded while unincorporated radioactivity should remain in the column.

Protocol 4. Hybridization after Northern transfer

Reagents

- Deionized formamide (see *Protocol 1*)
- 5 M NaCl
- 50% dextran sulphate
- 10% SDS
- 10 × SSC stock solution (1.5 M NaCl, 0.15 M trisodium citrate)
- 10 mg/ml single-stranded DNA. Dissolve salmon sperm DNA in water at 10 mg/ml, sonicate to a length of 200–500 bp. Denature the DNA by standing in a boiling water bath for 20 min, cool rapidly in ice, store at −20°C
- 3 × SSC (prepared from 10 × SSC stock)
- 1 × SSC, 1% SDS and 0.1 × SSC, 1% SDS solutions (each prepared from 10 × SSC and 10% SDS stocks)
- Denatured radiolabelled probe. The probe should preferably be a purified DNA restriction fragment labelled by the oligo-labelling method (*Protocol 2*) to about 10^9 c.p.m./μg. Denature the probe by heating at 100°C for 5 min and then immediately chilling on ice.

Method

1. Soak the filter in 3 × SSC for 30 min.
2. Place the filter in a heat-sealable plastic bag and add the following prehybridization solution:

	Volume (for 20 ml)	**Final concentration**
deionized formamide	10 ml	50%
10 mg/ml single-stranded DNA	400 μl	200 μg/ml
5 M NaCl	4 ml	1 M
50% dextran sulphate	4 ml	10%
10% SDS	2 ml	1%.

   Use about 1 ml per 10 cm^2 of filter area. Heat-seal the bag and incubate at 42°C with shaking for 3–6 h.
3. After prehybridization, carefully snip off a corner of the bag and add about 10^6 c.p.m./ml of the denatured, radioactively labelled probe.
4. Re-seal the bag and mix the probe thoroughly with the hybridization buffer by gently squeezing and inverting the bag. Incubate at 42°C with shaking for 20 h.
5. After hybridization, wash the filter:
   - twice in 100 ml 1 × SSC, 1% SDS, for 1 min at room temperature
   - twice in 200 ml 1 × SSC, 1% SDS, for 30 min at 42°C
   - once in 100 ml 0.1 × SSC, 1% SDS, for 30 min at 42°C.
6. Dry the filter. Wrap it in cling film while still moist and expose it to X-ray film at −70°C with an intensifying screen. Develop the X-ray film according to the supplier's instructions.

Protocol 4. *Continued*

7. To re-use the filter, remove the bound probe by placing the filter in a tray containing sterile distilled water at 100°C and leaving it for 5–10 min or until the water has cooled to room temperature.

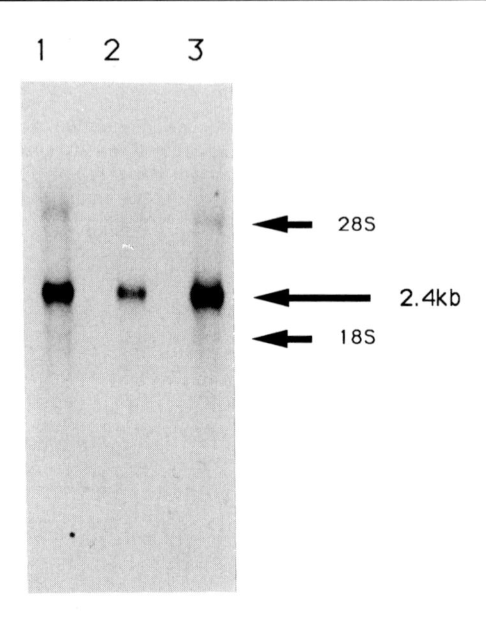

Figure 2. An example of Northern hybridization. For each sample, 20 μg total RNA was electrophoresed on a 1.2% agarose gel and subjected to the Northern hybridization procedure as described in *Protocols 1* and *4*. The probe was ^{32}P-human glucose-6-phosphate dehydrogenase cDNA labelled as in *Protocol 2*. The lanes were 1, HeLa cell RNA; 2, CV1 (a monkey kidney fibroblast cell line) cell RNA; 3, RNA from a CV1 cell line expressing the human glucose-6-phosphate dehydrogenase gene. The migration positions of the 28S and 18S rRNAs are indicated. The size (2.4 kb) of the hybridizing RNA was calculated from molecular weight markers run on the same gel as described in *Protocol 1*. Autoradiography was carried out for 16 h at −80°C using intensifying screens. Data kindly supplied by Colm Corcoran.

2.2 Quantitative Northern blots

In order to compare the amount of an RNA species in different samples it is necessary to ensure that equal amounts of RNA are loaded in each track. The best way to check that this has been achieved is to hybridize the blot with a probe detecting an RNA species that is expected to be at approximately the same level in all samples. RNA degradation is also detected by this procedure. Probes that detect actin mRNA are often used for this purpose since actin mRNAs are present at reasonably high levels in most cells. The signals from the test RNA are then normalized to those from the control RNA, either approximately by eye or more accurately after densitometry. Hybridization with the control probe can be carried out after the test probe has been stripped from the filter or, if the test and control RNAs are sufficiently different in size, both probes can be used at the same time.

2.3 Potential problems

The most common problem with Northern transfer is an unacceptable level of degradation of the RNA. A low level of degradation, which results in a 'shadow' migrating ahead of the intact RNA, is quite common. However, in extreme cases all the RNA will migrate as a diffuse smear of low molecular weight. This may arise because one of the solutions was contaminated with RNase in which case preparing fresh solutions may solve the problem. RNase is sometimes used to destroy RNA in 'mini-preparations' of bacterial plasmids before electrophoresis (e.g. ref. 11). We and others have found that an electrophoresis apparatus used for analysing such preparations retains sufficient RNase to degrade RNA if the apparatus is subsequently used for RNA electrophoresis. Hence it is advisable to reserve an apparatus solely for RNA electrophoresis and to treat it with 0.2% DEPC before use.

If neither of these measures proves effective, then it may be that the RNA was actually isolated in a degraded form. This can easily be determined by analysing an RNA sample which is known to be intact, in parallel with the suspect RNA. Indeed it is a good idea routinely to include a sample of intact cellular RNA during gel electrophoresis, which can be visualized subsequently by ethidium bromide staining of the corresponding segment of the gel followed by UV shadowing. To do this:

- cover a TLC plate (Merck Kieselgel 60F254) with cling film
- place the gel on top and view with incident UV light (254 nm) in the dark room.

RNA will appear dark on a fluorescent green background. If the RNA is intact, the large (28S) and small (18S) ribosomal RNA bands can be clearly seen and their position measured. This procedure also enables one to check that the amount of RNA loaded in each track is as expected.

3. RNA slot blots

3.1 Procedures

The principle of the method is that a known amount of sample RNA is spotted on to an inert support, such as nitrocellulose, and the amount of a specific RNA in the sample is determined by hybridization with a suitable radioactively labelled probe (12). The RNA can be applied as dots ('dot blots') or using an apparatus which applies the RNA in more elongated slots ('slot blots'). The technique is rapid, very sensitive, and can be made semiquantitative if sufficient radioactivity is hybridized to allow scintillation counting of the radioactive 'dots' or 'slots'. Alternatively, quantification can be achieved by densitometric scanning of autoradiographic images of the radioactive blots. For this application, 'slot blots' are preferred since their shape allows more accurate and reproducible scanning than 'dot blots'. Note that the technique provides no information as to the number or size of those RNA species hybridizing to the probe.

Two slot blot procedures are described here. In the first, cytoplasmic cell lysates are slotted directly on to nitrocellulose membranes. This procedure obviates the need for RNA purification and is particularly useful if only small numbers of cells are available. In the second procedure, purified RNA is applied to the membranes. *Protocol 5* describes the preparation of cytoplasmic lysates and purified RNA for slot blot analysis, while *Protocol 6* gives the procedure for application of these samples to the filter. Hybridization of the slot blots is carried out as in *Protocol 7*. Either DNA probes (as for Northern hybridization, *Protocol 4*) or RNA probes can be used. *Protocol 7* describes the method using RNA probes.

Protocol 5. Preparation of samples for slot blot analysis

Reagents

- Suspension buffer (10 mM Tris–HCl pH 7.0, 1 mM EDTA)
- Lysis solution (5% Nonidet-P40)
- Ribonuclease inhibitor; RNasin (10 U μl; Promega) or Nucleotide complex (BRL)
- 20 × SSC (3.0 M NaCl, 0.3 M trisodium citrate)
- 37% formaldehyde
- Phosphate-buffered saline pH 7.4 (PBS)
- Deionized formamide (see *Protocol 1*); for preparation of purified RNA only
- Trypan blue (Gibco/BRL).

Method

1. Harvest the cells using an appropriate procedure and pellet them by gentle centrifugation (approximately 1000 *g*) for 2–5 min. Resuspend the cells in a considerable excess of PBS (20–100 vol. of the original pellet).
2. Count the cells using a haemocytometer.
3. Pellet the cells by further centrifugation at 1000 *g* for 2–3 min.
4. Resuspend the cells in PBS at a concentration of 10^7 cells/ml. Transfer 1 ml of this cell suspension to a microcentrifuge tube and pellet by centrifugation for 2 min in a microcentrifuge. Remove the supernatant and place the cell pellet on ice.
5. Resuspend the cell pellet thoroughly in 45 μl of ice-cold suspension buffer taking care to keep the cells cold. To the evenly suspended cells, add 2 μl (10 U/μl) of ribonuclease inhibitor and mix well.

6. Lyse the cells by adding 5 μl of lysis solution, mixing well, and incubating on ice for 5 min. Add another 5 μl of lysis solution, mix, and incubate on ice for a further 5 min.[a]
7. Separate the lysate into nuclear and cytoplasmic components by centrifuging in a microcentrifuge at 4°C for 2.5 min.
8. Transfer 50 μl of the supernatant (contains cytoplasmic components including cytoplasmic RNA) to a fresh microcentrifuge tube into which 30 μl of 20 × SSC and 20 μl of 37% formaldehyde have just been added.
9. Denature the RNA by heating at 65°C for 15 min.
10. Either use immediately or store at −70°C for later use.

Purified RNA

1. Adjust the volume of the purified RNA sample to 10 μl with H_2O and keep on ice.
2. To a microcentrifuge tube add:

deionized formamide	20 μl
37% formaldehyde	7 μl
20 × SSC	2 μl.

   Mix the components and add the RNA sample (10 μl).
3. Denature the RNA by heating the sample at 65°C for 15 min, then transfer the denatured samples to ice.
4. Add 2 vol. of 20 × SSC. The samples are now ready to be applied to the nitrocellulose filter.

[a] Cells differ considerably in their susceptibility to lysis. The lysis procedure may be followed by staining reaction aliquots with either Trypan blue (mix equal volumes of cell suspension and 0.4% Trypan blue) and then examining them by light microscopy. Adjustments in either the amount of lysis solution added, the duration of the lysis steps, or the number of lysis steps performed may then be made accordingly to maximize the degree of lysis obtained.

Protocol 6. Application of samples

Equipment and reagents

- Schleicher & Schuell Minifold II slot blot system (or equivalent equipment)
- 15 × SSC (2.25 M NaCl, 0.225 M trisodium citrate)
- Nitrocellulose membrane filter (precut to correct dimensions)
- Filter paper (precut).

Method

Note: never handle nitrocellulose membranes without wearing disposable plastic gloves; grease from your hands or any other source will prevent even wetting of the membrane.

Protocol 6. *Continued*

1. Place a precut nitrocellulose membrane in water until it is evenly wetted. If the membrane does not wet completely, discard it in favour of a piece that does. Drain excess water from the membrane and transfer it to a dish of 15 × SSC.
2. Prepare 4–6 serial dilutions of each RNA sample to be applied; 5-fold dilutions generally work well. Make the dilutions using 15–20 × SSC taking care to keep the final SSC concentration greater than 12 × SSC since high salt concentrations are required for efficient binding of nucleic acid to nitrocellulose.
3. Assemble the slot blot apparatus according to the manufacturers instructions; first place two pieces of precut filter paper on to the vacuum manifold (base of unit) and then put the wetted nitrocellulose membrane on top. Attach the sample former (top of unit) and clamp the unit together. Connect the apparatus to a vacuum source.
4. Wash each well of the apparatus by flushing with 300 μl of 15 × SSC. Apply the samples. Wash each well with a further 300 μl of 15 × SSC.
5. Break the vacuum and dismantle the unit. Remove the nitrocellulose but take care that parts of the membrane do not contact each other at this stage as nucleic acid will be transferred at the point of contact. Place the nitrocellulose on to a piece of filter paper and allow to air dry.
6. When the nitrocellulose is dry, place it between two sheets of filter paper and bake in a vacuum oven at 80°C for 2 h.

Protocol 7. Hybridization of slot blots with RNA probes

Reagents

- 5 × SSC (0.75 M NaCl, 0.075 M trisodium citrate)
- Hybridization buffer [50% deionized formamide (see *Protocol 1*), 5 × SSC, 5 mM sodium phosphate buffer (pH 6.5), 0.1% SDS, 1 × Denhardt's solution (0.05% BSA, 0.05% Ficoll, 0.05% polyvinylpyrrolidine), 200 μg/ml denatured salmon sperm DNA (see *Protocol 4*)]
- Denatured radioactive RNA probe at about 10^9 c.p.m./μg (prepared as described in *Protocol 17*, diluted into hybridization buffer and denatured by heating at 90°C for 2 min then quenched immediately in ice)
- 2 × SSC, 0.1% SDS
- 0.1 × SSC, 0.1% SDS.

Method

1. Prepare fresh hybridization buffer and keep it on ice.
2. Wet the nitrocellulose membranes at room temperature in water, then transfer to 5 × SSC.

3. Place the wetted membrane in a plastic hybridization bag (see *Protocol 4*) and add the hybridization buffer (approximately 1–3 ml per membrane). Seal the bag and incubate in a shaking water bath at 65 °C for 1–16 h.[a] This is the prehybridization step.
4. After prehybridization, cut open the bag and discard the hybridization buffer. Add 1–3 ml of fresh hybridization buffer and 50–100 ng ($\sim 5 \times 10^7$ c.p.m.) of denatured radioactive RNA probe. Reseal the bag and incubate at 65 °C for a minimum of 18 h.
5. After hybridization, cut open the bag and remove the filter. Remove excess radioactive hybridization solution by washing the filter (by agitation) in a tray containing 300–500 ml of 5 × SSC.
6. Remove residual unbound probe and non-specifically bound probe by stringent washing at 65 °C in 2 × SSC, 0.1% SDS. Preheat the washing solution to 65 °C, use 100 ml per filter and wash for 20 min. Repeat this procedure four more times.
7. Perform a final wash using 100 ml of 0.1 × SSC, 0.1% SDS (preheated to 65 °C) for 20 min.
8. Drain excess wash solution from the membrane but do not let the membrane dry completely as this hampers probe removal prior to rehybridization with a new probe.
9. Wrap the filter in cling film and perform autoradiography at −70 °C (see *Protocol 4*).

[a] 1 h of prehybridization is usually sufficient. If high background is observed, try increasing the prehybridization time. An overnight prehybridization step is permissible and often convenient.

3.2 Quantification

Quantification may be achieved through scanning densitometry of the autoradiographs. If this method is used, it is important to be sure that all the signals measured fall within the linear range of the film response. Since this range is quite narrow and the signals obtained from different samples often vary considerably, it is normally necessary to scan several exposures of different duration.

A more reliable method for quantification uses direct counting of the radioactive emissions from the bound probe. This can be achieved easily and without destroying the membrane by using a purpose-built machine (suitable machines are available from Betagen, or Molecular Dynamics). If such a machine is not available then it is necessary to cut out the individual slots and count them in a scintillation counter. Compare the c.p.m. obtained for each of the serial dilutions of the samples. Ideally the c.p.m. should be directly proportional to the dilutions, but in practice the values will probably be proportional for only part of the dilution series employed. This is particularly so with slot blots of cytoplasmic lysates where, in the less diluted samples, there may be sufficient protein to block binding sites on the nitrocellulose and so render it inefficient in binding RNA. Clearly it is essential to use only those values that are proportional to the dilutions employed. Compare these values to those obtained from dilution series of known mRNA standards. If such standards are not available, they can be produced by *in vitro* transcription (using only unlabelled precursors) of sense RNA from cloned DNA templates. Alternatively single-stranded DNAs prepared from vectors containing an F1 origin may be used (make sure to generate the correct strand), although it should be remembered that DNA–RNA hybrids are more thermolabile than their RNA–RNA counterparts.

3.3 Potential problems

The major problem encountered in slot blot hybridization is one of non-specific or cross-hybridization. Cross-hybridization can be particularly problematic when analysing RNAs from multigene families, so in this case it is important to select probes from the most diverged regions of the mRNAs. Since it is impossible to know which RNAs in the sample are hybridizing to the probe, it is crucial to include a number of positive and negative controls in the analysis. If standard RNA samples are not available, then known concentrations of cloned DNAs can be used. If high non-specific background signals or undesired cross-hybridization are encountered, rewash the membranes at 65°C using 0.1 × SSC, 0.1% SDS containing 10–20% deionized formamide.

4. Analysis of specific RNA molecules by polymerase chain reaction (PCR)

4.1 Procedures

Providing that the nucleotide sequence of the RNA of interest is known, PCR (13, 14) can be used to detect that specific RNA molecule in a mixed population. A cDNA copy of the RNA population under test is synthesized using reverse transcriptase. This is then used as the template for PCR with the primers being two oligonucleotides specific for the RNA of interest. Ideally the primers should be 20–25mers whose sequences are separated in the RNA by 150–2000 bp. The PCR products of the reaction are analysed for the presence of an amplified DNA band of the predicted size on an agarose minigel. For detection of rare mRNAs (which may yield only a small amount of PCR product in a background of non-specific products) the gel can be blotted and probed with a specific radioactive probe.

The advantages of this combined reverse transcriptase and PCR analysis

(RT-PCR; also known as cDNA-PCR) are, firstly, that the process is extremely efficient, enabling the detection of RNAs that are present at less than one part per 10^8. Secondly, the PCR product can readily be subjected to cloning, direct sequencing, or restriction enzyme analysis (15). This may be useful in analysing the expression of closely related, but non-identical transcripts. It has been used to detect point mutations in RNAs in, for example, tissue samples from suspected carriers of human genetic diseases (15).

Either oligo-dT or a specific primer can be used to prime the reverse transcriptase reaction. If oligo-dT is used, the same cDNA can be used in many PCR reactions with different specific primers. If, however, the region of the RNA molecule being amplified is several kilobases from the poly(A) tail of the mRNA, a specific primer is normally used to ensure that reverse transcriptase will copy through the region of interest. *Protocol 8* describes the procedure for RT–PCR amplification of mRNA and *Figure 3* gives an example of its use.

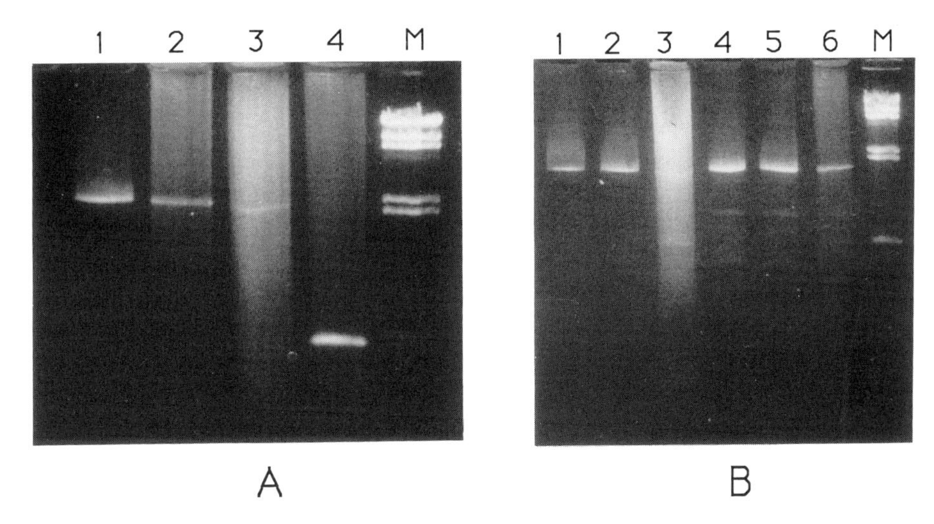

Figure 3. Examples of reverse transcription/PCR (RT/PCR). (A) RT/PCR was performed on total RNA from a cell line derived from a patient with acute myeloid leukaemia. Oligo-dT was used to prime the reverse transcription. The primers were designed to amplify fragments of 2157 bp (lanes 1–3) and 510 bp (lane 4) from the 3′ region of the human *c-fms* mRNA. Magnesium ion concentrations were: lane 1, 1.4 mM; lane 2, 1.6 mM; lane 3, 1.8 mM, and lane 4, 1.8 mM. (B) RT/PCR was performed on total RNA from bone marrow from two patients with acute myeloid leukaemia. Oligo-dT was used to prime the reverse transcription. The primers were designed to amplify fragments of 1732 bp from the 5′ region of the human *c-fms* mRNA. Lanes 1–3 represent PCR products of total RNA from one patient and lanes 4–6 from another patient. Magnesium ion concentrations were: lanes 1 and 4, 1.0 mM; lanes 2 and 5, 1.2 mM; lanes 3 and 6, 1.4 mM. Molecular weight markers (M) were a *Hin*dIII digest of phage λ DNA which yields DNA fragments of sizes 23.7, 9.5, 6.7, 4.3, 2.3, 1.96, and 0.5 kb. (Data kindly supplied by Nick Dibb.)

Protocol 8. Reverse transcription–PCR amplification of mRNA sequences (RT–PCR)

Equipment and reagents

- PCR temperature cycling apparatus (e.g. from Cetus)
- RNA sample (≤20 μg total RNA in 30 μl water)
- 1 mg/ml oligo-dT (Pharmacia) or specific 3′ primer
- Specific PCR primers
- 10 mM dNTP mixture (all four dNTPs; Pharmacia)
- 1 M Tris–HCl pH 8.3
- 1 M KCl
- 0.25 M $MgCl_2$
- 10 U/μl placental RNasin (Promega Biotech)
- 25 U/μl AMV reverse transcriptase (BRL)
- PCR buffer [50 mM KCl, Tris–HCl pH 8.3, 1–2 mM $MgCl_2$,[a] 0.001% gelatin, 0.05% NP40 (Sigma), 0.05% Tween 20 (Sigma), 0.3 mM dNTPs (all four dNTPs; Pharmacia), 1 μM of each of the PCR primers[b]]
- Taq polymerase (Cetus Corporation).

Method

1. Take 20 μg (or less) of total RNA in 30 μl sterile water and add:

1 M Tris–HCl pH 8.3	2.5 μl
1 M KCl	2.5 μl
0.25 M $MgCl_2$	2 μl
1 mg/ml oligo-dT or specific 3′ primer	5 μl
10 mM dNTP mixture	5 μl.

Mix and heat at 65 °C for 5 min. Cool to room temperature.

2. Add:

placental RNasin (10 U/μl)	2 μl
AMV reverse transcriptase (25 U/μl)	1.5 μl.

Incubate at 42 °C for 1 h.

3. Add 1 μl of the reverse transcription product to 50 μl PCR buffer containing 1 U of Taq polymerase. Mix, overlay with mineral oil, and heat at 93 °C for 3 min.

4. Set up the tubes in a PCR temperature cycling apparatus. Perform 30 cycles of, for example:

1 min denaturing at 94 °C
1 min annealing at 56 °C
2 min extension at 72 °C.

The extension time may need to be increased for long extensions (>500 bp).

5. After 30 cycles, hold at 72 °C for 10 min.

6. The reaction product is now ready for gel electrophoresis, restriction enzyme analysis, sequencing, or sub-cloning.

[a] The optimum Mg^{2+} concentration must be determined for each pair of primers and can vary between templates (see Section 4.3 and *Figure 1.3*).
[b] The 3′ primer could be the 3′ PCR primer.

4.2 Quantification

If only a low number of PCR cycles (<25) are performed, the intensity of the band on the agarose gel should be proportional to the amount of the specific RNA in the original sample. However, because of variability in effficiency of the PCR between samples, caution should be exercised in drawing quantitative conclusions. Nevertheless, the PCR reaction can be made quantitative (16) by constructing an artificial RNA template containing the same primer binding sites as the template under study but with a different distance between them, thereby giving rise to a different sized PCR product. Such a construct can be made by making a small deletion, or insertion in the DNA under test, and then cloning it into a plasmid adjacent to an RNA polymerase promoter (see Section 5.3.2). The RNA transcript from this plasmid is then purified and added, in different amounts, to the RNA populations under investigation. The extent to which this template competes with the authentic template reflects the amount of authentic template present. This gives a measure of the concentration of the required transcript that is independent of the overall efficiency of the reverse transcription and PCR reactions (*Figure 4*).

4.3 Potential problems

The most common problems that occur in PCR reactions, assuming intact RNA has been used as the template, are false positives due to contamination of one of the solutions with cloned cDNA or a PCR product. In order to avoid this, it is good practice to devote a section of the laboratory, and a set of micropipettes, exclusively to the setting up of PCR reactions.

It is important to establish the optimum magnesium ion concentration for each pair of primers. The PCR reaction is exquisitely sensitive to the magnesium concentration and this should be titrated at 0.2 mM intervals between 0.8–2.0 mM. If the magnesium concentration is too low, the primers will not form stable hybrids and no product will be seen. At too high a magnesium concentration, the primers will hybridize non-specifically and a ladder of fragments will be produced, which may or may not contain the required, specific fragment (see *Figure 3A*). The amount of $MgCl_2$ added may need to be altered for different samples since they may vary in their original content of magnesium ions or magnesium ion chelators (see *Figure 3B*).

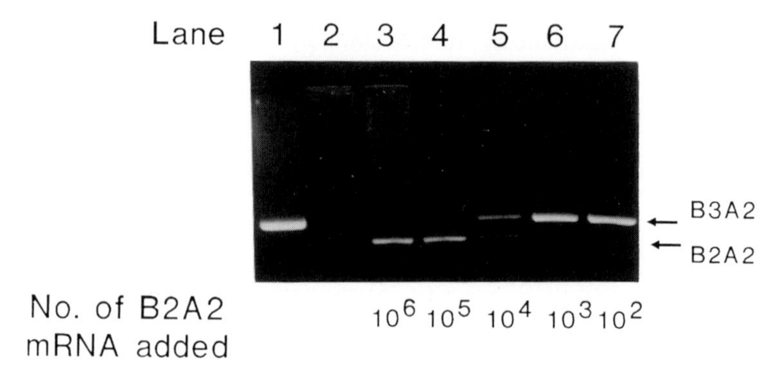

Figure 4. Quantitative PCR. RT/PCR was performed on RNA extracted from a mixed cell population of K562 and normal HL60 cells at a ratio of $1{:}10^4$. Reverse transcription was primed with the 3′ PCR primer. The primers used were designed to amplify a 379 bp fragment (B3A2) of the *bcr/abl* mRNA produced as a consequence of the 9:22 translocation found in K562 cells. The number of competing smaller synthetic RNA molecules (B2A2, 304 bp) added prior to reverse transcription is indicated below lanes 3–7. Lane 1, no added B2A2 RNA; lane 2 no RNA. Lane 5 shows amplification of both fragments indicating equivalence. (Data kindly supplied by Tim Hughes.)

If no specific PCR product of the expected size is present, or if the expected product is heavily diluted by non-specific fragments, a portion of the reaction (1/50–1/20th) can be subjected to a second round of PCR using a different pair of specific primers that hybridize within the region amplified in the first round. This use of 'nested' primers increases both the specificity and sensitivity of the technique.

5. Nuclease S1 mapping

The various methods of mapping RNA by nuclease protection share a common principle. A population of RNA is hybridized to a radioactively labelled probe that is complementary, along part or all of its length, to the specific RNA being analysed. Probe molecules that fail to hybridize, and those regions of the probe that are not annealed to the target RNA, are then removed by nuclease digestion. Finally, the digestion products are analysed by gel electrophoresis. The two commonly used methods, nuclease S1 mapping and RNase protection (sometimes called ribo-probe analysis), differ only in the nature of the probe and the nuclease that they employ. The rest of Section 5 describes nuclease S1 mapping, and RNase protection is described in Section 6.

5.1 An overview of nuclease S1 mapping

The enzyme nuclease S1 degrades single-stranded DNA and RNA but, at low temperature, and in high ionic strength buffers, it does not digest double-stranded nucleic acids at any appreciable rate. In S1 mapping, RNA is hybridized to a DNA molecule that is complementary to the RNA over only part of its length. After hybridization, the reaction mixture is incubated with nuclease S1 to degrade unhybridized segments of the DNA probe. The size of the nuclease-resistant DNA fragments represents the length of the nucleotide sequence over which there is perfect homology between the RNA and DNA. This method was used initially to determine the location and size of intervening sequences in eukaryotic mRNA (17). The DNA fragments were resolved by agarose gel electrophoresis and individual fragments were then identified by Southern transfer. A more sensitive and accurate procedure is to use radioactively labelled restriction fragments as probes (18) and so this technique is now much more commonly used than the original procedure.

A DNA fragment that spans the predicted RNA terminus or intron/exon splice point is selected and at its 5′ or 3′ end labelled with ^{32}P. The probe is denatured and hybridized with the RNA and the resulting RNA–DNA hybrids are treated with nuclease S1. The size of the labelled DNA strand in the protected hybrid is then determined by polyacrylamide or agarose gel electrophoresis under denaturing conditions. This length corresponds to the distance between the labelled end of the DNA restriction fragment and the RNA terminus or splice point (see *Figure 5*). Under conditions of DNA excess, the intensity of the autoradiographic signal from this labelled DNA fragment is directly proportional to the concentration of the hybridizing RNA species.

The most unambiguous, sensitive, and reproducible results are obtained using relatively short (100–600 nt) single-stranded DNA probes. The alternative is to use a double-stranded probe and perform the hybridization in 80% formamide at high temperature, conditions under which probe renaturation is suppressed but target-probe hybridization can occur (because RNA–DNA duplexes are more stable than their corresponding DNA–DNA duplexes). Both of these protocols are described below.

5.2 Preparation of single-stranded probes by strand separation and gel electrophoresis

5.2.1 Isolation of DNA fragment

The genomic DNA being studied by S1 mapping will already have been characterized by restriction enzyme analysis and, ideally, by DNA sequence analysis. The approximate locations of transcribed regions should have been

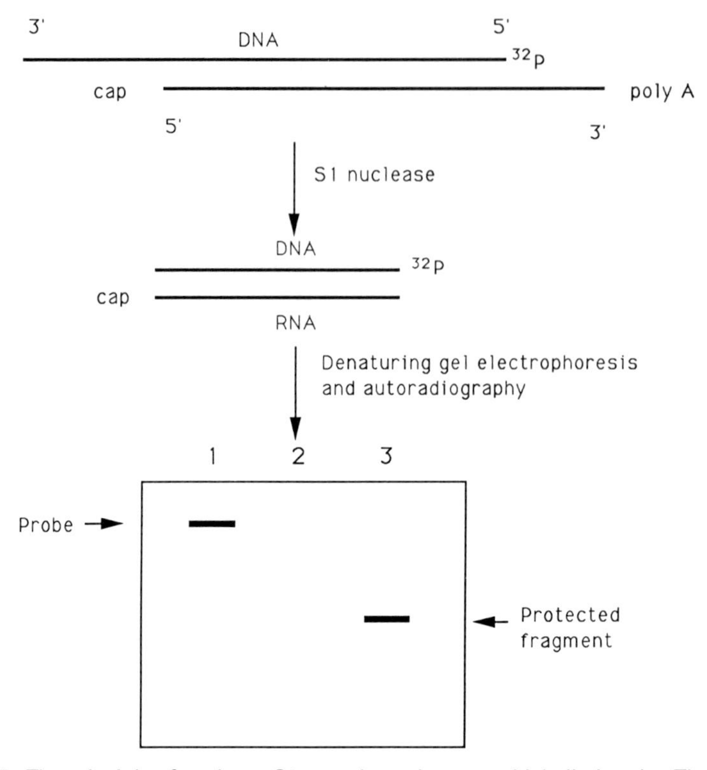

Figure 5. The principle of nuclease S1 mapping using an end-labelled probe. The diagram illustrates how the 5′ terminus of a poly(A)$^+$ mRNA might be determined using a single-stranded DNA probe. The schematic representation of the autoradiogram (lower part of figure) demonstrates the results expected from a typical experiment. Lane 1, the probe without nuclease S1 digestion; lane 2, probe incubated under hybridization conditions without complementary RNA and then digested with nuclease S1; lane 3, probe hybridized with complementary RNA and digested with nuclease S1 leaving the fragment protected from nuclease digestion by hybridized RNA.

deduced by Northern transfer, Southern blotting, or a combination of both techniques. The orientation of the gene may be deduced from comparison with a cDNA clone, or from Northern blotting with strand-specific probes. A probe is chosen which spans the RNA terminus or splice point of interest but which overlaps that point by at least 50 nt. The shorter the length of the hybridizing DNA, the greater the accuracy with which the RNA can be mapped.

When mapping a 5′ terminus, or an intron upstream of the labelling point, a 5′-labelled DNA fragment is used. Cleaner results will be obtained if an enzyme which leaves a projecting terminus after cleavage is used, because the label will then be removed by digestion during nuclease treatment. In contrast, when 3′-labelled probes are used, and if strand separation is not perfect, any renatured probe remains labelled after nuclease S1 treatment. Because of this problem, it is important to arrange that the region of complementarity between the DNA and the RNA is sufficiently smaller than the total length of the probe to ensure that the protected fragment is well separated from the probe during analytical gel electrophoresis.

Only in exceptional cases will the position of suitable restriction sites be such that the probe can be made by labelling a total restriction digest of a plasmid clone or subclone. In most cases, where there are multiple restriction sites, this would produce a bewildering array of labelled DNA molecules and greatly complicate the subsequent analysis. Therefore, whether using a double-stranded or single-stranded probe, it usually will be necessary to purify the restriction fragment before labelling. This can be achieved by preparative gel electrophoresis as described in *Protocol 9*.

Protocol 9. Isolation of DNA fragment by gel electrophoresis

Reagents

- 20–50 μg of the relevant cloned DNA
- Appropriate restriction nucleases and buffers
- Analytical polyacrylamide gels and agarose gels as appropriate, together with electrophoresis apparatus
- 0.5 × TBE buffer (45 mM Tris, 45 mM boric acid, 1.25 mM EDTA)
- Thin-layer chromatography plate (Merck 60F254, cat. no. 5554)
- Polymer wool (Interpet Ltd) as a plug in 1 ml disposable pipette tip
- 3 M sodium acetate pH 5.5
- TE buffer (10 mM Tris–HCl pH 8.0, 1 mM EDTA).

Method

1. Digest 20–50 μg of the cloned genomic DNA with the appropriate restriction enzymes. Check that the digestion is complete and the appropriate fragment is well isolated from other fragments on the gel by electrophoresis of an aliquot (0.5 μg) of the restricted DNA on a native *analytical* polyacrylamide gel (see Chapter 6, *Protocol 6* and ref. 8).
2. Run the rest of the restricted DNA on a 3-mm-thick *preparative* polyacrylamide gel, loading 10 μg DNA per lane. Electrophorese in 0.5 × TBE buffer.
3. After electrophoresis, place the gel on to a piece of cling film on a thin-layer chromatography plate. Illuminate the gel with incident UV light (254 nm) in a dark room to locate the DNA fragments. Expose the gel to UV for as short a time as possible since prolonged exposure will cleave the DNA. The DNA fragments will appear dark on a light background.
4. Using a clean razor blade, excise the gel slice containing the desired DNA fragment.

Protocol 9. *Continued*

5. Cut a length of dialysis tubing about 5 cm longer than the gel slice and seal it at one end with a plastic clip. Fill the bag with 0.5 × TBE buffer and place the gel slice in the bag. Gently extrude excess buffer and air bubbles then seal the bag with another clip.
6. Arrange the bag in a flat-bed electrophoresis tank with the slice to the cathodal side of the bag and at 90° to the current flow. Add sufficient 0.5 × TBE buffer to just cover the bag. Electrophorese at 100 V for sufficient time to electrophorese the DNA out of the gel slice (1 h is sufficient for fragments <1 kb long).
7. Reverse the direction of current flow for 30 sec to detach any DNA bound to the dialysis tubing.
8. Remove the liquid from the bag. Filter it through a plug of polymer wool (Interpet) in a 1 ml disposable pipette tip into a siliconized microcentrifuge tube.
9. Add 0.1 volume of 3 M sodium acetate pH 5.5 and then add 2.5 vol. ethanol. Mix well and place in an ethanol–dry ice bath for 10 min. Recover the DNA by centrifugation (12 000 *g*, 10 min).
10. Aspirate off the ethanol using a micropipette and add 200 μl 70% ethanol. Centrifuge at 12 000 *g* for 1 min.
11. Dry the DNA pellet by leaving the tube open for a few minutes. Dissolve the DNA in 20–50 μl of TE buffer.
12. Determine the concentration and purity of the eluted fragment by running an aliquot (~5%) on an analytical polyacrylamide or agarose gel with a known amount of a DNA molecular weight marker in a parallel track.
13. Store the rest of the DNA at −20°C.

5.2.2 Probe labelling

For mapping the 5′ terminus of an RNA or an acceptor/intron boundary (i.e. the 3′ end of an intron), the probe is labelled at the 5′ terminus (*Protocol 10*) with [γ-^{32}P]ATP by sequential reaction with calf intestine alkaline phosphatase (CIP) and T4 polynucleotide kinase. For mapping the 3′ end of an RNA or an intron/donor boundary (the 5′ end of an intron), the probe is labelled at the 3′ terminus by 'filling in' using the Klenow fragment of *E. coli* DNA polymerase I (*Protocol 11*). If maximal sensitivity is to be achieved the specific activity of the probe should be as high as possible (i.e. the labelled nucleotides should be used at as high a specific activity as possible).

Protocol 10. 5′ end labelling of DNA using alkaline phosphatase and T4 polynucleotide kinase

Reagents

- 10 × CIP buffer (0.1 M Tris–HCl pH 9.5, 10 mM spermidine, 1.0 mM EDTA)
- Calf intestinal alkaline phosphatase (CIP; Boehringer 108138)
- DNA fragment (1 μg in 15 μl of 1 × CIP buffer)
- 10 × CIP/kinase buffer [0.5 M Tris-HCl pH 7.5, 0.1 M $MgCl_2$, 50 mM dithiothreitol (DTT)]
- T4 polynucleotide kinase
- [γ-^{32}P]ATP (Amersham, >5000 Ci/mmol, 1 mCi/ml)
- TE buffer (10 mM Tris–HCl pH 8.0, 1 mM EDTA)
- Sephadex G50 spun columns (see *Protocol 3*)
- Reagents for phenol/chloroform extraction:
  Ultrapure phenol (BRL) equilibrated with 0.1 M Tris–HCl pH 8.0 according to the supplier's instructions
  Chloroform:isoamyl alcohol, 50:1.

Method

1. To 1 μg DNA fragment in 15 μl of 1 × CIP buffer, add 1 μl CIP (1 U). Incubate 1 h at 37°C.
2. **Either** incubate for 1 h at 70°C to inactivate the CIP **or** perform two phenol extractions and then one chloroform extraction (see *Protocol 16*, step 2) followed by a spun G50 column (see *Protocol 3*).
3. Heat at 100°C for 3 min to denature the DNA and then place immediately on ice.
4. Add 2 μl [γ-^{32}P]ATP, one tenth volume 10 × CIP/kinase buffer and 1 μl T4 polynucleotide kinase (PL-Biochemicals).
5. Incubate for 1 h at 37°C.
6. Adjust the volume to approx. 100 μl by the addition of TE buffer.
7. Extract the mixture by vortexing with 100 μl of phenol. Centrifuge briefly in a microcentrifuge to separate the phases, then remove and discard the organic phase. Repeat this phenol extraction on the aqueous phase.
8. Extract the aqueous layer by vortexing with 100 μl of chloroform. Centrifuge briefly as above to separate the phases then transfer the aqueous phase to a siliconized microcentrifuge tube.
9. Spin through a Sephadex G50 column in TE buffer (see *Protocol 3*).

Protocol 11. 3′ end-labelling of DNA using the Klenow fragment of DNA polymerase I (for 5′ protruding ends)

Reagents

- 10 × Klenow buffer stock (0.1 M Tris–HCl pH 7.5, 0.1 M $MgCl_2$, 0.5 M NaCl, 0.1 M 2-mercaptoethanol). Dilute 10-fold to give 1 × Klenow buffer for use
- DNA fragment to be labelled (0.1 μg/μl in 1 × Klenow buffer)
- [α-^{32}P]dNTP (>3000 Ci/mmol) e.g. [α-^{32}P]dCTP
- 1 mM dNTP solutions for all four dNTPs
- *E. coli* DNA polymerase I, Klenow fragment
- Sephadex G50 spun columns (see *Protocol 3*)
- Reagents for phenol/chloroform extraction (see *Protocol 10*).

Method

1. To 1–5 μg DNA in 20–50 μl of 1 × Klenow buffer,[a] add:

three unlabelled dNTPs (e.g. dATP, dGTP, and TTP)	to 0.1 mM final concentration each
[α-^{32}P] dNTP, e.g. [α-^{32}P]dCTP	20–50 μCi
E. coli DNA polymerase I (Klenow fragment)	1 unit/μg DNA.

2. Incubate for 15 min at room temperature.
3. If the labelled nucleotide is expected to be the 3′ terminal residue[b] after the reaction, proceed to step 5.
4. If the labelled triphosphate is not the 3′ terminal residue,[b] add unlabelled dNTP (of the nucleotide used as the label, e.g. dCTP) to 0.1 M and perform a 10 min 'chase' at room temperature.
5. Purify the labelled DNA by phenol and chloroform extraction (*Protocol 10*, steps 8 and 9) and G50 spun column chromatography (*Protocol 3*).

[a] The Klenow enzyme will also work in most restriction enzyme buffers.

[b] It is important to perform a chase when the label is internal to the 3′ end to prevent formation of a 'ragged' end which will give poor sequence data. On the other hand it is disastrous to perform a chase if the labelled nucleotide is the most 3′ nucleotide since the 3′ to 5′ exonuclease activity of the Klenow fragment will act to exchange the terminal nucleotide with its unlabelled counterpart and so reduce the labelling of the probe dramatically.

5.2.3 Strand separation

Strand separation is achieved by denaturing the DNA and resolving the two strands by gel electrophoresis under non-denaturing conditions using a vertical polyacrylamide slab gel. This procedure is normally successful but it relies on there being a sufficient difference in electrophoretic mobility for the two strands to be recovered separately from the gel. The difference in electrophoretic mobility results from the different secondary structure of the two strands. Hence it is important not to overheat the gel by running at too high a current as this will denature the DNA. *Protocol 12* describes a method that we have used to separate the strands of DNA fragments of between 130 and 600 bp in length. The two rinses of the DNA probe with 70% ethanol (step 3) prior to denaturation are designed to remove all traces of salt which would otherwise favour re-annealing. The gel contains a very low proportion of bis-acrylamide since this has been found empirically to increase resolution of the two strands.

Protocol 12. Separation of DNA strands by gel electrophoresis[a]

Reagents

- Radiolabelled DNA probe (see *Protocols 10* and *11*)
- 0.5 × TBE buffer (45 mM Tris, 45 mM boric acid, 1.25 mM EDTA)
- Appropriate polyacrylamide gel (40 cm long) for strand separation plus electrophoresis apparatus. (For DNA fragments of >200 bp use a 5% acrylamide–0.1% bis-acrylamide gel in 0.5 × TBE buffer. For DNA fragments of <200 bp use a 8% acrylamide–0.24% bisacrylamide gel in 0.5 × TBE buffer)
- 3 M sodium acetate pH 5.5
- Denaturation buffer [30% (w/v) dimethyl sulphoxide, 1 mM EDTA, 0.05% xylene cyanol].

Method

1. Pre-electrophorese the strand-separation gel at 5–10 V/cm for at least 1 h before use.
2. Meanwhile, recover the labelled probe in a siliconized tube by ethanol precipitation (see *Protocol 9*, step 9).
3. Rinse the DNA pellet *twice* with 70% ethanol and dry it under vacuum.
4. Dissolve the pellet in 10 μl denaturation buffer.
5. Heat the DNA mixture at 90°C for 3 min. Then chill the mixture quickly by plunging it into an ice/salt/water bath.
6. Load the DNA immediately on to the pre-electrophoresed strand-separation gel (step 1).
7. Continue electrophoresis at the same voltage (5–10 V/cm) for a time sufficient to run the double-stranded form of the DNA to the bottom of the gel. The xylene cyanol will co-electrophorese with a double-stranded DNA fragment of 260 bp in 5% gels and 160 bp in 8% gels.
8. Stop the electrophoresis and dismantle the apparatus leaving the gel adhering to one plate.
9. Cover the gel with cling film and make an asymmetric pattern of radioactive marks around the gel by sticking pieces of tape marked with ^{32}P on to the cling film. Expose to X-ray film for 5–30 min.
10. Use the autoradiogram to locate the labelled bands and cut out the slower-moving band or band and electroelute the DNA as in *Protocol 9*, steps 4–8.

Protocol 12. *Continued*

11. Recover the DNA by ethanol precipitation (see *Protocol 9*, step 9). *Note*: since the DNA is now single-stranded it is essential to use siliconized tubes to achieve good recovery of the DNA.

12. Dissolve the DNA in 10–50 μl sterile water at a concentration of 1 fmol/μl (150 pg of a 500 nucleotide fragment is 1 fmol).

[a] From ref. 19.

5.3 Preparation of single-stranded probes using single-stranded M13 DNA or 'phagemid' DNA

An alternative method of generating a single-stranded probe is to clone the restriction fragment to be used as a probe into the multiple cloning site of a suitable M13 DNA or 'phagemid' vector in an orientation such that single-stranded DNA present within the clone corresponds to the coding strand of the gene or mRNA under study. The single-stranded recombinant DNA then serves as the template for the synthesis of labelled DNA using the Klenow fragment of DNA polymerase I. These probes can be labelled at the 5′ end by using a polynucleotide kinase labelled oligonucleotide as the primer, or they can be uniformly labelled by incorporation of [α-^{32}P]dNTP. In both cases the probe is purified by cleaving the product of DNA synthesis with a restriction enzyme cutting distal to the cloned fragment and isolating the labelled single-stranded probe after electrophoresis in a denaturing polyacrylamide gel. For 5′ end-labelled probes, an excess of dNTPs can be added to the synthesis reaction since the specific activity of the final probe is determined by that of the primer. In the case of uniformly labelled probes, the specific activity of the final probe is determined by the specific activity of the [α-^{32}P]dNTPs in the reaction. If using labelled dCTP for example, the lower the amount of unlabelled dCTP that is added, the more radioactive the probe, but sufficient dNTPs must be present to ensure synthesis proceeds as far as the distal restriction site. A compromise can be achieved by adding a 5–10-fold excess of dCTP over what is needed to synthesize the probe, and including a brief chase with unlabelled dCTP.

It is important to use uniformly labelled probes as quickly as possible after synthesis (hours rather than days!) as they start to break down, owing to strand scission resulting from radioactive decay.

The primer used for 5′ end-labelled probes must be internal to the mRNA being mapped, while for uniformly labelled probes the primer may be internal or external to the sequence complementary to the mRNA. It is often convenient to use a 'universal' sequencing primer for this purpose (i.e. one outside the multicloning site). *Protocol 13* describes the preparation of both uniformly labelled and 5′ end-labelled probes using these techniques.

Protocol 13. Preparation of single-stranded probes using single-stranded M13 or 'phagemid' DNA

Reagents

- 0.1–1 μg/μl single-stranded DNA template
- 10 × Klenow buffer (see *Protocol 11*)
- Klenow fragment of DNA polymerase I (5 U/μl)
- Sephadex G50 spun columns (see *Protocol 3*)
- Appropriate restriction enzymes
- Formamide loading buffer [80% formamide, 0.5 × TBE buffer (see *Protocol 9*), 0.02% xylene cyanol, 0.1% bromophenol blue]
- TE buffer (10 mM Tris–HCl pH 8.0, 1 mM EDTA)
- Denaturing 5% polyacrylamide gel (see Chapter 6, *Protocol 12*)
- 3 M sodium acetate pH 5.5.

The following reagents are also required, depending on whether uniformly labelled or 5′ end-labelled probes are to be prepared.

For uniformly-labelled probes

- 10 ng/μl unlabelled oligonucleotide primer
- 10 mM dNTP-C (a mixture of dATP, dGTP, and dTTP each at 10 mM)
- 0.1 mM and 10 mM dCTP solutions
- [α-^{32}P]dCTP (>3000 Ci/mmol).

For 5′ end-labelled probes

- 10 ng/μl kinase-labelled oligonucleotide primer (see *Protocol 20*)
- 10 mM dNTP mixture (containing all four dNTPs, each at 10 mM).

A. *Preparation of uniformly labelled probe*

1. Mix together:

single-stranded DNA template	0.5 μg
10 × Klenow buffer	1 μl
unlabelled primer	2.5 pmol (=15 ng of a 20mer).

2. Heat at 65 °C for 5 min and allow to cool slowly to room temperature.

3. Add:

10 × Klenow buffer	1 μl
10 mM dNTP-C solution	1 μl
100 μM dCTP	1 μl
[α-^{32}P]dCTP	2 μl
H_2O	4 μl
Klenow enzyme	1 μl (5 units).

4. Incubate at room temperature for 30 min.

5. Add 1 μl 10 mM dCTP and incubate at room temperature for 5 min.

6. Heat at 70 °C for 10 min.

7. Add 10 units of the appropriate restriction enzyme and adjust the salt concentration of the reaction mixture if necessary.[a]

8. Incubate at 37 °C or other appropriate temperature for 1 h.

Protocol 13. *Continued*

9. Remove unincorporated nucleotides by applying the reaction mixture to a Sephadex G50 spun column (*Protocol 3*).
10. Recover the labelled DNA by ethanol precipitation (*Protocol 9*, step 9). Rinse the pellet with 70% ethanol and dissolve it in 5 μl TE buffer.
11. Add 5 μl formamide loading buffer. Heat the DNA at 90°C for 5 min and apply it to a denaturing 5% polyacrylamide gel (3–4 slots of a standard DNA sequencing gel).
12. Electrophorese until the probe is well separated from the template DNA (only 1–2 h at 1000 V will be required as the template DNA is 4–8 kb in length and therefore will migrate slowly).
13. Localize the labelled DNA fragment and recover the probe as described in *Protocol 12*, steps 8–11.

B. *Preparation of 5′ end-labelled probe*

The protocol is as above except:

- In step 1 use kinase-labelled primer.
- In step 3, add 1 μl of a 10 mM solution of dNTP mixture in place of the dCTP and the dNTP-C, and do not add either the 0.1 mM dCTP or the [α-^{32}P]dCTP.
- Omit step 5.

[a] Many restriction enzymes will work in this buffer, but if necessary the NaCl concentration can be increased with 1 M NaCl, or decreased by dilution. If using an enzyme such as *Sma*I which does not work in the presence of NaCl it will be necessary to purify the DNA using a Sephadex G50 spun column (*Protocol 3*) before the digestion step (step 7).

5.4 Hybridization and digestion with nuclease S1 using a single-stranded DNA probe

During the hybridization (*Protocol 14*) the probe should be in molar excess over the complementary RNA species. A simple way to check this is to perform parallel hybridizations with varying amounts of RNA or probe. If the probe is present in a sufficient excess, the signal obtained will be proportional to the RNA concentration but independent of the probe concentration. In addition, the probe should be at a concentration sufficient to drive the reaction to completion. Under the conditions given, 2 fmol of probe DNA in a reaction volume of 10 μl will drive the hybridization to completion in 6 h. We normally use 2–10 fmol of probe at a specific activity of 5×10^6 d.p.m./pmol with 5- or 10-fold excess of DNA over complementary RNA.

Protocol 14. Hybridization and digestion with Nuclease S1

Reagents

- Labelled single-stranded DNA probe (see *Protocols 10–12*)
- RNA samples
- 5 × hybridization buffer [2 M NaCl, 50 mM Pipes, pH 6.4 with NaOH]
- 10 × S1 digestion buffer (2 M NaCl, 20 mM $ZnSO_4$, 0.3 M sodium acetate pH 4.6). Store at −20°C
- nuclease S1 at 1000 U/μl from BRL
- 0.5 × TBE buffer (45 mM Tris, 45 mM boric acid, 1.25 mM EDTA)
- 3 M sodium acetate pH 5.5
- 1 mg/ml tRNA (prepare a 1 mg/ml solution in water after phenol/chloroform extraction and ethanol precipitation of yeast tRNA purchased from Sigma)
- Formamide dye mixture (80% formamide, 0.5 × TBE buffer, 0.02% xylene cyanol, 0.1% bromophenol blue)
- Glass microcapillaries (20–25 μl capacity)
- Denaturing (urea) 5% or 8% polyacrylamide gel (see Chapter 6, *Protocol 12*)
- Radioactively-labelled DNA size markers (e.g. a restriction digest of a common vector labelled as described in *Protocol 11*).

Method

1. Prepare the hybridization mixture in a siliconized microcentrifuge tube as follows:

   DNA probe: 2.5 μl containing 2.5 fmol of labelled probe (i.e. 400 pg of a 500 nucleotide fragment)

   RNA: up to 5.5 μl in sterile water containing 0.25–0.5 fmol of RNA complementary to the probe (i.e. 0.1–0.2 ng of a 1 kb mRNA)[a]

   2 μl of 5 × hybridization buffer

   water to 10 μl final volume.

2. Draw the mixture into a 20 μl or 25 μl glass microcapillary by touching one end of the capillary into the solution. Carefully invert the microcentrifuge tube while holding the capillary in the tube and allow the solution to move to the centre of the capillary.
3. Seal both ends of the capillary in a Bunsen flame and attach a piece of water-resistant tape labelled with water-resistant marker ink.
4. Incubate at 60–65°C[b] for at least 6 h by immersing the capillaries in a waterbath (an overnight incubation is permissible).
5. Just before use (in step 7 below), prepare 190 μl of nuclease S1 mixture for each hybridization reaction to be analysed. To do this, mix:

10 × S1 digestion buffer	20 μl
nuclease S1	0.1–0.5 μl (100–500 U)[c]
H_2O	170 μl.

6. Rinse and wipe the outside of each capillary tube. Snip off 0.5 cm at each end using a glass cutter.

Protocol 14. *Continued*

7. Expel the contents of each tube into 190 μl nuclease S1 mixture (prepared in step 5) in a siliconized 1.5 ml microcentrifuge tube. Incubate at 37 °C for 30 min.[c]
8. Add 5 μl tRNA (1 mg/ml) and recover the DNA by ethanol precipitation (*Protocol 9*, step 9).
9. Dry the DNA pellet under vacuum and dissolve it in 10 μl formamide dye mixture.
10. Heat at 90 °C for 2 min. Chill on ice.
11. Analyse on a denaturing 5% or 8% polyacrylamide gel containing 8 M urea and run in 0.5 × TBE buffer (see Chapter 6, *Protocol 12*). Also co-electrophorese radioactively labelled DNA size markers in parallel lanes.

[a] When the proportion of the complementary RNA in the total RNA population is not known, a range of RNA concentrations should be used.
[b] For a hybrid which is very short, or which has a high content of AT base pairs, it may be necessary to reduce the annealing temperature.
[c] The actual nuclease S1 concentration and time and temperature of incubation should be determined empirically for each probe.

5.5 Hybridization and digestion with nuclease S1 using a double-stranded probe

With some DNA fragments it is not possible to separate the strands by the simple procedure given in section 5.3.3. In this case, S1 mapping must be carried out using a double-stranded probe. In order to prevent the DNA probe re-annealing, the hybridization is performed in a high formamide concentration (*Protocol 15*). Under these conditions, DNA–RNA hybrids are more stable than DNA–DNA duplexes. Hence it is possible to select a hybridization temperature that allows hybridization of the probe to the RNA but which prevents re-annealing of the probe.

The optimum hybridization temperature depends on the length of the hybrid, its GC content and the primary sequence. Therefore when using a probe for the first time, it is necessary to try a range of temperatures around the expected optimum (see *Protocol 15*, footnote *b*). The precise conditions of hybridization (i.e. salt and formamide concentrations) are crucial. To prevent any change in these by evaporation, the hybridization is carried out in sealed capillaries. Rapid transfer after denaturation and rapid freezing after hybridization ensure that the reaction is not subjected to temperatures below the optimum, when probe re-annealing would be favoured.

Protocol 15. High formamide hybridization for double-stranded probes

Reagents

- Labelled double-stranded DNA probe (*Protocols 10* and *11*)
- RNA samples
- Formamide hybridization buffer (80% deionized formamide,[a] 0.4 M NaCl, 1 mM EDTA, 50 mM Pipes, pH 6.4)
- 10 × S1 digestion buffer (see *Protocol 14*)
- Nuclease S1 at 1000 U/μl (from BRL)
- 1 mg/ml tRNA
- Formamide dye mixture (see *Protocol 14*)
- Glass microcapillaries (20–25 μl capacity)
- Denaturing (urea) 5% or 8% polyacryalmide gel (see Chapter 6, *Protocol 12*)
- 0.5 × TBE buffer (45 mM Tris, 45 mM boric acid, 1.25 mM EDTA)
- Radioactively-labelled DNA size markers (see *Protocol 14*)
- 3 M sodium acetate pH 5.5.

Method

1. Into a siliconized microcentrifuge tube, place the radiolabelled double-stranded probe and the RNA in the amounts given in *Protocol 14*, step 1.
2. Co-precipitate them with ethanol (*Protocol 9*, step 9), wash the pellet with 70% ethanol and allow it to air dry.
3. Resuspend the pellet in 10 μl formamide hybridization buffer. Vigorous vortexing may be needed to dissolve the nucleic acids completely. Their dissolution can be checked using a Geiger counter.
4. Draw the solution up into a 20 μl or 25 μl glass capillary. Seal it and mark it as described in *Protocol 14* (step 3).
5. Immerse the capillaries in a 75°C water bath for 5 min to denature the DNA.
6. Transfer the capillaries rapidly to a water bath set at the hybridization temperature[b] and incubate for at least 6 h.
7. Snap freeze the reactions by transferring the capillaries rapidly to a tray of dry ice.
8. Remove the capillaries one by one from the dry ice. Snip off 0.5 cm at each end, wipe the outside and expel the contents into 190 μl of ice-cold nuclease S1 digestion mixture (prepared just before use as described in *Protocol 14*, steps 5–7). Vortex each tube and stand them in ice until the tubes are ready.
9. The rest of the procedure is identical to that described in *Protocol 14* (steps 8–11) for single-stranded probes.

[a] See *Protocol 1* for deionization of formamide.

[b] The optimal hybridization temperature varies with the base composition and length of the hybrid (for hybrids of <200 nucleotides). To determine the optimum, in an initial experiment vary the temperature in 2°C steps around the expected optimum of 50°C for a hybrid containing 50% GC base pairs.

5.6 Interpretation of results and potential problems

The size of any labelled bands in the gel is determined by comparison with the molecular weight markers run on the same gel. This is a measure of the distance between the labelled end of the probe and the terminus or splice junction of the RNA. Providing the hybridization was performed in probe excess, the intensity of the band will be proportional to the concentration of the complementary RNA species in the RNA population. Thus the technique can also be used to compare the concentrations of individual RNA species in different populations.

If the probe strands were not completely separated, or if the high formamide conditions were not sufficiently stringent to prevent probe re-annealing, some DNA–DNA duplexes will be present and these will be resistant to nuclease S1. Whether they will be detected on the final autoradiograph depends on the conditions of the experiment. It is not a problem for 5′ end-labelled probes, made by labelling a restriction site that leaves a 5′ protruding end using polynucleotide kinase, because nuclease S1 degrades the protruding nucleotides and thus removes the label from any re-annealed probe. For 3′-labelled probes, however, the label is not removed by the nuclease and the re-annealed probe can be seen on the final autoradiograph (see *Figure 6*).

Another problem to be aware of is that the nuclease S1 may cleave a double-stranded probe at AT-rich regions causing spurious bands lower down the gel. This problem can be diagnosed by running a 'minus RNA' control. If this is clean, and there is a band of the size of the original probe, then the probe has been completely protected by the RNA and therefore does not span the terminus or splice junction. A smear low down in the gel indicates that the RNA was degraded.

Sometimes, multiple bands are seen. Although this may be due to heterogeneity at the 5′ or 3′ ends of the RNA, it is more likely to be due to exonucleolytic cleavage of the DNA–RNA hybrid by nuclease S1 (often called 'nibbling'). Thus it is often necessary to perform the nuclease S1 digestion using different amounts of enzyme and different temperatures of incubation to determine the likely end point.

The experiment shown in *Figure 6* illustrates the 'nibbling' problem. The sequence in the area of the gene near the polyadenylation site for this mRNA contains a high proportion of AT base pairs. This leads to 'breathing' of the hybrid at this point and consequent 'nibbling' by nuclease S1. In an attempt to control for this, the nuclease S1 digestion (*Figure 6*) was performed under

---

Figure 6. Identification of the 3′ terminus of an mRNA by nuclease S1 mapping; the effect of varying digestion conditions. In this experiment a 316 nt fragment from the β1 globin gene of *Xenopus laevis* was hybridized to erythrocyte poly(A)$^+$ RNA. The probe was a 3′ end-labelled restriction fragment prepared by digestion with *Hin*dIII and *Hin*fI followed by strand separation on a native gel. There was a small amount of cross-contamination of the

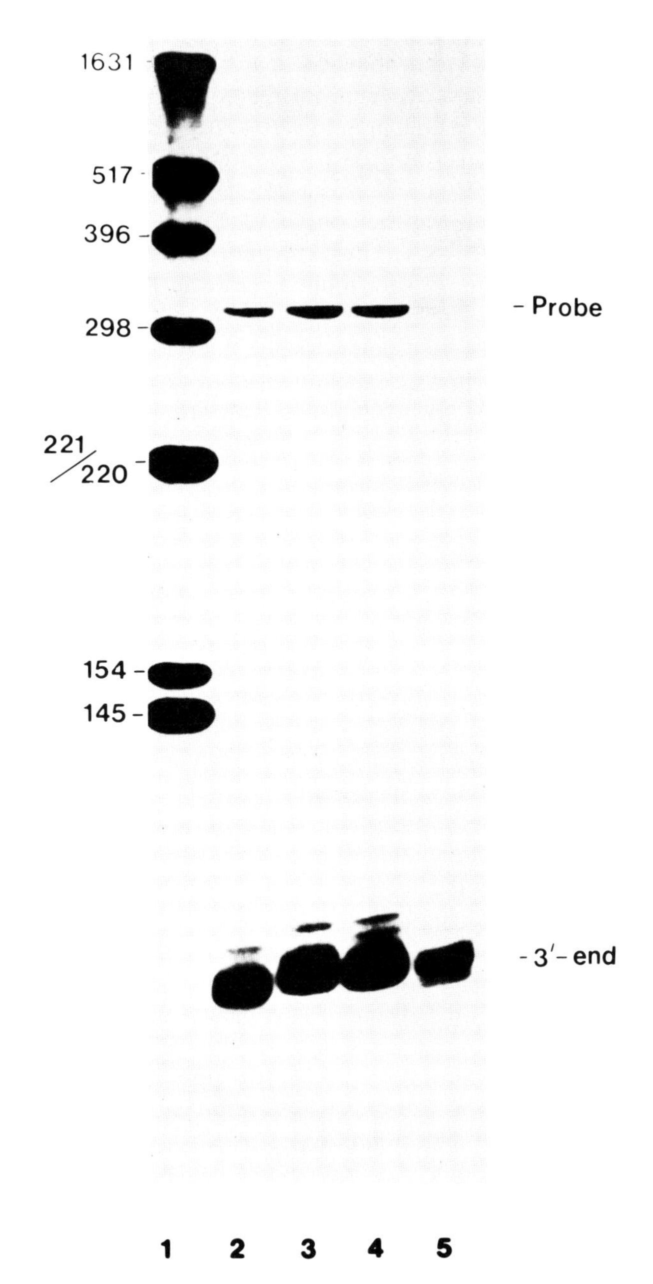

two strands which accounts for the presence of some renatured probe. Because a poly(A)$^+$ mRNA was analysed, the length of the protected fragment indicates the position of polyadenylation. Nuclease S1 digestion was performed at 37 °C (lane 2), 20 °C (lane 3), 30 °C (lane 4), and 42 °C (lane 5). Lane 1 was a *Hin*fI digest of pAT153 run as a size marker. From ref. 25.

the standard (37°C) conditions and at 20°C, 30°C, and 42°C. At the two lower temperatures, a higher proportion of the probe yields protected fragments which are *longer* than expected for the correct 3′ end, indicating that digestion by nuclease S1 is incomplete. At 42°C the proportion of longer transcripts is greatly reduced but the proportion of a fragment *smaller* than expected for the correct 3′ end is increased. At all temperatures, however, there are multiple fragments in the approximate position expected for DNA protected by the poly(A)$^+$ RNA. Thus, changing the temperature of nuclease S1 digestion has not reduced the 'nibbling' effect. This experiment nicely illustrates the limitation of nuclease S1 mapping, that is, it is normally very difficult to define a precise 'end point' for the nuclease S1 digestion. Almost inevitably, therefore, the precise position of an interruption or terminus can be defined only to within a few nucleotides. Thus if it is important to define a 5′ terminus precisely for an RNA, both nuclease S1 mapping and primer extension should be used since the artefacts observed with the two methods are quite different.

6. RNase mapping

6.1 An overview

The principle of RNase mapping (20, 21) is shown in *Figure 7*. In conceptual terms it is analogous to nuclease S1 mapping. The probe is a uniformly labelled, anti-sense, [^{32}P]RNA molecule that is complementary to the target RNA over only a portion of its length. Test RNA is then hybridized in solution to an excess of this [^{32}P]RNA probe in order to produce hybrid RNA duplexes, formed by the annealing of the probe RNA to its complementary (or 'target') RNA. After the annealing reaction is completed, the reaction mixture is treated with a combination of RNase A and RNase T1. These enzymes degrade single-stranded RNA with different sequence specificities (RNase A attacks pyrimidine nucleotides whereas RNase T1 attacks guanosine nucleotides) but leave double-stranded RNA intact. Thus, RNase treatment results in the degradation of unannealed (single-stranded) [^{32}P]RNA probe and unannealed (single-stranded) test RNA. In contrast, RNA–RNA duplexes, formed by the annealing of the [^{32}P]RNA probe and the target RNA with which it shares complementary homology, are resistant to RNase digestion in their double-stranded regions. The non-complementary regions of the hybridized [^{32}P]RNA–RNA duplexes remain single-stranded and are therefore unprotected from RNase digestion. The size of the remaining 'trimmed' duplexes reflects the length over which the probe and target RNA are completely homologous and may be determined by fractionation of the RNase-protected products on denaturing polyacrylamide gels. Depending on the probe used, the 5′ or 3′ termini, or the splice junctions, of the target RNA molecule may then be inferred.

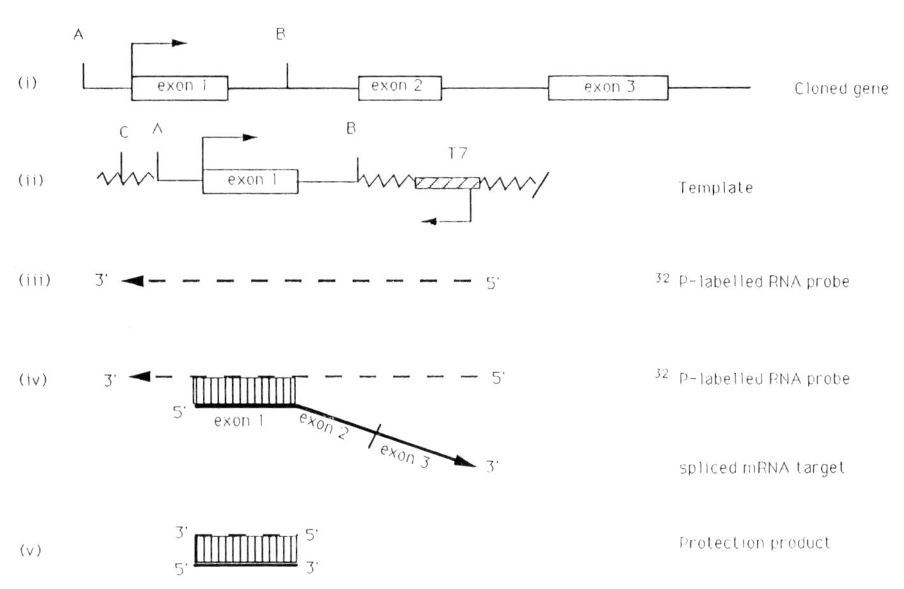

Figure 7. Schematic representation of RNase mapping. (i) From the cloned gene of interest, isolate the restriction fragment (AB) from which the probe is to be generated. (ii) Subclone the fragment into a riboprobe (or riboprobe-type) vector and linearize the template at restriction site A or at a suitable site (C) in the vector. (iii) Synthesize a run-off, anti-sense, ^{32}P-labelled probe, using an RNA polymerase appropriate to the vector (e.g. T7 RNA polymerase). (iv) Anneal the probe and test RNA to produce [^{32}P]RNA:RNA duplexes. (v) Trim the duplexes with RNase to reveal the protection products and analyse these by gel electrophoresis.

6.2 Subcloning a probe

Probes for RNase protection are prepared by *in vitro* transcription of cloned DNA templates. It is therefore necessary to subclone a DNA fragment, spanning the region from which the RNA to be probed is derived, into a plasmid vector which contains promoters for the viral polymerases (22, 23) used to produce the *in vitro* transcripts. Three polymerases are routinely used: T7, T3, and SP6. A number of vectors containing one or more of the promoters specific for these polymerases juxtaposed to multiple cloning sites are commercially available.

When preparing such a probe subclone, the following considerations should be borne in mind.

(a) The insert must be subcloned in an orientation such that transcription from the chosen promoter generates an anti-sense transcript, i.e. one that is complementary to the target RNA and can therefore anneal to it.

(b) The *in vitro* transcripts produced for RNase mapping are generally of the 'run-off' type where the 3′ end of the *in vitro* transcript is defined by cleavage of the DNA template with a suitable restriction endonuclease.

Thus when designing a subclone it is important to make sure that a suitably positioned restriction site is available in either the insert or the multiple cloning site. Note also that it has been reported that the ends produced by restriction enzymes which produce 3′ overhangs can serve as initiators of transcription for the viral polymerases routinely used to generate probes (24). End-initiated transcripts of this sort are undesirable as they would interfere with the RNase protection analysis. It is therefore preferable to avoid using restriction endonucleases of this type if at all possible. If this is unavoidable, the 3′ overhang may be converted to a blunt end using the 3′ to 5′ exonuclease activity of the Klenow fragment of DNA polymerase I (Chapter 2, *Protocol 8*).

(c) Attention should be paid to the length of both the probe to be synthesized, and the predicted protected product. Ideally the protected products should be less than 400 nt in length so as to allow their resolution on denaturing polyacrylamide gels. Probes should be designed such that the difference in size between the full length probe and the trimmed protection products can be resolved easily on denaturing polyacrylamide gels.

6.3 Probe preparation

Procedures for preparing a DNA template from the probe subclone and *in vitro* transcription to yield a suitable [^{32}P]RNA probe are described in *Protocols 16* and *17* respectively.

Protocol 16. Preparation of the DNA template for transcription

Equipment and reagents

- RNase-free microcentrifuge tubes (i.e. a stock of tubes kept separately and only handled with gloved hands)
- 10–20 μg probe sub-clone DNA
- Appropriate restriction nucleases and buffer
- 3 M sodium acetate pH 5.2
- Reagents for phenol/chloroform extraction
- Reagents and apparatus for agarose gel electrophoresis.

Method

1. Define the 3′ end of the probe to be made by restricting 10–20 μg of the template (probe subclone) DNA[a] with a suitable restriction endonuclease in a final volume of 50 μl.
2. After digestion add 5 μl 3 M sodium acetate pH 5.2 and extract by vortexing with 50 μl phenol. Centrifuge briefly in a microcentrifuge to separate the phases, then remove and discard the organic phase.
3. Extract the aqueous layer by vortexing with 50 μl chloroform. Centrifuge briefly as above to separate the phases, then transfer the aqueous layer to an RNase-free, siliconized microcentrifuge tube.
4. Add 2.5 vol. (125 μl) of absolute ethanol, and mix thoroughly.

5. Precipitate the cleaved DNA template by flash freezing the tube in liquid nitrogen. Thaw the tube and centrifuge for 5 min in a microcentrifuge. Remove and discard the supernatant.
6. Wash the DNA pellet by vortexing in 250 μl ice-cold 70% ethanol. Centrifuge for 5 min in a microcentrifuge to re-pellet the DNA. Remove and discard the supernatant.
7. Dry the pellet under vacuum, and resuspend it in RNase-free water at a concentration of approximately 1 μg/μl. Aliquot 1 μl portions into RNase-free, siliconized microcentrifuge tubes and store at −20°C.
8. Check that the cleavage reaction has gone to completion by subjecting one aliquot to analytical agarose gel electrophoresis.

[a] Although only 0.5–1.0 μg DNA is required for each probe transcription, it is normally more convenient to prepare considerably more cleaved template and store it as 1.0 μg aliquots for future use. Most procedures for plasmid DNA preparation produce templates of adequate quality for *in vitro* transcription.

Protocol 17. *In vitro* transcription of probes

Reagents

- Cleaved template DNA (see *Protocol 16*)
- 5 × transcription buffer (0.2 M Tris–HCl pH 7.5, 30 mM $MgCl_2$, 10 mM spermidine)
- 0.1 M DTT
- Separate 10 mM stocks of rATP, rCTP, and rGTP at pH 7.0 (Promega)
- [α-^{32}P]rUTP (3000 Ci/mmol; 10 mCi/ml)
- SP6, T7, or T3 RNA polymerases (15 U/μl; Promega, New England Biolabs) as appropriate for the cloning vector used
- Ribonuclease inhibitor (10 U/μl RNasin, Promega)
- RNase-free DNase (RQ1 DNase, Promega)
- 25 μg/μl yeast tRNA (Boehringer Mannheim)
- 10 M ammonium acetate
- Phenol/chloroform/isoamyl alcohol (25:24:1)
- Chloroform
- Hybridization buffer (80% deionized formamide,[a] 0.4 M NaCl, 40 mM Pipes pH 6.4, 1 mM EDTA).

Method

1. Thaw a single 1 μl aliquot of cleaved template DNA and then add the following reagents in order and at room temperature.[b]

5 × transcription buffer	2.5 μl
100 mM DTT	1.0 μl
RNase inhibitor	1.0 μl (10 U)
10 mM rATP	0.5 μl
10 mM rGTP	0.5 μl
10 mM rCTP	0.5 μl
[α-^{32}P]rUTP (10 mCi/ml)	5.0 μl
RNA polymerase (SP6, T7, or T3)	1.5 μl (10 U).

Protocol 17. *Continued*

Mix gently, then centrifuge briefly in a microcentrifuge to collect the contents to the bottom of the tube. Incubate at 37°C (for T7 or T3 RNA polymerase) or 40°C (for SP6 RNA polymerase) for 90 min.[c]

2. After the incubation, centrifuge briefly to collect the contents to the bottom of the tube. Add a further 1 μl (10 U) RNase inhibitor and then 1 μl (1 U) RNase-free DNase to destroy the template DNA. Mix gently, centrifuge briefly, and then incubate at 37°C for 15 min.

3. Next, add:

DEPC-treated water	64.5 μl
yeast tRNA (25 μg/μl)	1 μl
10 M ammonium acetate	20 μl.

Mix by vortexing and then centrifuge briefly.

4. Extract the mixture by vortexing with 200 μl phenol/chloroform/isoamyl alcohol (25:24:1). Centrifuge briefly to separate the phases, then remove and discard the organic phase.

5. Extract the aqueous phase by vortexing with 100 μl chloroform. Centrifuge briefly to separate the phases, then remove and discard the organic phase. The following steps (6–8) serve to remove unincorporated rNTPs by serial ethanol precipitation. An alternative procedure is chromatography using a Sephadex G50 spun column (*Protocol 3*), followed by one round of ethanol precipitation (*Protocol 9*, step 9).

6. Add 2 vol. (200 μl) absolute ethanol, mix by vortexing, and precipitate the RNA by flash freezing in liquid nitrogen. Thaw the tube and centrifuge for 5 min in a microcentrifuge. Remove all the supernatant and resuspend the pellet thoroughly in 80 μl RNase-free water. Centrifuge briefly and add 20 μl 10 M ammonium acetate.

7. Repeat step 6.

8. Add 2 vol. (200 μl) absolute ethanol, mix by vortexing, and precipitate the RNA by flash freezing in liquid nitrogen. Thaw the tube and centrifuge for 5 min in a microcentrifuge. Remove all the supernatant and wash the pellet by vortexing in 300 μl ice-cold 70% ethanol. Centrifuge for 5 min in a microcentrifuge to re-pellet the RNA and then carefully remove and discard the supernatant.

9. Dry the pellet under vacuum and resuspend it thoroughly in 50 μl hybridization buffer. If resuspension is difficult, heat briefly at 40°C, and vortex. Repeat the heat treatment as needed until resuspension is complete.

10. Add 1 μl of the resuspended probe to 1 ml of scintillation fluid and count in a scintillation counter. Adjust the probe concentration to 1×10^5 c.p.m./μl with hybridization buffer.

[a] See *Protocol 1* for preparation of deionized formamide.
[b] These components are mixed at room temperature to avoid the precipitation of template DNA by spermidine which can occur at 0°C.
[c] If the restriction enzyme used to cleave the template (*Protocol 16*) generates a 3′ protruding end, set up the transcription reaction but omit the rNTPs and the RNA polymerase. Add 5–10 units of DNA polymerase I (Klenow fragment) and incubate for 15 min at room temperature (22°C), before adding the rNTPs and RNA polymerase.

6.4 Hybridization and nuclease digestion

Hybridization of test RNA samples with the [^{32}P]RNA probe, followed by digestion with RNases A and T1 then analysis of the RNase-resistant fragments by gel electrophoresis are described in *Protocols 18* and *19*.

Protocol 18. Hybridization

Equipment and reagents

- RNase-free microcentrifuge tubes (see *Protocol 16*)
- Test RNA samples
- [^{23}P]RNA probe (see *Protocol 17*)
- Hybridization buffer (see *Protocol 17*).

Method

1. Thaw the test RNA on ice and transfer a portion to be analysed[a] into a fresh RNase-free microcentrifuge tube. Dry this under vacuum.[b]
2. To the desiccated RNA sample, add 10 μl [^{32}P]RNA probe in hybridization buffer. Resuspend by vortexing, centrifuge briefly, incubate at 37°C for 2–3 min and vortex again. Finally, centrifuge briefly to bring contents to the bottom of the tube. From this point on, handle the tube carefully to avoid disturbing the contents from the bottom.
3. Wrap the lid of the tube with parafilm and place it in a rack in a 90°C oven for 20 min to denature both the probe and target RNAs.
4. Transfer the tube from the 90°C oven or water bath to another oven or water bath at 45°C[c] without letting the tube cool. Incubate at 45°C for at least 8 h. Overnight incubation is permissible and usually more convenient.

[a] The amount of RNA to be analysed can be as little as a few ng or as much as 50 μg. If quantitative results are required, and the approximate proportion of the target RNA in the total test RNA is not known, then a range of RNA concentrations should be used to ensure that hybridization occurs under conditions of probe excess.
[b] If the sample volume is large and requires extended drying time, then it is recommended that drying be conducted at reduced temperature, or that the sample be precipitated with ethanol (using RNase-free carrier tRNA if necessary).
[c] This temperature is suggested as a rough guide and normally constitutes a lower limit. In practice the optimal temperature for any given probe may have to be determined empirically. In choosing a hybridization temperature remember that RNA:RNA duplexes are less thermolabile than their RNA:DNA counterparts.

Protocol 19. Nuclease digestion

Reagents

- RNase digestion buffer (10 mM Tris–HCl pH 7.5, 5 mM EDTA, 0.3 M NaCl)
- RNase A (Sigma type IIIA)
- RNase T1 (Sigma)
- Proteinase K stock at 3.4 μg/μl
- 13% SDS
- Phenol/chloroform/isoamyl alcohol (25:24:1)
- Loading buffer (80% deionized formamide,[a] 10 mM NaOH, 1 mM EDTA, 0.1% xylene cyanol, 0.1% bromophenol blue)
- Denaturing (urea) polyacryalmide gel (Chapter 6, *Protocol 12*) and electrophoresis apparatus.

Method

1. After hybridization (*Protocol 18*) is complete, remove the samples from the incubator and add 100 μl RNase digestion buffer to which RNase A and T1 have been freshly added to final concentrations of 40 μg/ml and 2 μg/ml, respectively. Mix by brief vortexing and centrifuge to collect the contents to the bottom of the tube. Incubate at room temperature for 30–40 min.
2. Stop the digestion by adding 5 μl of 13% SDS and 5 μl of 3.4 μg/μl Proteinase K to each sample. Mix thoroughly, centrifuge briefly to collect the contents and then incubate at 37°C for 20 min.
3. Extract the samples with 190 μl phenol/chloroform/isoamyl alcohol by vortexing. Centrifuge for 2 min in a microcentrifuge. Remove and discard the organic (lower) phase.[b]
4. Add 333 μl of absolute ethanol, mix well, and then precipitate the nucleic acids by flash freezing in liquid nitrogen. Thaw the mixture and then centrifuge for 5 min in a microcentrifuge. Remove and discard the supernatant.
5. Wash the pellet by vortexing in the presence of 0.5 ml of ice-cold 70% ethanol. Centrifuge for 5 min in a microcentrifuge to re-pellet. Remove and discard the supernatant.
6. Dry the pellet under vacuum and resuspend it in 10 μl loading buffer. Samples are often hard to resuspend. If this happens, samples should be repeatedly heated to 37°C and vortexed until they are completely resuspended, since unevenly resuspended samples will block the loading pipette. Resuspended samples should be spun briefly to collect the contents to the bottom of the tube.
7. Denature the samples by heating at 95°C for 2–3 min. Transfer the samples immediately to ice and analyse them without delay by denaturing

polyacrylamide gel electrophoresis (Chapter 6, *Protocol 12*). Co-electrophorese DNA size markers in parallel lines.

[a] See *Protocol 1* for deionization of formamide.
[b] If quantitative data are required, it is important not to remove any of the aqueous phase, even if this necessitates leaving some of the organic phase.

6.5 Interpretation and problems

The labelled protection products are sized by comparison with labelled DNA molecular weight markers run on the same gels (the difference in migration between single-stranded RNA and single-stranded DNA is normally only slight and can in most instances be ignored). As with nuclease S1 mapping (Section 5), if one end point of the homology between the probe and the target RNA is known, then the size of the protection product can be used to determine the distance between it and the unknown end point.

If the hybridization is performed under conditions of probe excess, then the intensity of the bands will be proportional to the concentration of the target RNA in the test RNA sample. That such conditions prevail is most easily checked by performing parallel hybridizations with varying amounts of RNA or probe. If many samples are being compared for relative concentrations of a target RNA, then it is advisable to control for variability in loading by co-probing the samples with a probe for an RNA whose concentration is not expected to vary between samples. The ratio of the intensities of the protection products produced by the two probes provides a measure of the relative concentration of the target RNA in the test sample that is not affected by the absolute amount of RNA in the hybridization reaction or by losses during subsequent manipulations.

RNase mapping is some 25–50-times more sensitive than nuclease S1 mapping with conventional end-labelled probes. This difference in sensitivity presumably results from the higher specific activity of the uniformly labelled probes and the increased thermostability of RNA:RNA versus RNA:DNA duplexes. However, as with most highly sensitive techniques, the major problem encountered in RNase mapping is an unacceptably high level of non-specific or background signals. In many cases this problem is sufficiently severe to prevent identification of the bona fide RNase protected products and it is often necessary to vary the ratio of probe to test RNA, as well as the conditions for hybridization and RNase digestion in order to reduce background and maximize sensitivity. Some of the common reasons for high background associated with the probe synthesis, hybridization, and RNase digestion steps of the RNase mapping procedure are listed below.

Probe problems

If the probe is not full length (i.e. contains molecules that fall short of the terminus of the target RNA being mapped) then these molecules give rise to a ladder of protection products that are smaller than the expected protection product. Probe molecules that are not full length may arise for a number of different reasons including incomplete synthesis, synthesis from a degraded template, degradation of the probe by contaminating RNases in the 'RNase-free' DNase and use of a probe that is too old, that is, one in which ^{32}P decay has caused internal breaks in the RNA chain. If the problem is that the probe is not full length, this is easily diagnosed by assessing the quality of the probe by denaturing polyacrylamide gel electrophoresis.

Hybridization problems

The problem here is the formation of nucleic acid duplexes other than those beween the probe and the target RNA. If the template DNA is not completely degraded by DNase I after the synthesis of the probe, then [^{32}P]RNA:DNA duplexes may form between the probe and complementary fragments of the undigested template DNA. These RNA:DNA hybrids will be resistant to degradation by RNase and will contribute to the background. If the template is completely intact, this will result in full length probe protection. However if the template has been partially degraded by DNase then a ladder of bands will be observed. RNA:DNA hybrids may also occur if the test RNA is contaminated by complementary DNA. This situation often arises in RNase protection analysis of transient transfection or gene expression assays using *in vitro* transcription. Regions of self complementarity within the RNA probe may also anneal resulting in internal [^{32}P]RNA:[^{32}P]RNA duplexes that are resistant to RNase degradation. That the protection products observed are actually dependent on input test RNA may be checked by analysing probe samples which have been hybridized and nuclease digested in the absence of test RNA. Duplexes of this nature are likely to be less thermostable than the bona fide protection products desired. Therefore, increasing the hybridization temperature should help reduce background caused in this way.

RNase digestion problems

Either over-digestion or under-digestion of the hybridization reaction products with RNase may give rise to high backgrounds. In the case of under-digestion, bands resulting from partially digested unhybridized probe fragments as well as partially digested [^{32}P]RNA–RNA duplexes will be seen. In the case of over-digestion, the [^{32}P]RNA–RNA hybrids may be internally degraded by RNase, giving rise to a ladder of bands smaller than the anticipated protection product. Background resulting from over-digestion can sometimes be ameliorated by running native as oppposed to denaturing polyacrylamide gels since over-digested products will tend to fragment less on native gels. However it is normally preferable to experiment with a range of RNase digestion conditions to allow selection of optimal conditions. Altering the time and temperature of the RNase incubation step is more likely to affect the background than altering the RNase concentration to which the digestion appears to be relatively insensitive.

7. Primer extension

7.1 Introduction

The principle of this technique is illustrated in *Figure 8*. It is the direct converse of nuclease mapping. A radioactively labelled probe derived entirely

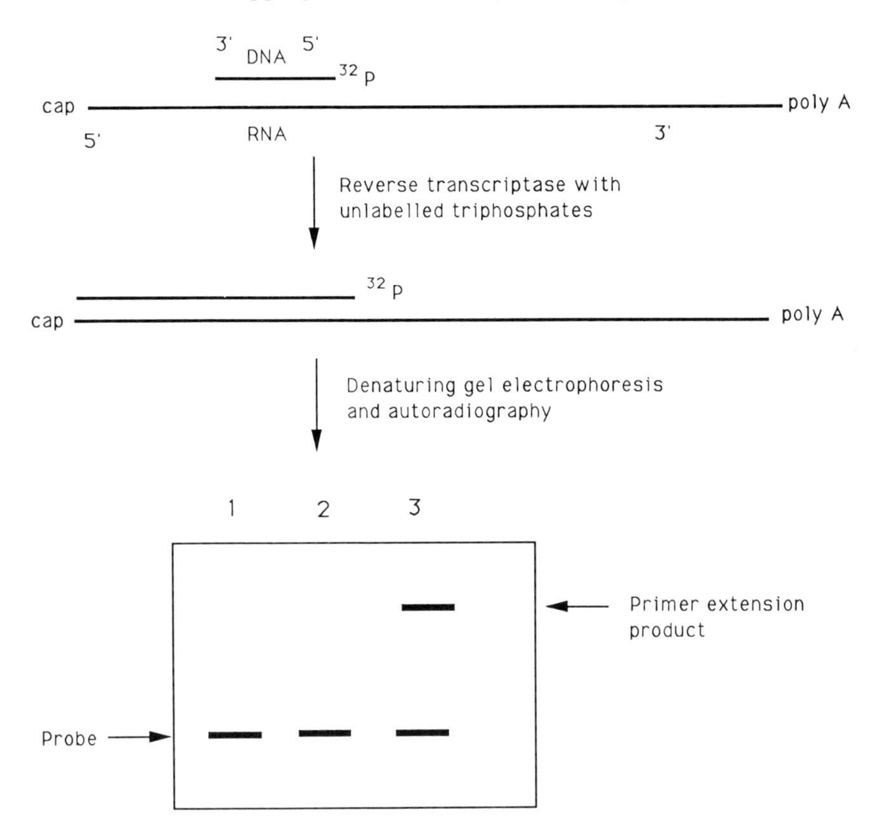

Figure 8. The principle of primer extension using an end-labelled probe. The diagram illustrates how the 5′ terminus of a poly(A)$^+$ mRNA might be determined using a single-stranded DNA probe (or oligonucleotide). Normally the probe would be a fragment derived from a cloned copy of the gene by cleavage with appropriate restriction enzymes or an oligonucleotide synthesized from the known cDNA sequence. The schematic representation of the autoradiogram (lower part of figure) demonstrates the results expected from a typical experiment. Lane 1, untreated probe; lanes 2 and 3, probe incubated under annealing conditions in the absence or presence of complementary RNA, respectively, and then incubated with reverse transcriptase in the presence of unlabelled nucleotide triphosphates. Only when complementary RNA is present to form a hybrid with the primer is an extension product synthesized (lane 3).

from within the gene is hybridized to complementry RNA and extended using the enzyme reverse transcriptase. The cloned probe is normally derived from a region near the 5′ end of the gene and the extension reaction terminates at the extreme 5′ end of the RNA. As with S1 mapping, this technique can be used to determine precisely the start point of transcription of an mRNA sequence.

The primer extension reaction is more sensitive and yields cleaner results using poly(A)$^+$ RNA. When using total RNA, primer extension yields a relatively large number of prematurely terminated transcripts (i.e. there are many primer extension products which are longer than the probe but which do not extend to the true 5′ terminus of the mRNA). These presumably result from a cellular inhibitor of reverse transcriptase which is removed by oligo(dT)-cellulose chromatography [the normal method of purification for poly(A)$^+$ RNA]. However, with a probe derived from near the 5′ end of the mRNA (i.e. within ~200 nucleotides), reasonable results can certainly be obtained using total RNA so poly(A)$^+$ RNA need be prepared only if the level of prematurely terminated reverse transcription is unacceptable.

7.2 Preparation of probe

Because only a small fragment of DNA is required as a primer, synthetic oligonucleotides are now almost exclusively used (although it is possible to use double-stranded DNA fragments under hybridization conditions that disfavour re-annealing, or to generate single-stranded primers by restriction digestion and gel electrophoresis; see ref. 25).

The protocol for labelling oligonucleotides is described in *Protocol 20*. It is important to know the specific activity of the probe, so that the correct amount of DNA can be used in the hybridization. This should be calculated immediately after the labelling reaction, by determining the number of Cerenkov counts which are excluded from the Sephadex G50 column used to remove unincorporated nucleoside triphosphates (*Protocol 20*, step 4).

Protocol 20. Oligonucleotide labelling by phosphorylation

Reagents

- Synthetic oligonucleotide to be labelled
- 10 × CIP/kinase buffer (0.5 M Tris–HCl pH 7.5, 0.1 M $MgCl_2$, 50 mM DTT)
- [γ-^{32}P]ATP (7000 Ci/mmol, 10 mCi/ml)
- T4 polynucleotide kinase
- SDS buffer (10 mm Tris–HCl pH 8.0, 1 mM EDTA, 0.1% SDS)
- Sephadex G50 spun columns equilibrated in SDS buffer (see *Protocol 3*)
- 3 M sodium acetate pH 5.5.

Method

1. To 100 ng of the oligonucleotide in a volume <10 μl add:

10 × CIP/kinase buffer	2.5 μl
[γ-^{32}P]ATP	5 μl (50 μCi)
H_2O	to 25 μl final volume.

2. Add 1 μl T4 polynucleotide kinase (10 U/μl) and vortex gently. Incubate for 1 h at 37°C.
3. Add 75 μl SDS buffer.
4. Centrifuge the oligonucleotide mixture through a Sephadex G50 spun column previously equilibrated in SDS buffer (see *Protocol 3*). Measure the radioactivity in the excluded probe using Cerenkov counting.
5. Recover the oligonucleotide by ethanol precipitation, wash the pellet with ethanol and let it dry (*Protocol 9*, steps 9–11).
6. Dissolve the radioactively labelled probe in water at a concentration of 1 fmol per μl (7.5 pg of a 25 nt primer is 1 fmol). The specific activity should be approximately 10^6 d.p.m./pmol of primer.

7.3 The hybridization and primer extension reactions

The procedures for hybridization and primer extension are described in *Protocol 21*. It is necessary to exercise care in optimizing the conditions under which the probe is hybridized to the RNA. The hybridization reaction should be performed under conditions of moderate (5- to 10-fold) excess of DNA over complementary RNA. While it is important to ensure that an adequate DNA excess is achieved, a massive excess of primer should not be used because this will lead to non-specific priming during the extension reaction. It is therefore wise to perform an RNA titration around the expected optimum.

Depending upon the GC content of the region of complementarity, and using a primer 20–30 nt, the optimum temperature for the hybridization reaction will normally be between 40°C and 60°C. However it is best to perform parallel annealing reactions at different temperatures around the expected optimum (from 10°C below to 10°C above, in 5°C steps). This will also aid interpretation of the data obtained. At the end of the annealing reaction, the hybridization mixture is diluted (*Protocol 21*, step 3) to lower the salt concentration and the primer is extended using reverse transcriptase and unlabelled nucleoside triphosphates (*Protocol 21*, step 4). This is performed at 42°C to minimize secondary structure in the mRNA, which can lead to premature termination of the reverse transcriptase reaction. The extension products are then separated on a high resolution urea-acrylamide gel.

Protocol 21. Procedure for hybridization and primer extension

Equipment and reagents

- Glass capillaries (20–25 μl capacity)
- Radiolabelled oligonucleotide probe (see *Protocol 20*)
- RNA samples
- 5 × hybridization buffer (2 M NaCl, 50 mM Pipes pH 6.4)
- 0.5 × TBE buffer (45 mM Tris, 45 mM boric acid, 1.25 mM EDTA)

Protocol 21. *Continued*

- Extension buffer
  Mix (per 90 μl): 5 μl M Tris pH 8.2, 5 μl 0.2 M DTT, 5 μl 0.12 M $MgCl_2$, 2.5 μl 1 mg/ml Actinomycin D in water,[a] 5 μl 10 mM dCTP, 5 μl 10 mM dATP, 5 μl 10 mM dGTP, 5 μl 10 mM dTTP, 10 U AMV reverse transcriptase,[b] and H_2O to 90 μl final volume.
- 3 M sodium acetate pH 5.5
- Formamide dye mixture (see *Protocol 14*)
- Denaturing (urea) polyacrylamide gel (see Chapter 6, *Protocol 12*); concentration suitable for fractionation of the fragments expected (see *Appendix 2*)
- Radioactively labelled DNA size markers (see *Protocol 14*).

Method

1. In a plastic microcentrifuge tube, mix:
   oligonucleotide probe: 2.5 μl containing 2.5 fmol primer (i.e. 20 pg of a 25 nucleotide primer)
   RNA: up to 5.5 μl in sterile water containing 0.25–0.5 fmol RNA complementary to the probe (i.e. 0.1–0.2 ng of a 1 kb mRNA[c])
   5 × hybridization buffer: 2 μl
   water to 10 μl final volume.
2. Seal the reaction mixture in a 20 or 25 μl glass capillary (see *Protocol 14*) and anneal for 6 h at an appropriate temperature,[d] e.g. 60°C.
3. Break open the capillaries (see *Protocol 14*) and expel the contents of each into 90 μl extension buffer.
4. Incubate the primer extension reaction for 1 h at 42°C.
5. Recover the nucleic acid by ethanol precipitation (see *Protocol 9*, steps 9–11).
6. Centrifuge at 12 000 *g* for 5 min.
7. Rinse the pellet with 0.5 ml 95% ethanol. Then leave it open to dry at room temperature.
8. Dissolve the dried pellet in 5 μl formamide dye mixture and electrophorese on a denaturing acrylamide gel alongside radiolabelled DNA size markers (see Chapter 6, *Protocol 12*).
9. Perform autoradiography of the gel using an intensifying screen. With an oligonucleotide at the lower end of the specific activity range specified in *Protocol 20*, an RNA constituting only 0.1% of the population can easily be detected in an overnight exposure of the gel.

[a] Actinomycin D is included to prevent self-copying of the primer by reverse transcriptase.
[b] Reverse transcriptase obtained from different manufacturers varies greatly in activity per unit. The figure of 10 units is based on enzyme obtained from Life Sciences Inc. When using a new batch of enzyme it is advisable to perform a titration in which replicate reactions are subjected to primer extension using various amounts of enzyme around the expected optimum.
[c] When the proportion of the complementary RNA in the total RNA population is not known, a range of RNA concentrations should be employed.
[d] See Section 7.3 for empirical determination of appropriate hybridization temperature.

7.4 Interpretation of primer extension data and potential problems

Because the hybridization goes to completion under the conditions specified in *Protocol 21* and because there is an excess of primer, it is possible to gauge the concentration of the complementary RNA by estimating the proportion of primer which has been extended. Therefore when deciding upon electrophoresis conditions, it is advisable to ensure that the unhybridized primer does not run off the end of the gel.

The primer extension reaction will often yield more than one transcript. There are several possible reasons for this.

(a) *'Cap effects'*. Generally a full-length reverse transcript of a eukaryotic mRNA sequence will also be accompanied by a minor reverse transcript which will migrate during electrophoresis with an apparent length about 1–2 nt shorter than the full length transcript. This is thought to be due to premature termination of reverse transcription at the methylated residue situated next to the cap site.

(b) *Multiple start sites of transcription*. Many genes which are transcribed by RNA polymerase II have multiple start sites for transcription which will yield mRNA sequences differing in length in their 5′ non-coding region.

(c) *Cross-hybridization to related RNA transcripts*. The different transcripts from multigene families may share sufficient homology over the region of the primer used to yield multiple reverse transcripts in the primer extension reaction, each derived from a different mRNA.

(d) *Premature termination of reverse transcription*. This can be a problem when using a primer which hybridizes to a region distant from the 5′ terminus and is exacerbated if total rather than poly(A)$^+$ RNA is utilized. It may be caused by regions of secondary structure in the RNA which prevent copying by reverse transcriptase or by cleavage of the RNA in particularly sensitive regions by contaminating RNase.

Figure 9 shows typical results obtained when the primer extension technique is used to analyse the transcripts of a multigene family. In this experiment a probe prepared from a conserved region of a *Dictyostelium* actin gene was hybridized to poly(A)$^+$ RNA. Seven extension products were reproducibly obtained at the optimal hybridization temperature (defined as the temperature at which the maximum number of reverse transcripts were obtained—in this case 60°C). The same result was also obtained with total RNA. Reduction of the hybridization temperature produced a uniform decrease in the amounts of the various reverse transcripts. However, as the temperature was increased, there was a differential loss of extension products (e.g. 1a, 1b, 2, 4) until at the highest temperature only three reverse transcripts were obtained (3a, 3b, and 3c). These three transcripts are derived from the same gene but

Figure 9. The identification of multiple transcripts derived from a gene family by primer extension; the effect of varying the hybridization temperature. In this experiment a 42 nt fragment of an actin gene from *Dictyostelium discoideum* was hybridized to *Dictyostelium* total poly(A)$^+$ RNA (26). The probe was a 3′ end-labelled restriction fragment prepared by sequential digestion with *Hin*dIII and *Hpa*II. The fragment was end-labelled with the

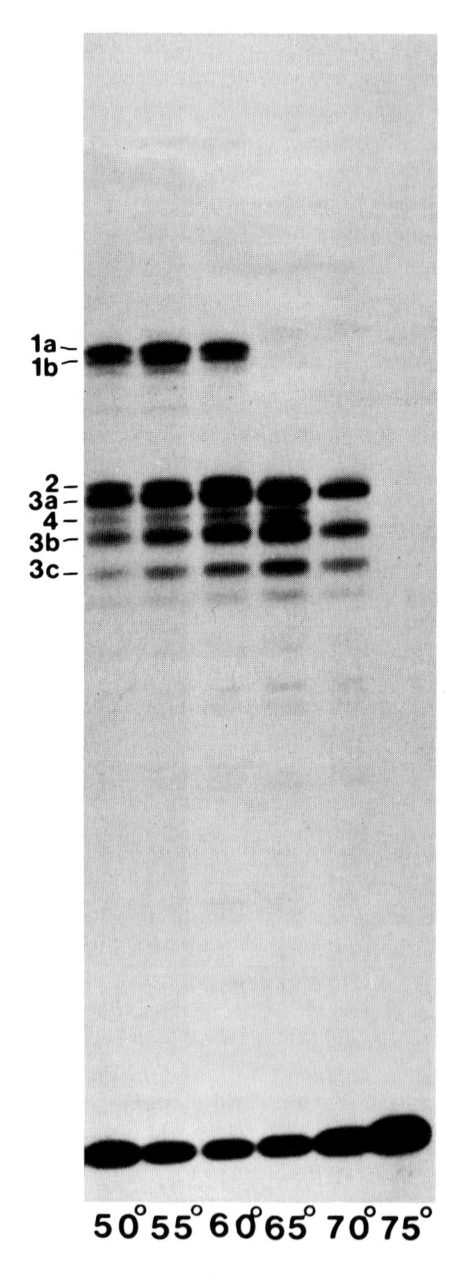

Klenow fragment of DNA polymerase I after restriction with *Hin*dIII but before *Hpa*II cleavage. This generated a 2 nt difference in the length of the coding and anti-coding strands permitting their separation on a denaturing polyacrylamide gel. The fragment was hybridized to the RNA at the temperatures indicated. The various reverse transcripts (1a, 1b, 2, 3a, 3b, 3c, and 4) are described in the text. From ref. 25.

they differ in the length of their 5′ non-coding regions. The reverse transcripts which were lost as the temperature was elevated are copied from other actin mRNA sequences which cross-hybridize to this probe. Nucleotide sequence analysis of preparative primer extension reaction products (using a 5′ end-labelled probe—ref. 25) showed that the seven transcripts are derived from four different actin genes (26). Two of the genes (1 and 3) display 5′-terminal heterogeneity, that is, they give rise to multiple transcripts (1a and 1b; 3a, 3b, and 3c) differing in length at their extreme 5′ ends because of heterogeneity in the precise position at which transcription is initiated.

The primer extension procedure described here has proved to be highly reproducible using RNA from diverse sources and purified by a number of different procedures. The only step which may give problems is the reverse transcriptase reaction, either because too little enzyme is used or because an inhibitor of reverse transcriptase has been co-purified with the primer or with the RNA. The simplest way to test for this problem is to determine whether a reasonable proportion of the DNA primer has hybridized to RNA in the hybridization step. This is described in *Protocol 22*. If the reverse transcriptase is working correctly then *all* of the primer which has formed a hybrid with the RNA should be extended. It may be difficult to quantify this precisely in situations where there is a very large excess of DNA primer over RNA, because only a small proportion of the primer will participate in hybrid formation. It may therefore be necessary to perform this analysis at several different ratios of DNA to RNA such that at the highest ratio of RNA a significant proportion of the primer forms a hybrid.

Protocol 22. Determination of the proportion of the primer in hybrid form using nuclease S1

Reagents

- 10 × S1 digestion buffer (see *Protocol 14*)
- Nuclease S1 (1000 U/μl from BRL)
- DEAE paper discs (DE81, Whatman)
- 5% Na_2HPO_4 (prepare by adding 50 g solid Na_2HPO_4 to 1 litre of water stirred during the addition with a magnetic stirrer bar).

Method

1. Just before use (step 2), prepare 210 μl nuclease S1 mixture for each hybridization reaction to be analysed. To do so, mix:

10 × S1 digestion buffer	21 μl
H_2O	199 μl.

Protocol 22. *Continued*

2. At the end of the hybridization reaction (*Protocol 21*, step 2), remove 2 μl of the mixture into 210 μl of nuclease S1 mixture. Mix by vortexing.
3. Place 4 × 50 μl aliquots into microcentrifuge tubes.
4. Add 300 units of nuclease S1 (3 μl at 10^5 units/ml; i.e. a 1/10 dilution of the enzyme in 1 × S1 digestion buffer) to only two of the aliquots.
5. Incubate all four tubes for 30 min at 42°C.
6. Remove 45 μl from each tube and spot it on to a disc of DEAE paper which has been labelled with a soft lead pencil.
7. Place the four discs in a 1 litre beaker containing 200 ml of 5% Na_2HPO_4.
8. Swirl the beaker gently for about 1 min. Pour off the buffer.
9. Repeat this washing step four times more.
10. Wash the filters twice with tap water, once with ethanol, and then air-dry the filters.
11. Count the filters in a suitable scintillant. Determine the proportion of the probe which is resistant to nuclease S1 by averaging the duplicate values and expressing the radioactivity of the samples with nuclease S1 as a percentage of the radioactivity of the samples with no nuclease S1.

8. *In situ* hybridization

8.1 Introduction

The analysis of gene expression by the various techniques described above gives useful information on the structure and general tissue distribution of transcripts. However, these methods do not allow analysis of the spatial control of gene expression at a cellular resolution, an important step in the characterization of many genes.

Immunocytochemical detection of protein gives the most precise and direct information on sites of gene expression, but raising specific antibodies is relatively slow and can be difficult. The detection of RNA by *in situ* hybridization is much faster, since specific probes can be produced from cloned DNA or by oligonucleotide synthesis. Single-stranded RNA probes have found general favour as they are very sensitive and RNase treatment can be used to reduce non-specific background (27). DNA oligonucleotides, although less sensitive, are also useful for certain studies, particularly since probes can be designed that distinguish transcripts with few sequence differences (28, 29). ^{35}S-labelled radioactive probes have been commonly used as they provide the best compromise in terms of the length of autoradiographic exposure required and the resolution of the signal. A method for the use of ^{35}S-labelled

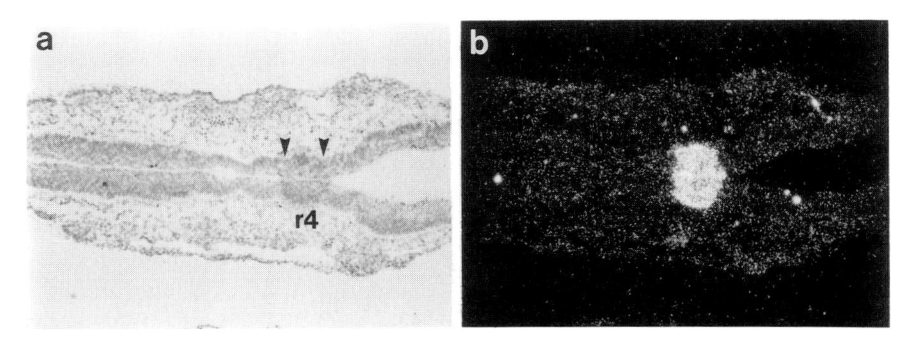

Figure 10. Example of *in situ* hybridization analysis of the pattern of gene expression. The pattern of expression of the *Hox*-2.9 gene in the hindbrain of a 9.5-day mouse embryo was analysed by the methods described here. (a) Bright-field photograph showing the tissue morphology. (b) Dark-field photograph showing the distribution of silver grains. In the coronal plane of section used, *Hox*-2.9 expression is seen to be restricted to a single morphological segment of the hindbrain, rhombomere 4 (r4). The arrows indicate boundaries of the tissue expressing *Hox*-2.9. From ref. 35.

RNA probes (30) is described below and an example of the use of these techniques is shown in *Figure 10*.

Recently developed methods (31, 32) based on the use of non-radioactive probes, although still less sensitive than ^{35}S-labelled probes, offer advantages of safety, speed of results, single-cell resolution, and the ability to visualize expression patterns in whole mounts. These latter methods should be considered if sensitivity is not a critical issue.

8.2 Preparation of single-stranded RNA probes for *in situ* hybridization

The synthesis of radiolabelled single-stranded RNA probes is carried out as described in *Protocols 16* and *17*, except that 10 µl [^{35}S]UTP (10 mCi/ml, 1000–1500 Ci/mmol) is used rather than ^{32}P-nucleotides. Probes of ~100 nt in length give optimal signals for certain tissues, and thus limited alkali hydrolysis is commonly used in order to reduce the average size of the probe (27). For other tissues, for example mouse embryos, little difference has been found in either the strength of the signal or non-specific background between probes degraded to an average size of 1 kb and shorter probes. Since the signal strength is proportional to sequence complexity, we recommend using probes of ~0.5–1.0 kb and the use of alkaline hydrolysis only for transcripts of >1 kb.

8.3 Preparation of tissue sections

Paraformaldehyde fixation and paraffin wax embedding are the methods of choice for the preparation of tissue sections for *in situ* hybridization (*Protocol 23*). This allows the accurate orientation of tissues, gives good tissue preservation, and allows easy collection of serial sections. It is important that the wax fully permeates into the tissue during the embedding, so >99.7% ethanol should be used in the final dehydration and the incubation times in *Protocol 23*, steps 3 and 4 should be increased for tissues of >3 mm thickness. The results of *in situ* hybridization analysis are often much easier to interpret in sections from accurately orientated embryos or tissues. This can be achieved by using glass embryo dishes as a mould, and orientating the tissues with a warmed needle before the wax has set. Sections are mounted on slides that have been subbed with TESPA (3-aminopropyltriethoxysilane) in order to retain tissues during subsequent steps (*Protocol 24*).

Protocol 23. Embedding and sectioning of embryos

Equipment and reagents

- Glass embryo dishes (Raymond Lamb)
- TESPA-subbed slides (see *Protocol 24*)
- PBS (0.14 M NaCl, 2.7 mM KCl, 10 mM Na_2PO_4, 1.8 mM KH_2PO_4); sterilize by autoclaving
- 4% paraformaldehyde in phosphate-buffered saline (PBS) (dissolve the paraformaldehyde in PBS at 65°C on the day of use. *Note*: paraformaldehyde is toxic and so inhalation or skin contact should be avoided)
- Saline (0.83% NaCl); sterilize by autoclaving
- Absolute ethanol and separate solutions of 70%, 85%, and 95% ethanol
- 1:1 saline:ethanol mixture
- Toluene
- Pastillated Fibrowax (BDH)
- 1:1 toluene:wax mixture.

Method

1. Dissect out the tissue in ice-cold PBS.
2. Place the tissue in 10 ml ice-cold 4% paraformaldehyde in PBS and leave at 4°C overnight.
3. Successively replace the solution with 10 ml of the following, each for at least 30 min and with occasional agitation: saline at 4°C (twice), 1:1 saline:ethanol mix at 4°C, 70% ethanol (twice), 85% ethanol, 95% ethanol, absolute ethanol (twice).
4. Replace the ethanol with 10 ml of the following, each for 20 min and with occasional agitation: toluene (three times), 1:1 toluene:wax at 60°C, pastillated Fibrowax at 60°C (three times).
5. Transfer the tissues to an embryo dish, orientate and allow to set.
6. Store at 4°C until required for sectioning.
7. Cut ribbons of 6 μm sections.
8. Float the sections on a bath of distilled water at 50°C until the creases disappear and then collect on TESPA-subbed slides.
9. Dry the slides at 37°C overnight and store desiccated at 4°C.

Protocol 24. Preparation of subbed slides

Reagents

- 10% HCl/70% ethanol
- 95% ethanol
- 2% TESPA (Sigma) in acetone
- Acetone.

Method

1. Dip glass slides in 10% HCl/70% ethanol, followed by distilled water and then 95% ethanol.
2. Dry in an oven at 100–150°C for 5 min and then allow to cool.
3. Dip the slides in 2% TESPA in acetone for 10 sec.
4. Wash the dipped slides twice with acetone, and then once with distilled water.
5. Dry at 37°C.

8.4 Pretreatment of sections

Sections are subjected to several pretreatment steps prior to the application of the hybridization probe in order to improve the signal and reduce background (*Protocol 24*). The digestion with Proteinase K (*Protocol 25*, step 6) increases the signal, presumably by improving the accessibility of the tissue to probe. The post-fixation after proteinase treatment is required in order to retain RNA and to prevent the disintegration of tissues. The treatment with acetic anhydride acetylates amino residues that might otherwise bind probe non-specifically.

Protocol 25. Pretreatment of sections

Reagents

- Tissue sections on TESPA-subbed slides (see *Protocol 23*)
- Xylene
- Absolute ethanol and separate solutions of 30%, 50%, 70%, 85%, and 95% ethanol
- Saline (0.83% NaCl); sterilize by autoclaving
- PBS (see *Protocol 23*)
- Fresh 4% paraformaldehyde in PBS (see *Protocol 23*)
- 20 μg/ml Proteinase K (dilute the Proteinase K from a 10 mg/ml stock just before use into 50 mM Tris–HCl, 5 mM EDTA pH 8.0)
- 0.1 M triethanolamine-HCl pH 8.0
- Acetic anhydride (*Note*: this reagent is toxic and volatile).

Method

1. De-wax the slides in 200 ml xylene, twice for 10 min, and then place in 200 ml of absolute ethanol for 2 min to remove most of the xylene.

Protocol 25. *Continued*

2. Transfer the slides quickly (several seconds each) through 200 ml absolute ethanol (twice), 95%, 85%, 70%, 50%, and then 30% ethanol.
3. Transfer the slides to 200 ml saline (0.83% NaCl), and then PBS for 5 min each.
4. Immerse the slides in 80 ml fresh 4% paraformaldehyde in PBS for 20 min.
5. Wash with 200 ml PBS, twice for 5 min.
6. Drain the slides, place them horizontally and overlay the sections with freshly prepared 20 μg/ml Proteinase K. Leave for 5 min.
7. Shake off excess liquid and wash the slides with 200 ml PBS for 5 min.
8. Repeat the fixation of step 4. The same solution can be used.
9. Quickly wash the slides in distilled water and place them in a container with 200 ml 0.1 M triethanolamine–HCl pH 8.0, set up with a rapidly rotating stir bar and in a fume hood.
10. Add 0.5 ml acetic anhydride (note: this is toxic and volatile) and leave for 10 min. The stirrer can be turned off when the acetic anhydride has fully dispersed.
11. Wash the slides with 200 ml PBS, and then saline for 5 min each.
12. Dehydrate the slides by passing them through 200 ml 30%, 50%, 70%, 85%, 95%, absolute ethanol (twice), and then allow to air-dry. To avoid salt deposits on the slides, they are left in 70% ethanol for 5 min; the other washes can be carried out quickly.

8.5 Hybridization and washing of sections

Whenever possible, it is important to include a thiol reducing agent, such as DTT, in the hybridization and washes (*Protocol 26*) since this prevents oxidation of the thio-substituted probe. However, RNase A, used to degrade non-specifically bound probe, is inhibited by thiol reducing agents and so it is important to wash out the DTT before RNase treatment. Studies of the relationship between the probe concentration and signal have indicated that ~0.3 ng probe/μl per kb of sequence complexity is optimal (27). However, if background problems are encountered, then they may be alleviated by reducing the concentration of probe and increasing the time period of autoradiography.

Protocol 26. Hybridization and washing of sections

Reagents

- Hybridization mixture (50% formamide, 0.3 M NaCl, 20 mM Tris–HCl pH 8.0, 5 mM EDTA, 10% dextran sulphate, 1 × Denhardt's solution, 0.5 mg/ml yeast tRNA)
- Pretreated tissue sections (see *Protocol 25*)
- 5 × SSC, 10 mM DTT
- 2 × SSC

- 0.1 × SSC
- 50% formamide, 5 × SSC
- 50% formamide, 2 × SSC, 10 mM DTT
- NTE buffer (0.5 M NaCl, 10 mM Tris–HCl pH 8.0, 5 mM EDTA)
- 20 μg/ml RNase A in NTE buffer
- Separate solutions of 30%, 60%, 80%, and 95% ethanol each containing 0.3 M ammonium acetate
- Absolute ethanol.

Method

The volumes indicated below are suitable for up to 20 slides, but can be varied according to the size of containers used. It is important to use different sets of containers for post-hybridization washing (in this protocol) and pre-treatment (*Protocol 25*) to avoid the latter becoming contaminated with RNase.

1. Apply the hybridization mixture including the probe (10^5 c.p.m./μl) to each slide adjacent to sections (~2.5 μl per square centimeter of coverslip is sufficient).
2. Gently lower a clean coverslip in such a way that the hybridization mixture is spread over the section.
3. Place all the slides horizontally in a plastic box, together with tissue paper soaked in 50% formamide, 5 × SSC, and seal the box to form a moist chamber.
4. Incubate overnight at 50°C.
5. Remove the slides and place them in a slide rack in 200 ml of 5 × SSC, 10 mM DTT at 50°C for 30–60 min for the coverslips to fall off. (This step is accelerated if the 5 × SSC, 10 mM DTT is prewarmed to 50°C.)
6. Place the slides in 40 ml 50% formamide, 2 × SSC, 10 mM DTT at 65°C for 30 min.
7. Wash the slides with 200 ml NTE buffer at 37°C, three times for 10 min each.
8. Treat with 40 ml 20 μg/ml RNase A in NTE buffer at 37°C for 30 min.
9. Wash with 200 ml NTE buffer at 37°C for 15 min.
10. Repeat step 6.
11. Wash in 250 ml of 2 × SSC, then 0.1 × SSC for 15 min each.
12. Dehydrate the slides by passing them quickly through 250 ml of 30%, 60%, 80%, and 95% ethanol (all including 0.3 M ammonium acetate), followed by absolute ethanol, twice.
13. Allow the slides to air dry, then set up each for autoradiography.

8.6 Autoradiography

Liquid photographic emulsions are the method of choice to obtain high-resolution images of the autoradiographic signal. The emulsions supplied by various manufacturers differ somewhat in their properties, in particular as to whether they tolerate repeated melting and solidification. The method given in *Protocol 27* is appropriate for Ilford K5 emulsion; the reader is advised to follow the manufacturer's recommendations for other emulsions. A typical exposure time for a moderately abundant RNA is 5–7 days, but if the level of non-specific background is sufficiently low then longer exposure times can be used. It is important not to over-expose the slides, since the signal/noise ratio will decline as the signal becomes saturating.

After autoradiography, the sections are stained with the nuclear stain, toluidine blue. It is also possible to counterstain the cytoplasm with, for example, eosin, but it is important not to overstain otherwise the signal can be obscured. The localization of silver grains over tissues can be visualised in several ways. Silver grains can be seen under bright-field illumination, particularly if the slides has been heavily exposed. However, dark-field illumination allows more sensitive visualization of silver grains.

Protocol 27. Autoradiography, staining, and mounting of slides

Reagents

- Hybridized, washed tissue sections (see *Protocol 26*)
- Photographic emulsion liquid (protocol given here is for Ilford K5 emulsion)
- Slide mailers for two slides (BDH Chemicals Ltd)
- 2% glycerol
- Developer (Kodak D19; 160 g/l)
- 1% glycerol, 1% acetic acid
- 30% sodium thiosulphate
- Absolute ethanol and 70% ethanol
- Xylene
- DPX mountant (BDH)
- 0.02% toluidine blue in water.

Method

1. Batches of emulsion ready for use are set up under safelight conditions (Kodak Wratten series II) as follows. Melt the emulsion shreds at 43 °C for 20–30 min. With a wide-mouthed pipette, transfer 6 ml aliquots of molten emulsion into slide mailers, each containing 6 ml 2% glycerol (pre-warmed to 43 °C) and then wrap each in foil. Invert each mailer several times to mix and store at 4 °C. When required, melt at 43 °C for 20 min.
2. Remove bubbles from the emulsion by dipping clean slides. Next, dip the experimental slides, allowing each to drain vertically for 2 sec, wiping the back of the slide and placing it horizontally in a light-tight (but not air-tight) box; suitable containers can be made by punching several holes in the lid of a plastic box, which is wrapped in two layers of foil after the slides have been placed in it.
3. Leave for 2 h then add a sachet of desiccant and leave for a further 2 h.

4. Transfer the slides to slide boxes containing a fresh sachet of desiccant, seal with tape, and place at 4°C to expose.
5. Before developing, allow each box of slides to warm to room temperature (>1 h).
6. Under safelight conditions in a darkroom, transfer the slides through Kodak D19 developer (160 g/l) for 2 min, 1% glycerol, 1% acetic acid for 1 min, and then 30% sodium thiosulphate for 2 min. The lights can now be turned on. At least 5 ml per slide should be used for each of these reagents.
7. Wash the slides twice in 200 ml distilled water for 10 min, quickly transfer through 70% ethanol then absolute ethanol and finally allow to air dry.
8. Stain in 0.02% toluidine blue for 1 min, then wash in water, 70% ethanol, and absolute ethanol. If the intensity of staining is not sufficient, air-dry the slides and repeat the staining, increasing the length of time in stain if necessary.
9. Transfer the slides from absolute ethanol to two changes of xylene. Place the slides horizontally and, while still wet, put several drops of DPX mountant on to each slide and lower a coverslip on top. Squeeze out any air bubbles and allow to dry overnight. These steps should be carried out in a fume hood.

8.7 Potential problems

The most common problem of *in situ* hybridization is the presence of non-specific background, especially since this limits the sensitivity of the technique and sites of low level gene expression might be missed. Different probes from the same gene can give different levels of background even under the high stringency washing conditions in the protocol described here. It is therefore worth testing several probes from different regions of the gene. Clearly, background must be distinguished from low levels of gene expression. Sense-strand RNA, or anti-sense probes from other genes with distinct expression patterns, have commonly been used as controls for the level of background, but this approach is limited by the variability in noise between different probes. A more direct method is to use unlabelled RNA identical to probe RNA, which will efficiently compete for specific, but not for non-specific, binding; inclusion of this competitor in the hybridization mix will drastically reduce specific signals but have no effect on background.

References

1. Wilkinson, M. (1991). *Essential molecular biology: a practical approach*, Vol. 1, (ed. T. A. Brown), p. 69. IRL Press, Oxford.

2. Alwine, J. C., Kemp, D. J., Parker, B. A., Reiser, J., Renart, J., Stark, G. R., and Wahl, G. M. (1979). In *Methods in enzymology*, Vol. 68, (ed. R. Wu), p. 220. Academic Press Inc., London and New York.
3. Thomas, P. S. (1980). *Proc. Natl Acad. Sci. USA,* **77,** 5201.
4. Lehrach, H., Diamond, D., Wozney, J. M., and Boedtker, H. (1977). *Biochemistry,* **16,** 4743.
5. Goldberg, D. A. (1980). *Proc. Natl Acad. Sci. USA,* **77,** 5794.
6. McMaster, G. K. and Carmichael, G. G. (1977). *Proc. Natl Acad. Sci. USA,* **74,** 4835.
7. Bailey, J. M. and Davidson, N. (1976). *Anal. Biochem.,* **70,** 75.
8. Wahl, G. M., Stern, M., and Stark, G. R. (1979). *Proc. Natl Acad. Sci. USA,* **76,** 3683.
9. Sealey, P. G. and Southern, E. M. (1982). In *Gel electrophoresis of nucleic acids: a practical approach* (ed. D. Rickwood and B. D. Hames), p. 39. IRL Press, Oxford.
10. Feinberg, A. P. and Vogelstein, B. (1983). *Anal. Biochem.,* **132,** 6 and (1984) Addendum in *Anal. Biochem.*, **137,** 266.
11. Holmes, D. S. and Quigley, M. (1981). *Anal. Biochem.,* **114,** 193.
12. Kafatos, F. C., Jones, W. C., and Efstratiadis, A. (1979). *Nucleic Acids Res.,* **7,** 1541.
13. Mullis, K. B. and Faloona, F. A. (1987). In *Methods in enzymology,* Vol. 155 (ed. R. Wu), p. 335. Academic Press, London and New York.
14. Saiki, R. K., Gelfand, D. H., Stoffel, S., Scharf, S. J., Higuchi, R., Horn, G. T., Mullis, K. B., and Erlich, H. A. (1988). *Science,* **239,** 487.
15. Gibbs, R. A., Nguyen, P. N., McBride, L. J., Keopf, S. M., and Caskey, C. T. (1989). *Proc. Natl Acad. Sci. USA,* **86,** 1919.
16. Becker-Andre, M. and Hahlbrock, K. (1989). *Nucleic Acids Res.,* **17,** 9437.
17. Berk, A. J. and Sharp, P. A. (1977). *Cell,* **12,** 721.
18. Weaver, R. F. and Weissman, C. (1979). *Nucleic Acids Res.,* **7,** 1175.
19. Maxam, A. and Gilbert, W. (1980). In *Methods in enzymology,* Vol. 65, (ed. L. Grossman and K. Moldave), p. 65. Academic Press, London and New York.
20. Zinn, K., DiMaio, D., and Maniatis, T. (1984). *Cell,* **34,** 865.
21. Melton, D. A., Krieg, P. A., Rebagliati, M. R., Maniatis, T., Zinn, K., and Green, M. R. (1984). *Nucleic Acids Res.*, **12,** 7035.
22. Davanloo, P., Rosenberg, A. H., Dunn, J. J., and Studier, F. W. (1984). *Proc. Natl Acad. Sci. USA,* **81,** 2035.
23. Butler, E. T. and Chamberlin, M. J. (1982). *J. Biol. Chem.,* **257,** 5772.
24. Schenborn, E. T. and Mierendorf, R. C. (1985). *Nucleic Acids Res.,* **12,** 7057.
25. Williams, J. G. and Mason, P. J. (1985). In *Nucleic acid hybridisation: a practical approach* (ed. B. D. Hames and S. J. Higgins), p. 139. IRL Press, Oxford.
26. Tsang, A. S., Mahbubani, H., and Williams, J. G. (1982). *Cell,* **31,** 375.
27. Cox, K. H., Deleon, D. V., Angerer, L. M., and Angerer, R. C. (1984). *Devel. Biol.,* **101,** 485.
28. Bradley, D. J., Young, W. S., III, and Weinberger, C. (1989). *Proc. Natl Acad. Sci. USA,* **86,** 7250.
29. Weiner, D. M., Levey, A. I., and Brann, M. R. (1990). *Proc. Natl Acad. Sci. USA,* **87,** 7050.
30. Wilkinson, D. G. and Green, J. (1990). In *Postimplantation mammalian embryos:*

a practical approach (ed. A. J. Copp and D. L. Cockroft), p. 155. IRL Press, Oxford.
31. Tautz, D. and Pfeifle, C. (1989). *Chromosoma,* **98,** 81.
32. Hemmati-Brivanlou, A., Frank, D., Bolce, M. E., Brown, B. D., Sive, H. L., and Harland, R. M. (1990). *Development,* **110,** 325–6.
34. Mason, P. J. and Williams, J. G. (1985). In *Nucleic acid hybridisation: a practical approach* (ed. B. D. Hames and S. J. Higgins), p. 113. IRL Press, Oxford.
35. Wilkinson, D. G., Bhatt, S., Cooke, M., Boncinelli, E., and Krumlauf, R. (1991). *Nature,* **341,** 405.

2

Transcription of exogenous genes in mammalian cells

KEVIN DOCHERTY and ANDREW R. CLARK

1. Introduction

The expression of exogenous genes in mammalian cells in culture has proved invaluable in a wide variety of applications. The advantages over the use of *E. coli* for the large-scale production of biologically active polypeptides stem from the ability of mammalian cells to undertake post-translational modification (e.g. proteolytic processing or glycosylation) of the gene product. The principal uses of mammalian cells for expressing foreign genes can therefore be summarized as follows:

- analysis of transcriptional regulatory sequences
- investigation of mechanisms involved in RNA processing and in the intracellular sorting and post-translational processing of secreted polypeptides
- expression cloning of cDNAs
- large-scale production of proteins from cloned cDNAs
- gene therapy.

In this chapter we describe strategies and methods for expressing exogenous DNA in mammalian cells in culture with emphasis on the use of these techniques in identifying transcriptional regulatory sequences.

DNA is very rarely introduced into cells as a genomic fragment. It is usually inserted into an appropriate vector which contains sequences which will:

- amplify the DNA copy number within the cell
- increase the efficiency of expression from the exogenous gene
- improve the stability of the transcript
- allow selection of cells in which the exogenous DNA has stably integrated into the genome.

There are a number of methods available for introducing DNA into mammalian cells in culture (see *Table 1*). The DNA can be taken up directly by the

Table 1. Methods for introducing DNA into mammalian cells

Method	Advantages/disadvantages	Efficiency	Ref.
DNA-mediated transfer methods			
Calcium phosphate co-precipitation	*Advantages* • Most widely used method • Simple and highly efficient	10–15%	(2, 3)
DEAE–dextran mediated transfection	*Advantages* • Simple and highly efficient *Disadvantages* • DNA subject to high mutation rate • Not recommended for stable transfections	Up to 80%	(4)
Electroporation	*Advantages* • Ease of operation • Reproducibility of conditions • Applicable to cells which grow either attached or in suspension • Stable and transient transfections • Control of copy number of transfected DNA molecules • Works on cells that are resistant to transfection by other methods *Disadvantages* • Optimal transfection occurs at a field strength resulting in cell death of 50% or more	Up to 50%	(6)
Polybrene-mediated transfection	*Advantages* • Highly efficient stable transfection of CHO cells *Disadvantages* • No improvement in stable transfection frequency over other methods for cells other than CHO	High in CHO cells	(34)
Liposome-mediated transfection	*Advantages* • Effective for stable and transient transfections • Applicable to cells which grow attached or in suspension *Disadvantages* • Complexity of liposome preparation techniques	10–15%	(35)

Protoplast fusion	*Advantages* • High efficiency transient and stable transfection *Disadvantages* • Manipulations are time consuming • Not possible to perform cotransfections	High	(36)
Lipofection	*Advantages* • Very high transfection efficiencies *Disadvantages* • Potential toxicity of the synthetic cationic lipid (DOTMA)	Very high	(37)
Red blood cell-mediated transfection	*Disadvantages* • Technical complexity in preparing erythrocyte ghosts	Moderate	(38)
DNA microinjection	*Advantages* • Particularly useful in production of transgenic mice • High transfection efficiency combined with control of copy number of integrated sequences *Disadvantages* • Requires use of sophisticated equipment and technical expertise	High	(39, 40)
Laser-mediated transfection	Sophisticated form of DNA microinjection. It allows more efficient treatment of a large number of cells	High	(41)
Microprojectile-mediated gene transfer	DNA coated on a microbead is introduced into cells by biolistic transfection. Originally developed to transfect plant cells, recent encouraging results indicate potential use in transfecting mammalian cells	–	(42, 43)
Virus-mediated transfer methods			
SV40 vectors	*Advantages* • Very high levels of expression in COS cells *Disadvantages* • Range of permissible cells limited to monkey cells • Viral infection results in cell lysis within 2–3 days • Only useful for transient transfection • Restriction on size of foreign DNA • DNA rearrangement frequently occurs during replication of SV40 viruses	High in COS cells	(17, 18)
Vaccinia virus	Viral infection shuts down host cell protein synthesis while retaining high levels of viral gene expression. This can be an advantage or a disadvantage depending on the design of the experiment.	100%	(9)

Table 1. *continued*

Method	Advantages/disadvantages	Efficiency	Ref.
Bovine papilloma virus	*Advantages* • Episomal vectors • Moderate levels of expression in wide variety of mammalian cells • Used for stable transfections *Disadvantages* • Biology of BPV not well understood • Vector prone to rearrangement	Moderate	(44)
Epstein–Barr virus	*Advantages* • Can incorporate relatively large DNA fragments • Wide variety of mammalian host cells • Used to establish cell lines with multiple episomal copies of the foreign gene	Moderate	(45)
Retroviral vectors	*Advantages* • Wide variety of host cells from variety of species • Stable lines • DNA introduced into 100% of cells (dependent on production of high titre virus stock) *Disadvantages* • Protein expression low due to problems associated with RNA splicing and mRNA translation • Limit on size of inserted foreign DNA	100%	(46)
Adenovirus	Foreign gene is inserted into a viral DNA segment within a plasmid vector and then rescued into full-length viral DNA *Advantages* • Virus easy to grow • Widespread use in oral administration of vaccines *Disadvantage* • Restriction on size of foreign DNA		(8)
Herpes simplex virus	*Advantages* • Can accommodate large (150 kb) DNA inserts • Particularly useful for transfecting neural cells		(47)

cell (transfection), introduced as a viral particle to the cell (infection) or microinjected directly into the cell. The choice of DNA transfer system depends on the cell line and the design of the experiment. For most experimental purposes, uptake and integration of the DNA into a small fraction (1 in 10^5 to 1 in 10^3) of the cell population is acceptable. This low efficiency can be achieved by using DNA transfection techniques. However, there are instances, such as transducing cells for gene therapy or related purposes, where high efficiency of DNA uptake and integration is imperative. In these instances retroviral vectors have proved invaluable, the efficiency of retroviral infection being limited only by the titre produced by the helper-free packaging cell line (1).

The choice of DNA transfer system also depends on the degree of expertise of the experimenter. Thus the transfection of naked DNA, often as a recombinant plasmid, is relatively simple, whilst the use of viral vectors or direct microinjection of DNA requires more specialized skills and facilities.

This chapter is limited to a description of some common DNA transfection procedures with which we have some experience.

2. Introduction of exogenous DNA into mammalian cells in culture

In all the techniques used to introduce recombinant plasmid DNA in mammalian cells, the quality of the DNA preparation is crucial. To obtain t highest transfection efficiencies, plasmid DNA should be purified by equili rium centrifugation in CsCl–ethidium bromide density gradients (*Protocol*

Protocol 1. Preparation of plasmid DNA by the alkali lysis method[a]

Equipment and reagents

- Sorvall RC5B Superspeed centrifuge and Sorvall G5A rotor or their equivalents
- 250 ml polycarbonate (wide mouth) centrifuge bottles and 50 ml sterile conical centrifuge tubes
- Beckman L8 ultracentrifuge, type Ti70 rotor, and 33 ml Beckman Quickseal polyallomer centrifuge tubes (or their equivalents)
- Superbroth [prepare this solution by dissolving 12 g bacto-tryptone, 24 g bacto-yeast extract and 4 ml glycerol in 900 ml water. Autoclave this solution and cool to 60°C before adding 100 ml of a sterile solution of 0.17 M KH_2PO_4 and 0.72 M K_2HPO_4. Once the salts have dissolved, adjust the volume to 1 litre with water and autoclave once more]
- Solution I (50 mM glucose, 25 mM Tris–HCl, 10 mM EDTA pH 8.0)
- Lysozyme (10 mg/ml in 10 mM Tris– pH 8.0)
- Solution II [0.2 M NaOH (freshly di from a 10 M solution), 1% (w/v) SDS]
- Solution III [Prepare this solution by combining (5 M potassium acetate, 11.5 ml glacial acid, 28.5 ml water. The final solution with respect to potassium and 5 M respect to acetate]
- TE buffer (10 mM Tris–HCl, 1 mM l pH 7.6)
- Ethidium bromide solution (10 mg/m buffer)
- Solid CsCl
- Water-saturated butan-1-ol
- Isopropanol.

Protocol 1. *Continued*

A. *Preparation of plasmid DNA from lysed bacteria*

1. Plate out the *E. coli* host cells containing the recombinant plasmid of interest to give single colonies on agar plates containing the appropriate antibiotic (usually 50 μg/ml ampicillin or 12.5 μg/ml tetracycline).
2. Pick a suitable colony and use it to inoculate 5 ml of Superbroth containing appropriate antibiotic.
3. Incubate overnight at 37°C with vigorous aeration.
4. Inoculate 1 ml of this fresh overnight culture into 400 ml of Superbroth containing the appropriate antibiotic. Incubate overnight at 37°C with vigorous aeration.
5. Transfer the overnight culture to two 250 ml plastic centrifuge bottles. Centrifuge at 2500 *g* for 10 min at 4°C in a Sorvall GSA rotor or its equivalent. Discard the supernatant.
6. Resuspend each pellet in 5 ml of solution I and combine these solutions in one 250 ml centrifuge bottle. Add 1 ml of lysozyme solution.[b]
7. Add 20 ml solution II, mix the contents by inverting gently several times, and leave the bottle on ice for 10 min.
8. Add 15 ml solution III and mix the contents by shaking the bottle several times. Leave the bottle on ice for 10 min.
9. Centrifuge the bacterial lysate at 2500 *g* for 15 min at 4°C in a Sorvall GSA rotor.
10. Filter the supernatant through four layers of cheesecloth into a 250 ml centrifuge bottle.
11. Add 0.6 vol. isopropanol, mix well and leave the bottle for 10 min at room temperature.
12. Centrifuge at 2500 *g* for 15 min at room temperature in a Sorvall GSA rotor.
13. Decant the supernatant carefully, and invert the bottle to allow the last drops of supernatant to drain away.

B. *Purification of plasmid DNA by equilibrium centrifugation in CsCl–ethidium bromide gradients*

1. Resuspend the nucleic acid pellet (from step 13 above) in 30 ml TE buffer.[c]
2. Add 30 g CsCl and mix the solution gently until the salt has dissolved.
3. Add 3 ml of ethidium bromide solution.
4. Transfer the solution to a Beckman Quickseal polyallomer 33 ml centrifuge tube. Seal the tube.

5. Centrifuge in a Beckman L8 centrifuge at 20°C using a type 70 Ti rotor and a programme which provides 250 000 *g* for 16 h followed by 150 000 *g* for 1 h.
6. Visualize the bands, using a UV lamp if necessary, and remove the lower of the two visible bands. To do this, puncture the top of the centrifuge tube with a 19-gauge hypodermic syringe needle. Then, using a syringe fitted with a 19-gauge needle, carefully puncture the side of the tube below the position of the lower band, and withdraw in a minimum volume the solution containing this band.

C. *Removal of ethidium bromide from plasmid DNA preparations*

1. Measure the volume of the plasmid solution from step 6 above (usually about 5 ml).
2. Add an equal volume of water-saturated butanol to the DNA solution in a sterile 50 ml conical centrifuge tube.
3. Mix the two phases by vortexing.
4. Centrifuge at 250 *g* for 3 min at room temperature in a bench centrifuge.
5. Remove and discard the upper organic phase.
6. Repeat the extraction (steps 1–5) 4–6 times until the pink colour disappears from both the aqueous and organic phases.
7. Add two volumes of water to the aqueous phase. Then add two volumes of ice-cold ethanol and store the tube on ice for 15 min. Under these conditions the CsCl should not precipitate along with the DNA.
8. Centrifuge at 250 *g* for 15 min in a bench centrifuge.
9. Resuspend the DNA pellet in an appropriate volume of TE buffer, pH 7.6. Quantify the DNA by diluting an aliquot of this solution into 1 ml of water and measuring the A_{260}. An A_{260} of 1.0 is equivalent to 50 μg/ml of DNA.

[a] From ref. 5.
[b] In our experience, for most of the bacterial strains in common use, the lysozyme digestion step can be omitted without affecting the recovery of plasmid DNA.
[c] We usually employ CsCl–ethidium bromide gradients in relatively large volumes. Alternatively, the crude nucleic acid pellet can be resuspended in 10 ml CsCl–ethidium bromide and divided into two smaller 'Quickseal' tubes. Also, other investigators often subject their plasmids to a second round of CsCl–ethidium bromide gradient centrifugation.

2.1 The calcium phosphate method

First described by Graham and van der Eb (2), calcium phosphate-mediated transfection involves mixing DNA directly with $CaCl_2$ and phosphate buffer to form a fine calcium phosphate precipitate containing the DNA which is then placed on the cell monolayer. The precipitate binds to the plasma membrane and is taken into the cell by endocytosis. This is one of the most commonly used techniques for introducing DNA into cells in culture and gives transfection efficiencies of about 10^3 (see *Protocol 2*). It is used for transfecting a wide range of cells for either transient expression or for the establishment of stable transformants (see Section 3). There are two common variations of the method:

- to leave the calcium phosphate-DNA precipitate in contact with the cells for approximately 6 h
- to leave the calcium phosphate-DNA precipitate in contact with the cells for up to 18 h.

Chen and Okayama (3) have described improvements to the basic procedure which can yield in the region of 10–50% stable transfectants. This involves optimizing the following parameters:

- growth temperature
- CO_2 concentration during growth
- DNA concentration
- physical form of the DNA.

Optimal conditions are reported (3) as being cell growth at 35°C in 2–4% CO_2 for 15–24 h with the DNA in circular form (which gives higher transfection efficiencies than linear DNA) and applied at a concentration of 20–30 μg/ml.

Protocol 2. Transfection by calcium phosphate

Reagents

- DNA for transfection (5–30 μg in approximately 50 μl TE buffer pH 7.6)
- 10 × HBS (Hepes-buffered saline) stock [8.18% (w/v) NaCl, 5.94% Hepes (w/v), 0.2% (w/v) Na_2HPO_4]
- 2 × HBS.
  Prepare this solution from 10 × HBS stock. Adjust the pH to 7.1 using 1 M NaOH
- 2.5 M $CaCl_2$
- DMEM medium (Gibco/BRL).

Method

1. Cell monolayers are transfected when at 50–70% confluence (i.e. whilst still rapidly growing). Approx. 4 h prior to transfection, feed the cells with 5 ml of fresh culture medium.[a]

2. Prepare solutions A and B as follows (volumes should be multiplied by the number of Petri dishes for each DNA sample):

A. 25 μl 2.5 M $CaCl_2$
   5–30 μg DNA
   H_2O to a final volume of 250 μl

B. 250 μl of 2 × HBS.

3. For each DNA sample, slowly add solution A to solution B with continuous mixing by vortexing. Store each mixture at room temperature for 30 min to allow a DNA calcium phosphate coprecipitate to form.
4. Add 0.5 ml of the precipitate to each Petri dish of cells with gentle swirling to distribute the precipitate evenly over the cells.
5. There are several variations of the next step. These include the following:

   Either (a) incubate the cells for 6 h at 37°C, then replace the medium with 1 ml 10% glycerol in serum-free DMEM medium for 1 min. Remove the glycerol solution and wash the cell monolayer twice with serum-free medium. Add 5 ml DMEM medium and incubate the cells as normal.

   Or (b) incubate the cells, overlaid with precipitate, overnight at 37°C. Remove the medium and wash the monolayer once with 10 ml serum-free DMEM medium and once with 10 ml PBS. Add fresh medium and replace the Petri dishes in the incubator.

[a] Avoid the use of culture medium such as RPMI-1640 which contains high concentrations of calcium, leading to the formation of dense precipitates and increased cell mortality. We usually perform transfections in DMEM medium, returning the cells to their preferred culture medium when the transfections are completed.

2.2 DEAE-dextran method

This is another highly efficient method for introducing DNA into a wide range of cells (4) and is based on the DNA binding to DEAE-dextran which is then applied to and endocytosed by the cultured cells (see *Protocol 3*). DEAE-dextran is preferred to the calcium phosphate method for analysing large numbers of DNA samples in transient transfection experiments since the sample preparation is less time-consuming. It is not recommended for the production of stable transformants.

Protocol 3. Transfection using DEAE-dextran

Reagents

- DNA for transfection (5–30 μg in approximately 50 μl TE buffer pH 7.6)
- 1 M Tris–HCl pH 7.3
- 0.5 M Hepes pH 7.2
- 100 mg/ml DEAE-dextran (M_r 5 × 10^5, chloride form from Sigma; make up in H_2O and sterilize by autoclaving)
- 25% (v/v) glycerol in serum-free culture medium
- Transfection solution; prepare just prior to use by adding the following to a 20 ml universal flask:
  - 0.5 ml 1 M Tris–HCl pH 7.3
  - 0.4 ml 0.5 M Hepes pH 7.2
  - 75 μl 100 mg/ml DEAE-dextran
  - serum-free medium to a final volume of 10 ml.

Protocol 3. *Continued*

Method

1. Add the required amount of DNA (10–30 μg) to 0.8 ml of the transfection solution and vortex.
2. Remove the medium from the cells in the Petri dish and wash the monolayer twice with serum-free medium.
3. Add the DNA solution (from step 1) to the monolayer and incubate at 37°C for 4 h.
4. Remove the DNA solution and add 1 ml 25% glycerol. Leave for 1 min.
5. Remove the glycerol and wash the monolayer twice with serum-free medium. Add fresh culture medium and return the Petri dish to the incubator.

2.3 Additional factors that increase transfection efficiency

The efficiency of transfection by calcium phosphate-DNA or DEAE dextran-DNA can be increased by additional treatments after exposure of the cells to the DNA. The most frequently used additional treatments involve incubation of the cells with reagents such as chloroquine, glycerol, dimethyl sulphoxide (DMSO), and sodium butyrate. However, it should be noted that these chemicals are toxic to cells. Thus, the optimal concentration and time of exposure must be empirically determined for each cell type.

(a) *Chloroquine* is generally applied to the cells simultaneously with the DNA. Chloroquine can be stored as a 2 mg/ml solution at 4°C in the dark for up to one week, and is used at a final concentration of 200 μg/ml. It is thought to enhance transfection by neutralizing the acidic pH within lysosomes thereby inhibiting the intracellular degradation of DNA during its transit to the nucleus.

(b) *Glycerol and DMSO solutions* [10–20% (v/v) in serum-free medium] are applied to the cells for about 1 min immediately following removal of the DNA precipitate. These reagents enhance uptake of DNA through the plasma membrane of the cell by a mechanism which is not well understood.

(c) *Sodium butyrate* is used after the glycerol or DMSO shock treatment (see above). Typically, 5 ml of a 10 mM sodium butyrate solution in Dulbecco's modified Eagle medium (DMEM) is added to the Petri dish, and the cell monolayer incubated in this solution for 16 h before the sodium butyrate is replaced with fresh growth medium.

The use of carrier DNA can also improve transfection efficiency in the calcium phosphate technique, presumably by improving the quality of the precipitate formed. The carrier DNA should be chosen with care since some plasmids (e.g. pUC) contain consensus binding sites for transcription factors which may then cause problems in subsequent transcriptional analyses. Eukaryotic DNA prepared in the laboratory (5) or commercially available DNA such as calf thymus or salmon sperm DNA can be used as carrier. The concentration of the carrier DNA to use should be optimized in pilot experiments.

2.4 Electroporation

In this method (see *Protocol 4*) the cells and DNA are mixed in suspension and exposed to a high voltage electric shock (6). This apparently creates pores in the plasma membrane of the cells, which allow the DNA to enter and become integrated into the genome. The reclosing of the membrane is delayed by incubation at 0°C, thereby increasing the efficiency of DNA uptake. Electroporation has found widespread use since it appears to work on virtually any cell type, even those which are resistant to transfection by calcium phosphate or DEAE-dextran. Several types of electroporation equipment are available commercially. We use the Bio-Rad Gene Pulser machine, and find that it provides adequate control of the voltage and duration of the pulse.

The pulse conditions must be set empirically for each cell type. However, a useful rule of thumb is that optimal transfection occurs at a field strength that results in cell death of around 50%. Therefore when establishing electroporation conditions for a new cell line, we routinely determine the extent of cell death by Trypan Blue staining. For this, cells are resuspended at a concentration of approximately 5×10^5 cells/ml in phosphate-buffered saline (PBS), and 0.5 ml is mixed with 0.5 ml of 0.4% (w/v) Trypan Blue solution in PBS. This mixture is allowed to stand for 5–15 min. A drop of the Trypan-Blue cell suspension is then transferred to a haemocytometer and the number of stained and unstained cells counted. The viability is expressed as the percentage of unstained cells relative to the total number of stained and unstained cells.

Protocol 4. Transfection by electroporation

Equipment and reagents

- Electroporation apparatus (e.g. BioRad Gene Pulser™ with Pulse Controller and Capacitance Extender)
- DNA for transfection (5–30 μg in approximately 50 μl TE buffer pH 7.6)
- PBS (phosphate-buffered saline) [to prepare, dissolve 8 g NaCl, 0.2 g KCl, 1.44 g Na_2PO_4, and 0.24 g KH_2PO_4 in 800 ml H_2O. Adjust pH to 7.4 with HCl and add H_2O to 1 litre final volume].

Protocol 4. *Continued*

Method

1. Harvest the cells by gentle trypsination. Centrifuge and resuspend in ice-cold PBS at approximately 10^6 cells per ml.
2. Place 0.8 ml of the cell suspension in a sterile electroporation cuvette. Add DNA (10–30 μg), mix by pipetting, and store on ice for 10 min.
3. Transfer the cuvette to the chamber of the electroporation apparatus and deliver a pulse at the appropriate capacitance and voltage settings as determined by prior experimentation (see Section 2.4).
4. Place the cuvette on ice for 10 min.
5. Dilute the cell suspension into 10 ml culture medium and incubate overnight at 37°C. Remove the medium, which contains dead cells, and replace with fresh medium.

2.5 Other methods

Other less widely used transfection methods are listed in *Table 1*.

3. Transient and stable expression

Following transfection, DNA enters a large proportion of the cells in culture. However, of these, only a small proportion of cells transfer the DNA into the nucleus where it may be transiently expressed for a few days. In an even smaller proportion of cells, the DNA is integrated into the genome and is stably expressed in future generations of this cell population. The identification of these rare stable transfectants requires the inclusion of a drug-selectable marker either in the recombinant plasmid or on a separate plasmid which is co-transfected into the cells with the plasmid bearing the DNA under investigation (see Section 4.1.4). This presents a major drawback of stable transfection in that the drug-selection protocol, followed by the isolation and propagation of stably transfected clonal cell lines, can take weeks and often months to complete. This effectively rules out stable transfection for many purposes, for example as a method for analysing numerous modified DNA sequences for transcriptional regulatory properties.

4. Vectors for introducing DNA into mammalian cells in culture

Many types of vectors have been utilized for introducing DNA into cultured mammalian cells. We describe mammalian plasmid expression vectors which are widely used. Readers are directed elsewhere for a description of vaccinia virus, adenovirus and retroviral DNA transfer systems (7–9).

4.1 Components of mammalian plasmid expression vectors

4.1.1 Prokaryotic plasmid sequences

Almost all mammalian expression vectors contain sequences derived from the plasmid pBR322. These sequences include a replicon that permits propagation and growth of the vector in *E. coli,* a gene encoding antibiotic resistance to facilitate selection of recombinant plasmids, and a number of unique restriction endonuclease sites for insertion of foreign DNA sequences.

4.1.2 The transcriptional unit

The transcriptional unit comprises:

- sequences encoding the foreign protein
- sequences responsible for driving expression of the gene of interest
- intron sequences
- mRNA polyadenylation signals.

In the simplest vector, these sequences are provided by a genomic fragment containing a complete copy of the gene of interest. However, since many mammalian transcriptional control sequences (promoters and enhancers) are to some extent tissue-specific, expression will be possible only in a limited number of cell types. More commonly the foreign protein is expressed from a cDNA inserted into a vector containing the other required components of the transcriptional unit. The following sections describe these vector elements.

Promoter elements

The promoter elements commonly used in mammalian expression vectors include the SV40 early promoter, the Rous sarcoma virus (RSV) promoter, the adenovirus major late promoter, or the human cytomegalovirus (CMV) immediate early promoter. Since the inclusion of enhancer sequences can increase the transcriptional activity of the promoter by 10- to 100-fold, most expression vectors include a strong enhancer such as those derived from SV40, RSV, or CMV, which are active in a wide variety of cell types from many species.

Inducible promoters

Inducible promoters can be used to express a protein which is potentially cytotoxic. Several of the inducible promoters which have been used for expression of foreign genes in mammalian cells are listed in *Table 2*. In our studies, the metallothionein-1 promoter in pMt.neo.1 (see *Figure 1a*) provided an almost 20-fold stimulation of insulin expression in stably transfected mouse pituitary (AtT20) cells (see *Figure 1b*). It is important to determine the potential toxic effect of inducing agent on the cell line. Maximal expression

Table 2. Inducible promoters

Inducible promoter	**Inducing agent/treatment**	**Comment**	**Ref.**
Interferon promoter	1. Virus infection 2. Double-stranded RNA	1. Pretreatment of cells with interferon improves induction	(48)
Heat shock promoter	1. Heat shock	1. 100-fold induction achieved when CHO cells harbouring an hsp70 promoter fused to c-myc were incubated at 43°C; 2. Potential problem if heat shock is detrimental to the cell	(49)
Metallothionein promoter	1. Cadmium or zinc 2. Glucocorticoids 3. Interferon	1. High basal level of expression 2. Cadmium and zinc can be cytotoxic	(50)
Glucocorticoid response element (MMTV)	1. Glucocorticoids	1. Vectors available containing the MMTV long terminal repeat sequences (LTR)	(51)

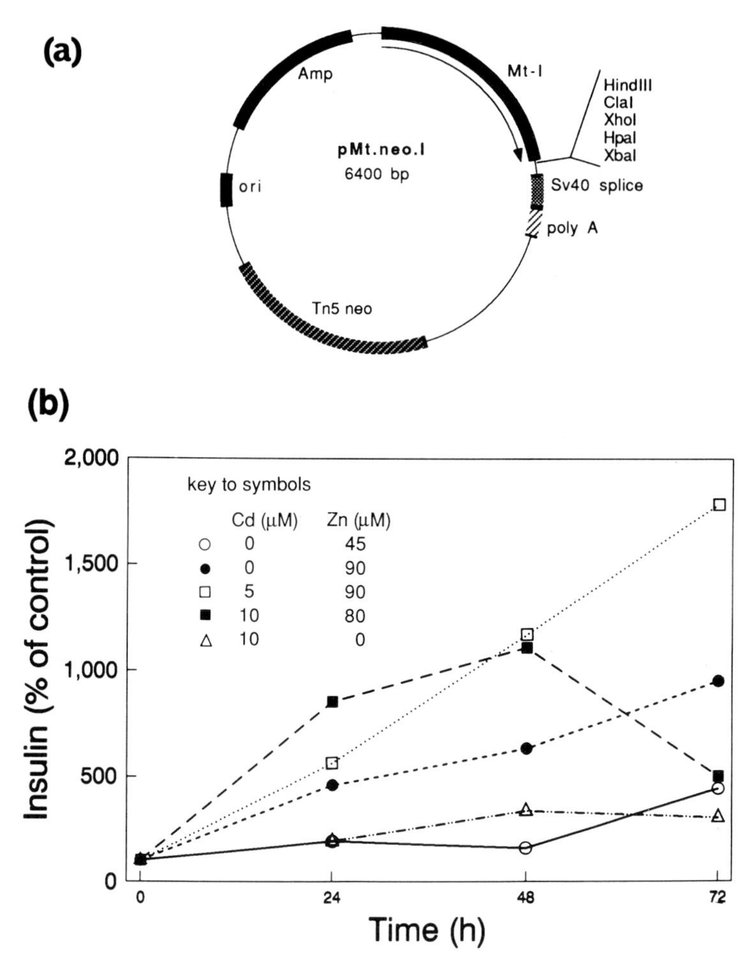

Figure 1. (a) Plasmid map of an inducible expression vector (pMt.Neo.1) for stable transfection of cell lines. The plasmid vector contains: an origin of replication (ori) and ampicillin resistance gene (Amp) for growth and selection of recombinants in *E. coli*; a heavy metal ion inducible promoter, the metallothionein promoter (MT-1); a multiple cloning region (MCR) with *Hin*dIII, *Cla*I, *Xho*I, *Hpa*I, and *Xba*I sites for insertion of foreign DNA sequences; an intron and polyadenylation signal [SV40 splice and poly(A)] from the SV40 small T antigen; and a gene coding for aminoglycoside 3′ phosphotransferase (Tn5 neo) which confers neomycin resistance. This vector was constructed by Dr K. Peden (NIH, Maryland). (b) Inducible expression of insulin in a mouse anterior pituitary cell line (AtT20) stably transfected with plasmid pMt.neo.1 bearing a human preproinsulin cDNA inserted at the *Xho*I site. The results are expressed as released insulin (as measured by radioimmunoassay) as a percentage of that in medium from cells incubated in the absence of Zn or Cd. This experiment shows that maximal stimulation of insulin expression was produced by 90 μM Zn in the presence of 5 μM Cd. The data also show that Cd concentrations above 5 μM proved toxic to the cells (data of N. A. Taylor).

from the metallothionein promoter was achieved with 5 μM Cd^{2+} and 90 μM Zn^{2+} (*Figure 1b*); higher Cd^{2+} concentrations proved toxic to the cells.

Introns

Although many cDNAs are efficiently expressed from vectors lacking splicing signals, introns are normally built into expression vectors since they very often enhance the efficiency of expression. The intervening sequences from SV40 small T antigen (10) or a hybrid intron containing adenovirus and immunoglobulin sequences (11) are commonly used. There appears to be a strong requirement for introns for expression of foreign genes in transgenic mice (12).

Polyadenylation signals

The stability and efficiency of translation of the expressed transcript is dependent on the addition of poly(A) to the 3′ end of the mRNA. Two sequences are important for polyadenylation: a hexanucleotide AAUAAA located 11 to 30 nucleotides upstream of the polyadenylation site and U-rich sequences downstream of the site. These signals are usually provided by inclusion of sequences from the SV40 early transcription unit, the hepatitis B surface antigen gene or from the mouse β-globin gene (13–15).

cDNA sequences

The minimum amount of foreign DNA encoding the protein under study should include an initiation code in the correct context (i.e. a purine at position −3 or a G at +4 where the A of the AUG is position +1) (16) and a termination codon. Other extraneous sequences such as pieces of plasmid DNA or polylinker, which are often carried over from the cDNA library or during subcloning stages, are usually removed by digestion with *Bal*31 or exonuclease III (see Section 6.1).

4.1.3 Viral replicons

Replicons are viral sequences that promote extrachromosomal (episomal) replication of the viral DNA in permissive cells that express *trans*-acting factors which interact with the replicon. Vectors have been constructed bearing replicons from SV40, polyoma virus, bovine papilloma virus, and Epstein-Barr virus.

Vectors bearing the SV40 virus origin of replication, replicate to extremely high copy number (approximately 10^5 copies per cell) in cells that express the SV40 large T antigen. In order to improve the use of such vectors, Gluzman (17) developed the COS cell line. COS cells were derived from simian (CV1) cells that were genetically engineered to produce the SV40 large T antigen. A high efficiency expression vector, p91023(B) (ref. 18) for use with COS cells is shown in *Figure 2a*. Transfection of p91023(B) into COS cells results in runaway replication of the recombinant plasmid and very high levels of

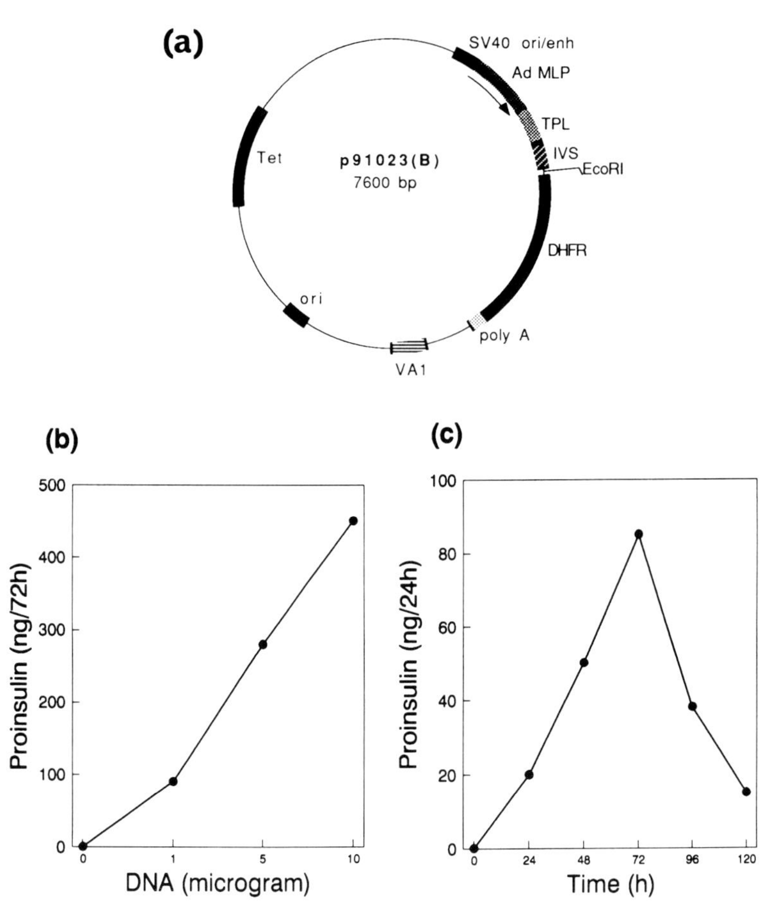

Figure 2. Transient expression in COS cells. (a) Structure of expression vector p91023(B). This plasmid is designed for high efficiency expression in COS cells. It contains: the pBR322 origin of replication (ori) and tetracycline resistance gene (Tet); the adenovirus major late promoter (Ad MLP); the adenovirus tripartite leader sequence (TPL) and VA1 gene (VA1); the SV40 origin of replication (SV40 ori); SV40 early polyadenylation signal [poly (A)]; a 5′ splice site from the adenovirus first late leader and an introduced 3′ splice site (IVS); and the mouse dihydrofolate folate reductase (DHFR) coding sequences. Foreign DNA is inserted at the *Eco*RI site. Expression is driven by the major late promoter, generating an mRNA containing the foreign sequence flanked by the tripartite leader at the 5′ end and the *dhfr* on the 3′ end. The *dhfr* sequence stabilizes the mRNA while the tripartite leader sequence interacts with the products of the VA genes to increase translation. When introduced into COS cells, the endogenous SV40 large T antigen binds to the SV40 origin and activates replication of the plasmid (vector constructed by Dr R. J. Kaufman, Genetics Institute, Boston, USA). (b) Expression of proinsulin in COS cells using the vector p91023(B). The human preproinsulin cDNA was inserted into the *Eco*RI site of p91023(B) and transfected into COS cells. This experiment shows that almost 0.5 μg of proinsulin is released into the medium 72 h after transfection with 10 μg of the recombinant plasmid DNA. (c) Expression of proinsulin in COS cells using p91023(B) is transient, peaking 72 h after transfection, then rapidly falling off as the cells lyse (19).

expression of exogenously introduced genes carried on the vector (about 0.5 mg/ml encoded protein; see *Figure 2b*). Expression begins shortly after transfection (peaking within approximately 72 h), and then rapidly disappears as the cells lyse (*Figure 2c*).

SV40-based vectors (e.g. pCDM8) have also been designed for cloning of secreted or membrane proteins following expression of cDNA libraries in COS cells (20). A cDNA library is constructed in the SV40-based expression vector and used to transform *E. coli*. The transformed *E. coli* are amplified, and the amplified library divided into pools composed of sub-populations of the library. Plasmid DNA is prepared from each pool and transfected into COS cells. For membrane proteins the transfected cells are plated in a tissue-culture dish coated with antibody or an appropriate affinity ligand. The cells expressing the cDNA of interest are selectively bound to the culture dish. Plasmid DNA is then extracted from the bound cells, and further rounds of expression/selection results in a single bacterial clone containing the required cDNA.

Vectors bearing the polyoma virus origin also replicate to very high copy number in permissive cells. Thus, SV40- and polyoma-virus-based vectors can be used for transient expression. On the other hand, vectors containing bovine papilloma virus and Epstein–Barr virus origins of replication are generally used to isolate stably transformed cell lines expressing low levels of exogenous genes since they are replicated at low copy number and do not cause cell death.

4.1.4 Selectable markers

As mentioned above, stably transfected cell lines are isolated by introducing into the cells, along with the gene of interest, a second gene that encodes an enzyme activity which confers resistance to an antibiotic or drug. Generally, the cells are allowed to recover after transfection for 24 h to allow efficient expression of the selectable gene before the appropriate drug is added to the culture medium. Transfected cells are then grown in selective medium for 2–3 weeks, with frequent changes of medium to eliminate dead cells and cellular debris, until distinct colonies are visible. These colonies are then trypsinized and transferred to microtitre wells for further culture in the presence of the selective medium. Since transfected DNA can be relatively unstable in the host genome, and reversion rates of transfected cells are high, the cells, once selected, should be cultured continuously in the presence of the selectable drug.

There is a broad and expanding range of selectable markers, and four are described below.

Thymidine kinase

A number of cell lines have been isolated which lack the thymidine kinase gene and so lack the ability to convert thymidine to dIMP. Selective pressure to maintain the tk^- phenotype is exerted by adding bromodeoxyuridine to the growth medium: cells that are tk^- cannot metabolize this lethal analogue and remain viable. Cells transfected with a plasmid bearing a functional *tk* gene are selected in medium containing hypoxanthine, aminopterin, and thymidine (HAT medium).

Xanthine-guanine phosphoribosyl transferase

The enzyme xanthine–guanine phosphoribosyl transferase (XGPRT) is a bacterial enzyme encoded by the *gpt* gene, with no mammalian equivalent, that converts xanthine to xanthosine monophosphate. Following transfection, vectors expressing the *gpt* gene (e.g. pMSG, see *Figure 3A*) permit mammalian cells to grow in medium containing adenine, xanthine, mycophenolic acid, and aminopterin.

Resistance to aminoglycoside antibiotics

The bacterial gene for aminoglycoside 3′-phosphotransferase (APH) is the most widely used dominant selection system. Two APH bacterial transposons Tn5 and Tn601, each encoding an *aph* gene, are commonly used and confer resistance to aminoglycoside antibiotics such as kanamycin, neomycin, and geneticin (G418). *Figure 3B* shows an example of a plasmid containing one of these *aph* genes, pRSVneo, which is commonly used in transfections. Unfortunately, however, a major disadvantage of the *aph* selectable marker is that G418 is very expensive compared to other selectable drugs.

Resistance to hygromycin B

Vectors containing the gene for hygromycin B phosphotransferase have been used to prevent the killing of cells by the antibiotic hygromycin B, an inhibitor of protein synthesis.

Procedures for the use of these selectable markers are given in *Protocol 5*.

Protocol 5. Selection of stably transfected cells

Reagents

A. *For thymidine kinase (TK) selection*

- HAT stock
  [Dissolve 15 g hypoxanthine and 1 mg aminopterin in 8 ml 0.1 M NaOH. Adjust to pH 7.0 with 1 M HCl. Add 5 mg thymidine and make the volume to 10 ml with water]
- 1 × HAT
  [Prepare by diluting HAT stock 100-fold in culture medium].

B. *For xanthine–guanine phosphoribosyl transferase (XGPRT) selection*

- MPA stock
  [Dissolve 250 mg mycophenolic acid (MPA) in 9 ml 0.1 M NaOH, adjust the pH to 7.0 with 1 M HCl and add water to 10 ml]. Filter sterilize and store at −20°C.
- XAT stock
  [Dissolve 1.875 g xanthine, 189 mg adenosine, 15 mg aminopterin, 1 g glutamine, and 75 mg thymidine in 450 ml 0.1 M NaOH]. Filter sterilize and store at 4°C.

Protocol 5. *Continued*

- Working solution
  [Add 0.1 ml MPA stock and 12 ml XAT stock to 200 ml culture medium].

C. *For selection by resistance to geneticin (G418)*

- G418 stock
  [Prepare a 40 mg/ml stock solution of G418 sulphate (Gibco/BRL) in 100 mM Hepes pH 7.3]
- G418 working solution
  [Add 100 μl G418 stock solution to 10 ml of culture medium to give a final concentration of 400 μg/ml].

Note, however, that there are large variations in the potency of various batches of G418 and in the cellular susceptibility to G418. The exact required concentration should be determined empirically by testing the effect of G418 over the range 0.2–1.0 mg/ml on each untransfected cell line.

D. *For selection by resistance to hygromycin B*

- Hygromycin B stock
  [Prepare a solution of hygromycin B (100 mg/ml) in water, filter sterilize it, and store at −20°C]
- Hygromycin B working solution
  [Use in culture media at 10–400 μg/ml].

Method

The following applies to all the above selectable drugs.

1. 48 h after transfection, split the cultures to a 1:5 or greater dilution and plate in the drug-containing medium.
2. Feed the cell cultures with drug-containing medium every 3 days, examining the cultures under a microscope to determine the efficiency of drug selection.

4.1.5 Amplification

The level of expression of a foreign gene in stably transfected cells depends on the promoter/enhancer sequences present and the efficiency with which the transcript is translated. However, expression levels can also be increased by drug-induced amplification of DNA sequences. An example of this is *dhfr* gene amplification in response to methotrexate. When cells, stably transfected with a vector containing the *dhfr* gene and a second gene under study, are treated with increasing concentrations of methotrexate, clonal lines can be selected which contain amplified copies of the *dhfr* and flanking sequences containing the gene of interest. Gene amplification is achieved by step-wise increases in the concentration of the selective agent.

Other drug amplification systems are shown in *Table 3* (see ref. 21 for detailed protocols and strategies).

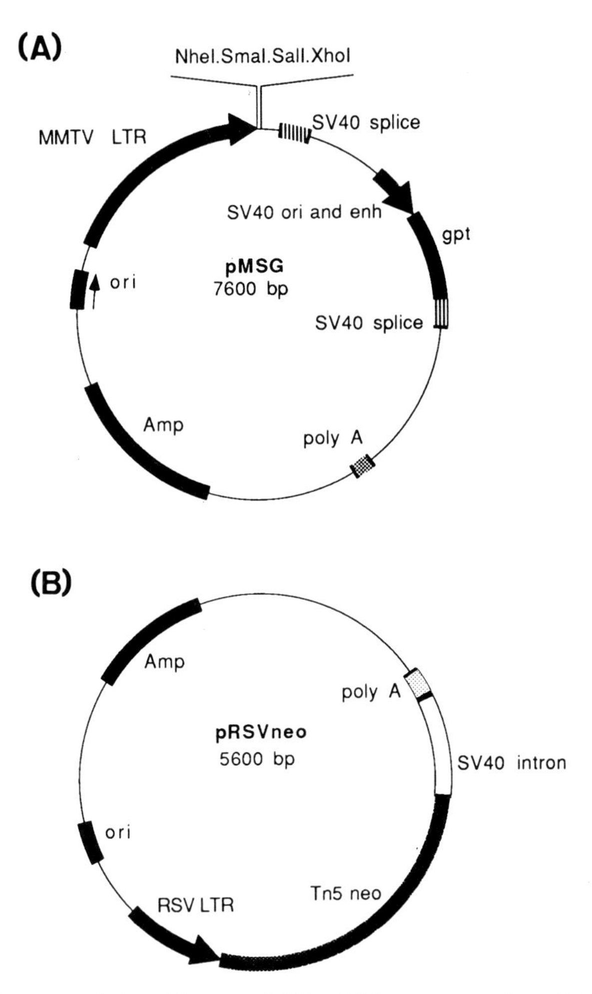

Figure 3. (A) Structure of plasmid vector pMSG. cDNA sequences cloned into the multiple cloning site (MCR; shown with its *Nhe*I, *Sma*I, *Sal*I and *Xho*I sites) are driven by the glucocorticoid-inducible mouse mammary tumour LTR promoter (MMTV LTR). The *E. coli* xanthine guanine phosphoribosyl transferase gene (gpt), which permits selection of stable transformants, is expressed from the SV40 promoter located in the SV40 origin (SV40 ori). Copies of the SV40 small T intron (SV40 splice) and polyadenylation signal (poly A) are located downstream from both the MCR and the gpt gene. Amp and ori are an ampicillin resistance gene and a bacterial origin of replication, respectively. (B) Structure of plasmid vector pRSVneo. The aminoglycoside phosphotransferase gene (Tn5 neo) is driven by the Rous sarcoma virus LTR promoter (RSV LTR). The SV40 small T intron (SV40 intron) and polyadenylation signal (poly A) are located downstream of the Tn5 neo. Also present in the vector is a bacterial origin of replication (ori) and an ampicillin resistance gene (Amp). This vector is used in cotransfections to select for G418-resistant transformants.

Table 3. Gene amplification systems[a]

Gene	Selection	Comment	Ref.
Dihydrofolate reductase (DHFR)	Methotrexate	Amplification normally performed with DHFR-deficient CHO cells	(52)
CAD[b]	PALA (*N*-phosphonacetyl-L-aspartate)	CAD gene product possesses three enzymatic activities; carbamyl phosphate synthetase, aspartate transcarbamylase and dihydroorotase	(53)
Adenosine deaminase	Xyl-A (9-D-xylofuranosyl adenine) and dCF (2′-deoxycoformycin)	Dominant selectable/amplifiable gene	(54)
P-glycoprotein 17	Multiple drugs	Dominant selectable gene. Selection normally in medium containing colchicine	(55)
Ornithine decarboxylase	D-difluoromethylornithine (DFMO)	Select in medium containing putrescine and DFMO	(56)
Asparagine synthetase	β-aspartyl hydroxamate (β-AH) and albizziin	In AS^- cells amplify in medium minus aspargine supplemented with glutamine and β-AH. In AS^+ cells add albizziin	(57)

[a] For further details and detailed protocols see refs. 21, 58, and 59.
[b] CAD is carbanoyl-phosphate synthase–aspartate transcarbamylase-dihydroorotase.

5. The characterization of transcriptional regulatory sequences

The regulation of eukaryotic gene transcription is dependent on the interaction of *cis*-acting regulatory sequences with gene regulatory proteins within the nucleus. The identification of these sequences and an understanding of the underlying mechanisms by which genes are expressed in a tissue-specific and temporal manner or respond to extracellular signals has developed principally from an ability to transfer cloned genes into cells in culture. The general scheme for such studies usually involves modification of the native sequence either by deletions or by single or multiple point mutations. The effect of these changes on the expression of the gene can be tested by introducing the modified gene into eukaryotic cells in culture and examining transient expression.

Introduction of a mutant gene into a cell which contains a functional copy of the same gene causes practical problems in differentiating between the product of the exogenous gene and that of the endogenous gene. To overcome this problem, putative regulatory sequences or their modified derivatives are usually joined to a reporter gene (that is a gene which is normally expressed in a very low or undetectable levels in the cell of interest) and the expressed reporter gene product is measured in extracts prepared from the transfected cells. In recent years, this approach has led to the identification of distinct promoter, enhancer, and silencer elements in many eukaryotic genes and has characterized sequences which mediate cell-specific and inducible expression (22).

The next section describes a variety of expression vectors containing suitable reporter genes and is followed by a detailed consideration of the practical procedures required in experiments of this kind. Much of this work involves manipulating DNA and moving it from plasmid to plasmid (subcloning). In our experience, many of the most elaborate strategies falter through difficulties incurred in subcloning fragments of DNA from plasmid to plasmid. This is usually because of either a failure to trim the ends of the DNA or fragment properly, or loss of the DNA during a gel purification step. In this context, the following protocols are provided to help with these important DNA manipulation steps.

5.1 Vectors containing reporter genes

The most commonly used reporter genes are those which encode enzymes such as alkaline phosphatase, bacterial β-galactosidase, firefly luciferase, and chloramphenicol acetyl transferase (CAT). Non-enzymic reporters include SV40 virus T antigen, globin, and growth hormone. The most widely used reporter gene is CAT, particularly since this has a low background activity in mammalian cells.

5.1.1 Novel CAT expression vectors

The original CAT vector, pSVOCAT (23), had several disadvantages, notably a lack of suitable cloning sites and a low copy number in *E. coli*. We therefore constructed the pBC series of CAT vectors (*Figure 4*) (24, 25) which have several advantages over the original pSVOCAT. Plasmid pBCO is designed for assaying the promoter activity of fragments cloned 5′ to the CAT. It contains a convenient multiple cloning region (MCR), and the CAT gene upstream of the SV40 small T intron and polyadenylation signal. Plasmids pBCSVp and pBCTKp contain the SV40 virus early promoter and

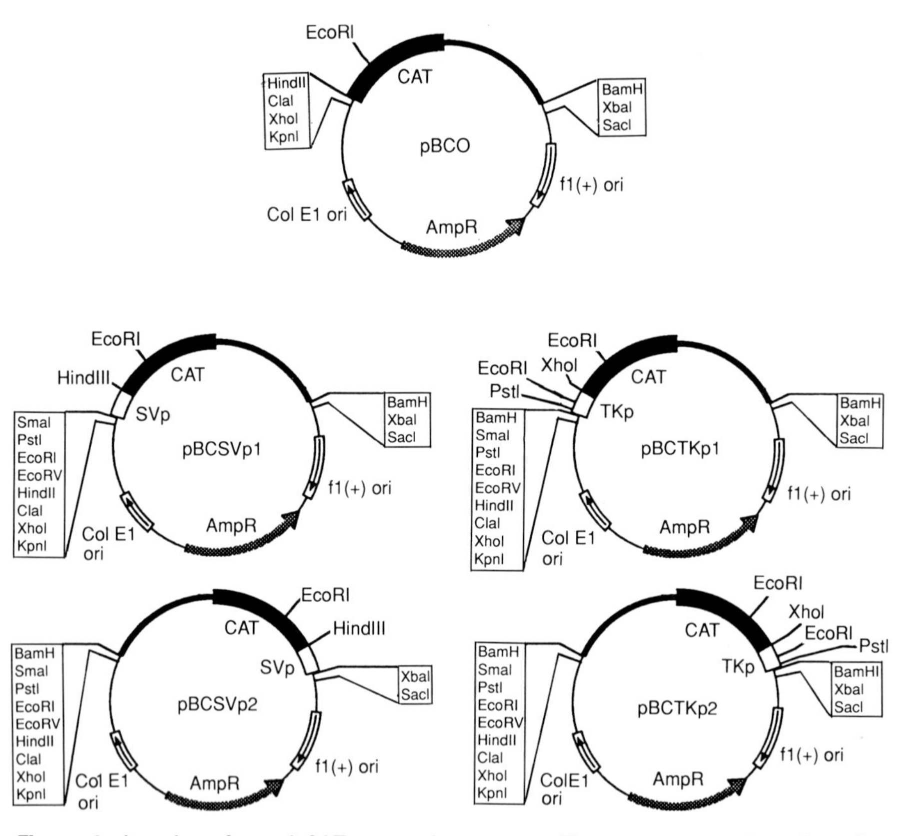

Figure 4. A series of novel CAT expression vectors. These vectors are based on the Bluescript plasmid. pBC0 contains the multiple cloning region (MCR) for the introduction of test sequences 5′ to the CAT gene whereas in pBCSVp1 and pBCTkp1 the MCR is located 5′ to the SV40 virus early promoter (SVp) and the herpes simplex virus thymidine kinase promoter (TKp) respectively. In plasmids pBCSVp2 and pBCTKp2, the MCR is located 3′ to the CAT gene. Other abbreviations used are as described in the legend to *Figure 3* except F1(+) ori which is an origin of replication from filamentous bacteriophage F1 which permits rescue of single-stranded plasmid. Further details are provided in the text.

herpes simplex virus thymidine kinase promoter respectively. These plasmids are designed for assaying enhancer or silencer activities of fragments cloned 5′ to the promoter. Plasmids pBCSVp and pBCTKp are also available with the MCR located 3′ of the CAT gene. These plasmids are used to test the effect of putative regulatory sequences when placed 3′ of the reporter gene. We have observed that the TK and SV promoters have different activities in a given cell type. Both vectors should therefore be tested in preliminary experiments before embarking on a detailed enhancer mapping exercise.

These novel CAT expression vectors were constructed by subcloning appropriate fragments of DNA in the Bluescript vector. They therefore possess all the advantages of Bluescript:

- high copy-number Col E1 origin of replication
- F′ origin of replication, making possible the rescue of single-stranded plasmid for use in sequencing and site-directed mutagenesis
- a large polylinker with nested 5′ and 3′ overhang restriction sites, designed for the synthesis of deletions by the exonuclease III/mung bean nuclease technique (25). This polylinker also makes cloning and orientation of fragments easier
- T3 and T7 promoters either side of the polylinker (not shown in *Figure 4*) which can be used for the generation of labelled RNA transcripts for RNase mapping.

5.1.2 Luciferase expression vectors

Firefly luciferase has become increasingly popular as a reporter gene. The principle advantage is that the luciferase assay is approximately 100-fold more sensitive than assays for CAT activity. A series of luciferase reporter plasmids (GeneLight^{TM}, pGL) are available from Promega. The pGL2-basic vector lacks promoter or enhancer sequences; pGL2-enhancer contains an SV40 enhancer; pGL2-promoter contains an SV40 promoter; and pGL2-control contains SV40 promoter and enhancer sequences.

5.2 Subcloning DNA fragments or oligonucleotides into reporter plasmids

The putative regulatory sequence to be studied in a transfection experiment may be a fragment derived from a particular gene or a synthetic oligonucleotide corresponding to a genomic sequence. Whatever the source, one of the first stages in the study is to subclone the fragment into the reporter plasmid.

5.2.1 Preparation of the DNA fragment for ligation into the vector

With luck it may be possible to isolate the DNA as a fragment with overhanging ends which can simply be annealed into appropriate sites of the vector

MCR and ligated. However, more likely than not, it will be necessary to trim the DNA ends prior to cloning into a blunt-ended site e.g. *Sma*I. The first step is to digest the cloned gene with appropriate restriction endonucleases and purify the desired fragment by electrophoresis in an agarose gel (*Protocol 6*) or polyacrylamide gel (*Protocol 7*). The fragment is then rendered blunt-ended by filling in 5′ overhangs with DNA polymerase I (Klenow fragment) (*Protocol 8*), or digesting 3′ overhangs using the exonuclease activity of T4 DNA polymerase (*Protocol 9*). An alternative approach is to use mung bean nuclease to digest single-stranded tails (*Protocol 10*). This enzyme acts on both 3′ and 5′ overhangs. Mung bean nuclease is preferable to nuclease S1 for this purpose due to its very low intrinsic activity on double-stranded DNA.

Protocol 6. Purification of DNA fragments and plasmid vectors from agarose gels

Reagents

- DNA to be fractionated
- 6 × gel-loading buffer [0.25% (w/v) bromophenol blue, 0.25% (w/v) xylene cyanol FF, 30% (v/v) glycerol in water]. Store at 4°C.
- Ethidium bromide stock (10 mg/ml in water). *Caution*—ethidium bromide is a powerful mutagen
- Buffer-equilibrated phenol
- Phenol:chloroform:isoamyl alcohol (25:24:1)
- 0.5 × Tris/borate/EDTA (TBE) (0.045 M Tris-borate, 1 mM EDTA, pH 8.0) [Prepare this solution as a 5 × stock by adding 54 g Tris and 27.5 g boric acid to 900 ml water. Stir on a heated stirring plate, then add 20 ml 0.5 M EDTA pH 8.0, and make up to 1 litre with water. This stock is usually made up in 5 litre batches and stored at room temp. in a container with a plastic stopper.]
- Agarose (BRL Ultrapure).

Method

1. Knowing the size of the DNA fragment(s) to be purified, prepare an appropriate agarose gel (0.8–2.0%, see *Appendix 2*). To do this, dissolve the agarose in 0.5 × TBE buffer, place in a microwave oven for a few minutes and then allow the solution to cool to about 50°C before adding ethidium bromide to a final concentration of 0.5 μg/ml. Pour the agarose solution into a horizontal gel mould.
2. After suitable restriction endonuclease digestion, electrophorese the DNA sample on the agarose gel, if necessary using a wide-toothed comb to allow large volumes to be loaded. Use 0.5 × TBE as the electrophoresis buffer.
3. Visualize the fractionated DNA using a UV transilluminator, and excise a portion of the gel containing the desired fragment using a scalpel blade.
4. Chop the gel slice into small pieces and place these into a microcentrifuge tube. Add an approximately equal volume of buffer-equilibrated phenol. Vortex the microcentrifuge tube for 1 min, and then place the tube in a dry ice–methanol bath for 15 min.

5. Centrifuge the tube for 15 min in a microcentrifuge at room temp.
6. Collect the aqueous phase and place this into a fresh microcentrifuge tube.
7. Re-extract the phenol phase with 200 μl distilled water, and combine this with the previous aqueous phase.
8. Extract the combined aqueous phases with phenol:chloroform:isoamyl alcohol and then precipitate the DNA by adding 2 vol. ethanol and leaving on ice for 15 min.
9. Recover the DNA precipitate by centrifugation for 10 min in a microcentrifuge and dissolve in the buffer required for storage or the next procedure.

Protocol 7. Purification of DNA fragments by polyacrylamide gel electrophoresis

Reagents

- DNA to be fractionated
- 6 × gel-loading buffer (see *Protocol 6*)
- 40% (w/v) acrylamide (38:2, acrylamide: bis-acrylamide)
- 5 × TBE stock (see *Protocol 6*)
- TEMED (*N,N,N′,N′*-tetramethylethylenediamine)
- 20% (w/v) ammonium persulphate [Prepare a fresh solution weekly]
- 1 × TBE (diluted from 5 × TBE, see *Protocol 6*)
- Ethidium bromide stock solution (see *Protocol 6*)
- X-ray film (Fuji RX)
- High salt elution buffer (10 mM Tris–HCl pH 7.5, 1 mM EDTA, 0.3 M NaCl pH 8.0)
- TE buffer pH 8.0 (10 mM Tris–HCl, 1 mM EDTA pH 8.0)
- 1 ml Sephadex G50 spun column previously equilibrated in TE buffer pH 8.0.[a]

Method

1. Knowing the size of the DNA fragment(s) to be purified, prepare a polyacrylamide gel of the required concentration (see *Appendix 2*) as follows. Add to a flask in order:

40% (w/v) acrylamide	volume appropriate to chosen gel concentration
5 × TBE	10 ml
water	to 49.8 ml
TEMED	75 μl
20% (w/v) ammonium persulphate	150 μl.

   Mix and pour the gel. Allow to polymerize undisturbed.
2. Load the DNA samples and electrophorese in 1 × TBE for approximately 4 h at 150 V.
3. After electrophoresis, for unlabelled DNA fragments, stain the gel with

Protocol 7. *Continued*

0.5 μg/ml ethidium bromide and observe the position of the required DNA fragments using an UV transilluminator. For radiolabelled DNA fragments wrap the gel in Saran wrap and expose it to X-ray film (Fuji RX) for 5–60 min. Identify the position of the fragment(s) from the resultant autoradiograph.

4. Place the gel behind a small perspex shield, and excise that portion of the gel containing the fragment using a scalpel blade. Cut the gel into small pieces and place these into a microcentrifuge tube.

5. Add 1 ml of high salt elution buffer and rotate the tube on a revolving wheel overnight at room temperature.

6. Collect the elution buffer and centrifuge through a 1 ml Sephadex G50 spun column,[a] previously equilibrated in TE pH 8.0. The eluate contains the DNA, free of polyacrylamide.

[a] To make this column, plug a 1 ml syringe with siliconized and autoclaved glass wool, and fill with a slurry of Sephadex G-50 (M) sterilized by autoclaving. Place the syringe in a 15 ml plastic centrifuge tube and centrifuge in a bench top centrifuge at 250 *g* for 5 min. Add 0.5 ml of water or a suitable buffer (for example, TE buffer) and centrifuge again. Then cut off the cap from a 1.5 ml microcentrifuge tube and place this in the centrifuge tube to collect the column eluate. Apply the DNA sample in a volume of 50–100 μl to the syringe and centrifuge exactly as before. The DNA can be collected from the microcentrifuge tube.

Protocol 8. Blunt-ending DNA fragments by filling in 5′ overhangs

Reagents

- DNA fragment with 5′ overhang(s); usually less than 1 μg per reaction
- 10 × 5′ overhang buffer [0.5 M Tris–HCl pH 7.2, 0.1 M $MgSO_4$, 1 mM dithiothreitol (DTT), 500 μg/ml bovine serum albumin (BSA)]
- 2 mM dNTP mixture (2 mM each of dATP, dCTP, dGTP and dTTP in 20 mM Tris–HCl pH 7.5)
- DNA polymerase I (Klenow fragment); 1 unit/μl.

Method

1. Set up the following reaction in a microcentrifuge tube.

DNA fragment	2 μl
10 × 5′ overhang buffer	2 μl
DNA polymerase I (Klenow fragment)[a]	2 units
H_2O	to 20 μl final volume.

2. Incubate for 15 min at room temperature.

3. Inactivate the enzyme by incubating at 70°C for 5 min. Add an equal volume of TE buffer (pH 7.6) and extract the solution with phenol:chloroform.

4. Collect the DNA by ethanol precipitation as follows. Add 0.1 vol. of 3 M sodium acetate (pH 5.2)[b] and 2 vol. of ice-cold ethanol.
5. Store the tube on ice for 15 min, then centrifuge for 10 min in a microcentrifuge.
6. Carefully remove the supernatant and add 1 ml of 70% (v/v) ethanol.
7. Centrifuge in a microcentrifuge for 2 min.
8. Repeat steps 6 and 7 then leave the open tube at room temperature for a few minutes to dry off all traces of ethanol.

[a] The Klenow fragment of DNA polymerase I works well in almost all buffers used for digestion of DNA with restriction enzymes.
[b] 2.0–2.5 M ammonium acetate is frequently used instead of sodium acetate to reduce co-precipitation of dNTPs.

Protocol 9. Blunt-ending DNA fragments by removing 3′ overhangs

Reagents

- DNA fragment with 3′ overhang(s); usually less than 1 μg per reaction
- 2 mM dNTP mix (see *Protocol 8*)
- T4 DNA polymerase
- 10 × 3′ overhang buffer[a] (0.3 M Tris-acetate pH 7.9, 0.66 M potassium acetate, 0.1 M magnesium acetate, 5 mM DTT, 1 μg/μl BSA).

Method

1. Set up the following reaction in a microcentrifuge tube:

DNA fragment	2 μl
10 × 3′ overhang buffer	2 μl
2 mM dNTP mixture	2 μl
T4 DNA polymerase	2 units
H_2O	to 20 μl final volume.

2. Incubate at 12°C for 15 min.[b]
3. Inactivate the enzyme by incubating at 70°C for 5 min.
4. Add an equal volume of TE buffer (pH 7.6) and extract the solution with phenol:chloroform.
5. Collect the DNA by precipitation with 2 vol. ethanol (see *Protocol 8*, steps 4–8).

[a] T4 DNA polymerase is active in all buffers commonly used for digestion of DNA with restriction enzymes. The filling reaction can therefore be carried out by adding the polymerase directly to the digestion mixture together with the four dNTPs.
[b] At 37°C the 3′ exonucleolytic activity of T4 DNA polymerase is threefold greater than its polymerizing activity, while at 12°C there is a smaller difference between the two activities. Thus when the enzyme trims back to the double-stranded region, there is a balance between further removal of nucleotides and incorporation of dNTPs when the reaction is performed at 12°C.

Protocol 10. Blunt-ending DNA fragments using mung bean nuclease

Reagents

- 10 × mung bean nuclease buffer (0.3 M sodium acetate pH 5.0, 0.5 M NaCl, 10 mM $ZnCl_2$, 50% glycerol)
- Mung bean nuclease dilution buffer [10 mM sodium acetate pH 5.0, 0.1 mM zinc acetate, 1 mM cysteine, 0.1% (v/v) Triton X-100, 50% (v/v) glycerol]
- DNA fragment (approx. 1 μg)
- Mung bean nuclease; this is supplied at high concentration (NBL) and immediately before use is diluted to 5 U/μl in dilution buffer.

Method

1. Set up the following reaction in a microcentrifuge tube:

DNA fragment	1 μg (approx.)
10 × mung bean nuclease buffer	2 μl
mung bean nuclease	5 units
H_2O	to 20 μl final volume.

2. Incubate at room temperature for 30 min.
3. Inactivate the mung bean nuclease by incubating at 70°C for 10 min.
4. Add an equal volume of TE buffer (pH 7.6) and extract the solution with phenol:chloroform.
5. Collect the DNA by precipitation with 2 vol. ethanol (see *Protocol 8*, steps 4–8).

5.2.2 Preparation of the vector

Another crucial stage in subcloning DNA fragments is preparation of the plasmid vector. The plasmid vector should be completely linearized with a restriction endonuclease and treated with calf intestinal phosphatase (CIP) (*Protocol 11*) to prevent recircularization and so reduce the background of non-recombinants. Since this stage is so important, we routinely repurify the linearized plasmid DNA from agarose gels using a simple 'phenol squeeze' method (see *Protocol 6*). This ensures that only fully linearized vector is used in ligations and also removes all trace of CIP.

Protocol 11. Preparation of dephosphorylated linearized plasmid vectors

Materials

- Plasmid DNA
- Appropriate restriction endonucleases
- Agarose gel electrophoresis reagents (see *Protocol 6*)
- TE buffer pH 8.0 (10 mM Tris–HCl, 1 mM EDTA pH 8.0)
- Calf intestinal phosphatase (CIP) (1 unit/μl)
- CIP buffer (10 mM $ZnCl_2$, 10 mM $MgCl_2$, 100 mM Tris–HCl pH 8.3)
- Phenol:chloroform:isoamyl alcohol (25:24:1).

Method

1. Linearize the plasmid vector by digestion with the appropriate restriction endonuclease(s). Monitor the progress of digestion by agarose gel electrophoresis (see *Protocol 6*).
2. When digestion is complete, extract the digest solution with phenol: chloroform:isoamyl alcohol and collect the DNA following precipitation with 2 vol. ethanol (see *Protocol 8*, steps 4–8). Resuspend the DNA in 43 μl TE buffer pH 8.0.
3. Add 5 μl CIP buffer and 2 μl (2 units) CIP. Incubate at 37°C for 1 h.
4. For blunt-ended plasmids, after incubating the reaction at 37°C for 1 h, add another 2 units of CIP and incubate the reaction mix at 56°C for a further hour.
5. Extract twice with phenol:chloroform:isoamyl alcohol and then ethanol precipitate (see *Protocol 8*, steps 4–8).
6. Purify the phosphatased linearized plasmid by electrophoresis on an agarose gel and recover it using the procedure described in *Protocol 6*.

5.2.3 Cloning oligonucleotides

Oligonucleotides can be synthesized with appropriate sticky ends to facilitate cloning. Before ligation into the plasmid, 5′ phosphate groups must be added to the oligonucleotide using T4 polynucleotide kinase (*Protocol 12*).

Protocol 12. Preparation of double-stranded oligonucleotides for cloning

Reagents

- Two complementary oligonucleotides (100 pmol of each)
- 20 × annealing buffer (0.2 M Tris–HCl pH 7.9, 40 mM $MgCl_2$, 1 M NaCl, 20 mM EDTA)
- T4 polynucleotide kinase (1 unit/μl)
- 10 × polynucleotide kinase buffer[a] (0.5 M Tris–HCl pH 7.6, 0.1 M $MgCl_2$, 50 mM DTT, 1 mM spermidine, 1 mM EDTA)
- 10 mM ATP
- 0.1 M DTT.

Method

1. In a microcentrifuge tube, mix 100 pmol each of two complementary oligonucleotides in a total volume of 20 μl containing 1 μl of 20 × annealing buffer.
2. Incubate the mixture at 90°C for 5 min then allow it to cool slowly to room temperature.

Protocol 12. *Continued*

3. Set up the following reaction mixture:

annealed oligonucleotides	10 μl
10 × polynucleotide kinase buffer	2 μl
10 mM ATP	1 μl
0.1 M DTT	1 μl
T4 polynucleotide kinase	10 units
H_2O	to 20 μl final volume.

4. Incubate at 37°C for 30 min.

5. Inactivate the enzyme by incubating at 70°C for 10 min.

6. Add an equal volume of TE buffer (pH 7.6) and extract the solution with phenol:chloroform.

7. Collect the DNA by precipitation from 2 vol. ethanol (see *Protocol 8*, steps 4–8).

[a] This buffer can also be made up containing ATP and DTT. However, we prefer to add these separately to the reaction mixture since they are less stable than the other components during prolonged storage.

5.2.4 Ligations and transformations

The ligation of DNA fragment into the MCR of the CIP-treated linearized vector is described in *Protocol 13*. When the recombinant vector has been produced, the next step is to prepare and transform competent *E. coli* host cells. Methods for this can be found in ref. 5.

Protocol 13. Ligation of DNA fragments or oligonucleotides to linearized vector

Reagents

- Dephosphorylated linearized plasmid vector (see *Protocol 11*); approx. 0.1–1.0 μg
- DNA fragment (or phosphorylated double-stranded oligonucleotide; see *Protocol 12*)
- 5 × ligase buffer (250 mM Tris–HCl pH 7.6, 50 mM $MgCl_2$, 5 mM ATP, 5 mM DTT, 25% (w/v) polyethylene glycol 8000)[a]
- T4 DNA ligase (2 Weiss units/μl).

Method

1. Use 0.1–1.0 μg of vector DNA per ligation reaction, with an approximately equimolar quantity of DNA insert for cohesive ends, or an approximately 10-fold excess of insert to vector for ligation of blunt-ended DNA or double-stranded oligonucleotides.

2. For each insert, set up the following reaction mixture in a microcentrifuge tube:

vector DNA	see step 1
insert DNA or oligonucleotide	see step 1

5 × ligase buffer	4 μl
T4 DNA ligase	2 Weiss units
H_2O	to 20 μl final volume.

3. Incubate overnight at 4°C for blunt-ended ligations, or at 16°C for cohesive end ligations. A control ligation should be performed with dephosphorylated linearized vector in the absence of insert to determine the extent of vector-only ligation occurring.

[a] This buffer contains polyethylene glycol-8000 as a condensing agent which increases the efficiency of ligation, and is suitable for all types (blunt and sticky-ended) of ligation.

5.3 Transfection of reporter constructs into cultured cells

DNA for transfections should be prepared from large-scale plasmid preparations and purified by equilibrium centrifugation in a CsCl gradient containing ethidium bromide. The ethidium bromide is then removed by extracting with butan-1-ol (*Protocol 1*). We then routinely use the calcium phosphate precipitation method for transfecting reporter gene constructs into cells in culture (*Protocol 2*).

5.4 Analysis of reporter gene products in transfected cells

Approximately 48 h after transfection, the cells are harvested and a cytoplasmic extract is prepared for assay of reporter gene activity. *Protocol 14* describes a suitable procedure for the preparation of cell extracts and the following sections provide details of reporter gene assays.

Protocol 14. Preparation of a cytoplasmic extract from transfected cells

Reagents

- Transfected cell monolayer
- PBS (see *Protocol 4*)
- 0.25 M Tris–HCl, pH 8.0.

Method

1. Approximately 48 h after transfection, remove the growth medium, and wash the cell monolayer with 10 ml PBS. Remove the wash and discard.
2. Add 1 ml PBS and dislodge the cells from the plate by scraping with a plastic spatula (rubber policeman).
3. Transfer the resuspended cells to a microcentrifuge tube and centrifuge for 1 min.

Protocol 14. *Continued*

4. Remove the supernatant and resuspend the cells by vortexing in 250 μl 0.25 M Tris–HCl pH 8.0.
5. Lyse the cells by repeated (three or four) cycles of freezing and thawing. Each cycle consists of 5 min in a dry ice–methanol bath and 5 min in a water bath at 37°C.
6. Centrifuge for 1 min, and transfer the supernatant to a fresh tube. If necessary, the extract can be stored at −20°C.

5.4.1 Assay for CAT activity

The role of CAT in bacteria is to modify chloramphenicol by mono- and diacetylation. CAT activity can therefore be measured by following the conversion of [^{14}C]chloramphenicol to its 1-acetyl and 3-acetyl derivatives (*Protocol 15*). The mono- and diacetyl derivatives of chloramphenicol are separated from the unmodified compound by thin-layer chromatography on silica gels. The silica gel is then exposed to X-ray film for autoradiography. The assay is relatively simple, but is not ideally suited for the large number of samples which can be generated in systematic investigations of putative regulatory sequences. For valid comparison of different samples, the protein content in individual cell extracts must be measured and the CAT assay performed using a similar amount of protein in each extract. A simple and quick protein assay kit is supplied by BioRad. An alternative is to use a β-galactosidase internal control (see Section 5.4.3).

If transcription factors are present in limiting amounts, the observed CAT activity may be very sensitive to the number of plasmids which enter a transfected cell, and therefore be affected by transfection efficiency. Since transfection efficiency can vary between experiments, it is important that the recombinant plasmids to be compared are transfected at the same time and an adequate number of replicates used.

The assay itself becomes non-linear as the substrate is exhausted, at which time comparison between samples is no longer meaningful. Therefore, reactions should be stopped before significant conversion to the 1,3-diacetyl form has occurred. Assays should be run for different times (e.g. 15 min, 1 h, 4 h) in order to determine the optimum time-course for the quantitative assays.

Protocol 15. Chloramphenicol acetyl transferase (CAT) assay

Reagents

- Cytoplasmic extract (see *Protocol 14*)
- 0.5 M Tris–HCl pH 7.8
- D-*threo*-[dichloroacetyl-1-^{14}C]chloramphenicol; (Amersham; 57 mCi/mmol, 25 μCi/ml)
- 8 mM acetyl-coenzyme A (Sigma)
- Ethyl acetate (BDH)
- Silica gel thin-layer chromatography (TLC) plate (Merck, type 553)

- Chloroform:methanol [95:5 (v/v)] (both from BDH; chromatography grade)
- X-ray film.

Method

1. Set up the following reaction mixture in a microcentrifuge tube by adding in order:

cytoplasmic extract[a]	100 μl
0.5 M Tris–HCl pH 7.8	70 μl
H_2O	90 μl
D-*threo*-[dichloroacetyl-1-^{14}C]chloramphenicol	20 μl
8 mM acetyl-coenzyme A	20 μl.

2. Incubate at 37°C for the optimum time period determined by prior experimentation (see Section 5.4.1). For particularly long incubations, add further aliquots of 20 μl acetyl-coenzyme A at 2 h intervals.
3. Extract by vortexing the reaction mixture with 1 ml of ethyl acetate for 30 sec.
4. Centrifuge in a microcentrifuge for 30 sec and then transfer the upper (organic) phase to a fresh tube.
5. Dry down the ethyl acetate in a vacuum evaporator (Savant Speed Vac) and resuspend the residue in 20 μl ethyl acetate.
6. Apply the sample to a silica gel TLC plate and subject the plate to ascending chromatography in a chamber containing 100 ml of a 95:5 (v/v) mixture of chloroform and methanol. When the solvent front is approximately 1 cm from the top of the plate, remove the plate from the tank, air-dry, and expose to X-ray film overnight at room temperature.
7. For purposes of quantification, use the autoradiograph to determine the position of acetylated and non-acetylated ^{14}C-chloramphenicol on the TLC plate. For each sample cut out a square corresponding to the monoacetylated form from the silica plate, add it to 5 ml scintillation fluid and count in a scintillation spectrometer.

[a] Quantities of cytoplasmic extract added to CAT assays should be normalized, against either β-galactosidase activity (see Section 5.4.3) or against protein concentration (see Section 5.4.1). Make up the volume to 100 μl with 0.25 M Tris–HCl pH 8.0. Incubate the extract at 65°C for 10 min to inactivate deacetylases.

5.4.2 Luciferase

The assay for luciferase is based on the oxidation of beetle luciferin with concomitant production of photons of light (*Protocol 16*).

Protocol 16. Luciferase assay[a]

Equipment and reagents

- Luminometer (Berthold Instruments model LB 9501 or similar instrument)
- 5 ml Sarstedt vials
- Cell extraction buffer (0.1 M potassium phosphate, 1 mM EDTA, pH 7.8)
- 30 mM glycylglycine (pH 7.8)
- Luciferase (10 μg/ml)
- Luciferase reaction buffer [30 mM glycylglycine (pH 7.8), 2 mM ATP, 15 mM $MgSO_4$]. Add solid $MgSO_4$ and ATP (18 mg $MgSO_4$ and 12 mg ATP per 10 ml) to 30 mM glycylglycine and adjust the pH to 7.8 immediately.
- 10 mM luciferin (Sigma L-9504; make up this solution in 30 mM glycylglycine (pH 7.8).

Method

1. Harvest the transfected cells. Pellet 5×10^6 cells by centrifugation and resuspend them in 100 μl extraction buffer. Prepare a cell extract by repeated cycles of freezing and thawing (see *Protocol 14*[b]).
2. Aliquot 350 μl of the luciferin reaction buffer into 5 ml Sarstedt vials and place the vials on ice. Prepare sufficient reaction vials for the number of samples plus a positive control.
3. Add 10 μl luciferase solution to the positive control vial and aliquots of cell extract to the sample vials.
4. Prepare the luciferin substrate solution by adding a volume of 30 mM glycylglycine (pH 7.8) equivalent to 150 μl per sample to be analysed to a 5 ml Sarstedt vial. Then add 0.1 vol. luciferin solution.[c]
5. Place a reaction vial in the tube holder of the luminometer and add 10 μl of the luciferin substrate solution. Record the light emission over approximately 10 sec.

[a] Method of de Wet *et al.* (60) as adapted by Dr R. Henschler, Paterson Institute of Cancer Research, Manchester, UK.
[b] The luciferase activity in the extract decreases rapidly with storage, even at −70°C.
[c] The final luciferin concentration of this solution can be reduced to 0.5 mM without affecting the sensitivity of the assay.

5.4.3 β-Galactosidase assay

When measuring the effect of potential transcriptional regulatory sequences on reporter gene activity, an internal control is usually included that will distinguish differences in transcriptional activity from differences in transfection efficiency. This is achieved by co-transfecting the cells with the reporter gene construct (e.g. CAT or luciferase) and a vector that contains sequences encoding a second enzyme that can be measured in the cell extract along with the vector of interest. A vector commonly used for co-transfection with CAT constructs is RSVβgal, which contains the coding sequences for bacterial β-galactosidase downstream of the broad host range Rous sarcoma virus (RSV) promoter. By performing the assay (*Protocol 17*) at the optimal pH of the bacterial enzyme, pH 7.3, there is little contribution to hydrolysis of the substrate by the endogenous mammalian lysosomal β-galactosidase (pH optimum between 3.0 and 6.0). For each assay a positive control should be included using commercially available *E. coli* β-galactosidase.

Protocol 17. β-Galactosidase assay

Reagents

- Cytoplasmic extract (see *Protocol 14*)
- 100 × magnesium solution (0.1 M $MgCl_2$, 5 M 2-mercaptoethanol)
- 0.1 M sodium phosphate buffer, pH 7.3
- 4 mg/ml CPRG [chlorophenol red-β-D-galactopyranoside (Boehringer Mannheim)] in 0.1 M sodium phosphate buffer
- 1.0 M Na_2CO_3.

Method

1. Set up the following reaction mixture in a microcentrifuge tube:

100 × magnesium solution	3 μl
0.1 M sodium phosphate buffer	0.2 ml
4 mg/ml CPRG	66 μl
cytoplasmic extract	31 μl.

2. Incubate at 37°C for 30–60 min.
3. Add 0.7 ml 1.0 M Na_2CO_3 to stop the reaction, and vortex the tubes.
4. Measure the A_{574} against an appropriate reaction blank. The reaction blank has the same composition as the reaction test mixture (step 1) but contains 31 μl 0.25 M Tris–HCl pH 8.0 in place of the cytoplasmic extract.

One should be aware of potential problems in the use of internal controls. If the amount of control plasmid is similar to the amount of reporter plasmid, competition for limiting transcription factors can occur. The omission of a control increases variation, but at least avoids the possibility of 'observer effect'. Experimenters are divided on whether to omit or include a transfection control. Without a control, multiple replicate samples must be analysed to provide meaningful results.

5.5 Analysis of transcripts by primer extension

In addition to measuring the reporter gene enzyme activity (e.g. CAT), it is important to map the precise location of the 5′ terminus of the RNA produced in the transfected cells to monitor for the potential expression of cryptic promoters located within the DNA under investigation or within the vector. Several methods have been used to map the 5′ ends of transcripts:

digestion of DNA–DNA hybrids with nuclease S1, digestion of RNA–RNA hybrids with RNase, and primer extension analysis. The most commonly used method is primer extension analysis. In this method an oligonucleotide which will hybridize close to the 5′ end of the transcript is synthesized and end-labelled with ^{32}P. Total RNA is prepared from each dish of transfected cells (*Protocol 18*). The end-labelled oligonucleotide is then allowed to anneal to the RNA and acts as primer for reverse transcriptase which will then extend the primer sequence until the 5′ end of the RNA transcript is reached (*Protocol 19*). The extended radiolabelled primer is then resolved from residual primer by electrophoresis on a denaturing (urea) polyacrylamide gel and the transcription start point is precisely located from the size of the extended product.

Protocol 18. Guanidinium isothiocyanate/caesium chloride centrifugation method for the preparation of total RNA from transfected cells

Equipment and reagents

- Beckman SW50.1 centrifuge tubes and rotor (or their equivalent)
- Transfected cell monolayer
- PBS (see *Protocol 4*)
- Solid CsCl
- 5.7 M CsCl, 4 mM EDTA
- STE buffer [10 mM Tris–HCl (pH 7.4), 5 mM EDTA, 0.1% SDS]
- 4 M guanidinium isothiocyanate. [Prepare this solution by dissolving 50 g guanidinium isothiocyanate in 10 ml 1 M Tris–HCl (pH 7.5) and add water to 100 ml. Filter through Whatman 1MM paper and store at room temperature (stable indefinitely). Just before use, add 2-mercaptoethanol to 1%]
- 10% (w/v) sodium lauryl sarcosine stock.

Method

1. Harvest the transfected cells by scraping with a plastic spatula and wash them three times in PBS. Resuspend the cells in 4 ml 4 M guanidinium isothiocyanate and vortex for 20 sec. This should be sufficient to completely disrupt the cells.
2. Add sodium lauryl sarcosine to 0.5% (v/v) and mix.
3. Centrifuge at 5000 × *g* for 5 min at room temperature to remove insoluble debris.
4. Add solid CsCl to the supernatant to 0.5 g/ml final concentration.
5. Layer 3.7 ml of this solution on to a 1.2 ml cushion of 5.7 M CsCl, 4 mM EDTA in a Beckman SW50.1 centrifuge tube.
6. Centrifuge at 100 000 *g* for 12–15 h at 20°C. After centrifugation the RNA will appear as a translucent pellet.

7. Carefully remove the supernatant and dissolve the RNA pellet in 0.4 ml STE buffer.
8. Add 0.1 vol. 3 M sodium acetate (pH 5.2) and 2.5 vol. ethanol and leave at −20°C overnight or in a dry ice–ethanol bath for 40 min.
9. Centrifuge for 15 min in a microcentrifuge and wash the pellet in 70% ethanol.
10. Briefly dry the pellet in a vacuum desiccator or vacuum evaporator (Savant Speedvac) and resuspend the pellet in 0.5–1.0 ml water.
11. Quantify the RNA by diluting 10 μl into 1 ml H_2O and measuring the absorbance of this solution at 260 nm. An A_{260} of 1.0 is equivalent to 40 μg/ml RNA.

Protocol 19. Primer extension analysis

Reagents

- Total RNA from transfected cells (see *Protocol 18*); 2.5 μg/μl
- Oligonucleotide primer (30 pmol/μl) complementary to the transcript being examined
- 10 × polynucleotide kinase buffer (see *Protocol 12*)
- [γ-^{32}P]ATP (10 μCi/μl, 3000 Ci/mmol)
- T4 polynucleotide kinase (10 units/μl)
- TE buffer pH 8.0 (10 mM Tris–HCl, 1 mM EDTA pH 8.0)
- 1 ml Sephadex G50 spun column equilibrated in TE buffer (see *Protocol 7*, footnote *a*).
- 5 × annealing buffer (125 mM Pipes, 2 M NaCl, 5 mM EDTA pH 6.8)
- Reverse transcriptase buffer [1 M Tris–HCl pH 8.3, 0.5 M KCl, 0.1 M $MgCl_2$]
- 10 mM dNTP
- 0.1 M DTT
- 1 mM actinomycin D
- AMV reverse transcriptase (50 units/μl)
- Formamide dye mixture [10 mM NaOH, 1 mM EDTA, 0.1% (w/v) xylene cyanol, 0.1% (w/v) bromophenol blue, 80% (v/v) deionized formamide].[a]

A. *End-labelling the oligonucleotide primer*

1. Add 12.5 μl (125 μCi) [γ-^{32}P]ATP to a microcentrifuge tube and vacuum evaporate (Savant Speedvac). Then add to the tube the following:

oligonucleotide primer	1 μl (30 pmol)
polynucleotide kinase buffer	5 μl
T4 polynucleotide kinase	2 μl (20 units)
H_2O	to 50 μl final volume.

2. Incubate at 37°C for 30 min.
3. Add 50 μl TE buffer pH 8.0 and centrifuge through a 1 ml Sephadex G50 spun column previously equilibrated in TE buffer. Store the eluate (labelled oligonucleotide) at −20°C.

Protocol 19. *Continued*

B. *Annealing reaction*

1. Set up the following mixture in a microcentrifuge tube:

RNA	4 μl (10 μg total RNA)[b]
5 × annealing buffer	2 μl
end-labelled oligonucleotide primer	2 μl
H_2O	to 10 μl final volume.

2. Close the microcentrifuge cap tightly and incubate at 70°C for 3 min in a heating block. Allow the heating block to slowly cool to 35°C and incubate the reaction overnight at 35°C.

C. *Primer extension reaction*

1. Set up the following mixture in a microcentrifuge tube:

annealed reaction mixture (see above)	10 μl
reverse transcriptase buffer	10 μl
0.1 M DTT	10 μl
10 mM dNTP	5 μl
1 mM actinomycin D	5 μl
AMV reverse transcriptase	10 units
H_2O	to 100 μl final volume.

2. Incubate at 42°C for 1 h.
3. Stop the reaction by adding 5 μl formamide dye mixture, and incubating the tube at 95°C for 5 min.
4. Separate the primer and extended products by electrophoresis on a 6% polyacrylamide sequencing gel.
5. Expose the gel to X-ray film to determine the size of the extended product.

[a] Deionize formamide as follows. Mix 100 ml of formamide with 5 g of mixed bed resin AG 501-X8 (Bio-Rad), stir for 30 min at room temperature, and then filter to remove the resin.
[b] Alternatively, use 0.5–1.0 μg poly(A)$^+$ RNA.

6. Mutagenesis of reporter gene constructs

The crucial criterion for the identification of a transcriptional regulatory sequence is that deletion or mutagenesis of the identified sequence should abolish the transcriptional activity of the reporter gene in transfected cells. The various levels of mutagenesis include:

- deletion mutants to provide a coarse map of putative regulatory sequences
- linker scanning and systematic block replacement mutagenesis, which can map to within approx. 10 bp

- oligonucleotide site-directed mutagenesis to map specific nucleotides within the regulatory sequences.

6.1 Generation of nested sets of deletion mutants

Nested sets of deletions can be generated by progressively removing sequences from one end of the target DNA using *Bal*31 or exonuclease III.

6.1.1 *Bal*31

*Bal*31 progressively degrades double-stranded DNA by liberating single nucleotide residues from the 3′ end. It also contains a weaker single-stranded endonucleolytic activity. To generate mutants, the recombinant plasmid is linearized with a restriction enzyme that cleaves at one end of the target sequence and is then digested with *Bal*31 for different time periods (*Protocol 20*). The overhanging termini are then repaired and the plasmid recircularized using T4 DNA ligase in the presence of linkers to faciliate efficient recovery of the truncated insert.

Protocol 20. Generation of deletions using exonuclease *Bal*31

Reagents

- Recombinant plasmid DNA (20 μg)
- Appropriate restriction endonuclease(s)
- Appropriate restriction enzyme linkers (phosphorylated)
- Phenol:chloroform:isoamyl alcohol (25:24:1)
- 2 × running buffer (40 mM Tris–HCl pH 8.0, 24 mM $CaCl_2$, 24 mM $MgCl_2$, 0.4 M NaCl, 2 mM EDTA)
- Buffer-equilibrated phenol
- TE-50 buffer (50 mM Tris–HCl pH 8.0, 50 mM EDTA)
- *Bal*31.
  [Immediately prior to use, dilute the *Bal*31 to 0.2 U/μl in *Bal*31 storage buffer (20 mM Tris–HCl pH 6.8, 100 mM NaCl, 5 ml $MgCl_2$, 5 mM $CaCl_2$)]
- TE buffer pH 8.0 (10 mM Tris–HCl, 1 mM EDTA pH 8.0)
- Agarose gel electrophoresis reagents (see *Protocol 6*).

A. *Generation of deletions*

1. Linearize 20 μg of the recombinant plasmid to completion using an appropriate restriction endonuclease(s).
2. Extract the DNA with phenol:chloroform. Ethanol precipitate the DNA (see *Protocol 8*, steps 4–8) and resuspend it in 125 μl H_2O.
3. Add 125 μl of 2 × running buffer, and incubate the mixture at 30°C.
4. In each of six microcentrifuge tubes, marked 1 to 6, place 50 μl phenol and 20 μl TE-50 buffer.
5. Add 30 μl of the linearized DNA solution (step 2) to tube number 1 (this is the zero time point for the *Bal*31 time course). Immediately vortex the tube.

Protocol 20. *Continued*

6. At time zero, add 2 μl of the (diluted) *Bal*31 to the remaining linearized DNA (step 2). At set intervals thereafter (normally 2–5 min apart) remove 30 μl aliquots from the *Bal*31/DNA solution and add to tubes 2–6 (step 4). Immediately vortex the tubes.
7. Centrifuge all of the tubes 1–6 in a microcentrifuge for 5 min.
8. Collect the aqueous phases separately. Extract each once with chloroform and then ethanol precipitate. Resuspend each DNA sample in 20 μl TE buffer.
9. In order to determine the approximate extent of deletion, digest a 5 μl aliquot of DNA from each time point with an appropriate second restriction endonuclease and the analyse the products by agarose gel electrophoresis (*Protocol 6*).

B. *Recircularization of deletion plasmid constructs*

1. To recircularize the plasmid, fill in the ends of the digested DNA (5 μl aliquot from each time point) using DNA polymerase I (Klenow fragment) (*Protocol 8*).
2. Ligate the plasmid DNA (*Protocol 13*) in the presence of 250 ng of appropriate phosphorylated restriction enzyme linkers. The inclusion of linkers in the ligation mixture ensures that the deletion fragments can be recovered in future, as needed, by restriction endonuclease digestion from the recircularized plasmid.

C. *Isolation and sequencing of individual deletions*

1. Following transformation into appropriate strains of *E. coli*, pick a number of colonies for each construct, and confirm the deletion by agarose gel electrophoretic analysis (*Protocol 6*) of restriction enzyme digests of plasmid minipreparations.
2. Map the precise end point of each deletion by DNA sequencing.
3. End-label DNA from the plasmid minipreparations using DNA polymerase I (Klenow fragment) and [α-^{32}P]dNTP (see *Protocol 21*), and sequence using the Maxam and Gilbert chemical method (32).
4. Amplify appropriate colonies and perform large scale plasmid preparations (see *Protocol 1*).

Because *Bal*31 degrades both ends of the linearized plasmid, both the target DNA and vector sequences are degraded. It is therefore necessary to purify the truncated fragments by gel electrophoresis and to reclone them into appropriate vectors. It is also essential to ascertain the exact extent of the deletions generated. This is performed by sequencing the fragment before recloning it into the reporter gene vector. The DNA is end-labelled using DNA polymerase I (Klenow fragment) (*Protocol 21*) and sequenced using the Maxam and Gilbert chemical sequencing method (32).

Protocol 21. End-labelling DNA using DNA polymerase I (Klenow fragment)

Reagents

- Linearized plasmid DNA (approx. 1 μg)
- 10 × 5′ overhang buffer (see *Protocol 8*)
- [α-^{32}P]dNTP (Amersham, 3000 Ci/mmol, 10 μCi/μl)
- 2 mM dNTP labelling mixture; contains 2 mM concentration of each of the three dNTPs excluding the dNTP to be supplied in ^{32}P-labelled form
- DNA polymerase I (Klenow fragment)
- An appropriate restriction enzyme which cuts the linearized DNA to leave a 5′ overhang
- A second restriction enzyme to release the labelled fragment from the plasmid
- Agarose gel electrophoresis reagents (*Protocol 6*) or polyacrylamide gel electrophoresis reagents (see *Protocol 7*) depending on the size of the fragment to be purified (see *Appendix 2*).

Method

In order to label the DNA selectively at one terminus the plasmid is linearized with a restriction enzyme which generates a 5′ overhang. The 5′ overhang is then filled in by DNA polymerase I (Klenow fragment) using dNTPs one of which is radiolabelled. Finally, the labelled DNA fragment is released from the plasmid by a second restriction enzyme digestion.

1. Set up the following reaction in a microcentrifuge tube

linearized plasmid DNA	(approx. 1 μg)
10 × 5′ overhang buffer	2 μl
[α-^{32}P]dNTP	1–5 μl
2 mM dNTP labelling mixture	1 μl
DNA polymerase I (Klenow fragment)	2 units
H_2O	to 20 μl final volume.

2. Incubate at room temperature for 15 min.

3. Incubate at 70°C for 10 min to inactivate the enzyme.

4. For digestion with the second restriction enzyme, make up the volume to 50 μl with appropriate restriction enzyme buffer, then add the restriction enzyme and incubate to digest the DNA.

5. Purify the end-labelled DNA fragment by agarose gel electrophoresis (*Protocol 6*) or polyacrylamide gel electrophoresis (*Protocol 7*), monitoring the location of the separated fragments by autoradiography.

6.1.2 Exonuclease III

The original exonuclease deletion procedure involved digesting the DNA with two restriction enzymes, one to leave a 3′ overhang and the other to leave a 5′ overhang or blunt end. Exonuclease III was reported not to digest from a 3′ single-stranded overhang so that when this linearized recombinant DNA was challenged with exonuclease III, digestion proceeded only from the 5′ overhang or blunt end, not from the 3′ overhang. However, in our experience and that of others, exonuclease III has sufficient activity with 3′ overhangs to present substantial problems in trying to generate deletions according to the original protocol (26). The amended protocol (*Protocol 22*) uses a restriction enzyme which linearizes the recombinant plasmid DNA to leave a 5′ overhang which is then filled in with thio-dNTPs. The resulting thio-dNTP end is blocked from exonuclease III digestion. A second digestion with another enzyme then generates an end from which exonuclease digestion can occur, now unidirectionally. When the exonuclease III reaction has finished, a 3′ overhang is present on each DNA molecule, the extent of which depends on the degree to which the enzyme has been allowed to digest that DNA. Since a highly staggered end cannot serve as an efficient ligation substrate, the overhang is eliminated by digestion with a single-strand-specific nuclease; usually nuclease S1 or mung bean nuclease (see *Protocol 10*).

Protocol 22. Generation of an exonuclease III deletion series[a]

Reagents

- Recombinant plasmid DNA (5 μg per deletion time point)
- 2 × exo III buffer (0.1 M Tris–HCl pH 8.0, 10 mM $MgCl_2$, 20 μg/ml tRNA)
- 1 × mung bean nuclease dilution buffer (10 mM sodium acetate pH 5.0, 0.1 mM zinc acetate, 1 mM cysteine, 0.1% Triton X-100, 50% glycerol)
- 10 × ligation buffer (0.5 M Tris–HCl pH 7.5, 70 mM $MgCl_2$, 10 mM DTT)
- Stop buffer
  [For each deletion time point, prepare this by mixing 20 μl 0.3 M sodium acetate pH 5.0, 0.5 M NaCl, 10 mM $ZnCl_2$, 50% glycerol, and 155 μl H_2O]
- Thio-dNTP mixture; a mixture of 1 mM of each of the four 2′-deoxynucleoside-5′*O*-1-thiotriphosphates (dNTPαS) (e.g. Stratagene, Pharmacia)
- Exonuclease III[b] (specific activity varies depending on the commercial source)
- *E. coli* DNA polymerase I (Klenow fragment) (e.g. Pharmacia)
- Mung bean nuclease (Pharmacia; ~130 units/μl)
  [Dilute this enzyme in 1 × mung bean nuclease dilution buffer to the working concentration only just before use][c]
- TE buffer (10 mM Tris–HCl pH 8.0, 1 mM EDTA)
- 5 mM ATP
- T4 DNA ligase (1–5 U/μl)
- 0.1 M 2-mercaptoethanol
- 1.0 M Tris–HCl pH 9.5
- 8.0 M LiCl
- 20% SDS
- 3.5 M sodium acetate pH 7.5
- 5 mg/ml glycogen (molecular biology grade; BCL)
- Appropriate restriction enzymes and buffers.

Method

1. Digest the recombinant plasmid with a restriction enzyme that cuts upstream from the site from which deletions are desired and which generates a 5′ overhang. Digest sufficient plasmid to yield 5 μg linearized plasmid for each deletion time point in step 10. Digest at a concentration of 5 μg plasmid per 20 μl reaction volume.
2. Check a 5 μl aliquot of the total volume of restricted DNA by agarose gel electrophoresis; at least 95% of the DNA must have been linearized otherwise confusing data will be obtained during gel electrophoretic analysis of the deletions.
3. Incubate the reaction mixture at 70°C for 10 min to inactivate the restriction enzyme.
4. For every 20 μl of original reaction mixture (step 1), add:

thio-dNTP mixture	1 μl
E. coli DNA polymerase (Klenow fragment)	2 units.

   Incubate at room temperature for 30 min to allow the polymerase to fill in the 5′ overhang.
5. Phenol extract the DNA and precipitate it with ethanol (see *Protocol 8*, steps 4–8).
6. Check the fill-in reaction by incubating 2 μg DNA with a large excess of exonuclease III (40 units). Incubate at 37°C for 15 min and analyse the resulting products by agarose gel electrophoresis. Co-electrophorese the original linearized DNA (step 1) and filled-in plasmid DNA (step 5) which was not incubated with exonuclease III. At least 95% resistance to exonuclease III is required, as indicated by no detectable change in the size of the DNA fragment.
7. Carry out the second restriction digest, making sure that the second enzyme site lies between the first restriction site and the site from where the deletions will be made. Use an excess of restriction enzyme and incubate at 37°C for 2 h. In choosing this enzyme, bear in mind that its recognition site should not lie too close to that of the first enzyme recognition site since some restriction enzymes require a minimum extension of DNA on either side of the site in order to cut efficiently.
8. Phenol extract the reaction mixture and ethanol precipitate the DNA (see *Protocol 8*, steps 4–8).
9. Wash the pellet in 70% ethanol and then dry the pellet.
10. Prepare a single reaction mixture in a microcentrifuge tube for generating the exonuclease III deletion series. The total volume will depend on

Protocol 22. *Continued*

the number of deletion time points to be taken. For each deletion time point (sample volume 25 μl), mix the following:

double-digested DNA	5 μg
2 × exo III buffer	12.5 μl
0.1 M 2-mercaptoethanol	2.5 μl
exonuclease III[b]	100 units
H_2O	to 25 μl final volume.

11. At suitable time points, remove 25 μl of the exonuclease III deletion mixture (step 10) and add it to 175 μl of stop buffer in a microcentrifuge tube on dry ice.

12. When all the deletion time point samples have been taken, heat all the samples at 68°C for 10 min and then cool them on ice.[d]

13. Add 15 units of mung bean nuclease to each sample time point mixture and incubate at 30°C for 30 min.[c]

14. It is of the utmost importance to remove all of the mung bean nuclease before proceeding to the ligation step. To do this, add to each reaction mixture:

1.0 M Tris–HCl pH 9.5	10 μl
8.0 M LiCl	20 μl
20% SDS	4 μl
buffer-equilibrated phenol	250 μl.

Mix by vortexing and spin in a microcentrifuge for 3 min.

15. Remove the upper aqueous phase to a clean microcentrifuge tube and re-extract with chloroform.

16. Add 25 μl 3.5 M sodium acetate pH 7.5, 5 μl 5 mg/ml glycogen (as carrier) and 2.5 vol. ethanol. Keep at −70°C for 30 min and then centrifuge as before.

17. Drain each pellet, wash it in 70% ethanol and then dry it.

18. Resuspend each pellet in 15 μl TE buffer.

19. Set up each reaction mixture for ligation by mixing:

exonuclease III treated DNA	1 μl
10 × ligation buffer	2 μl
5 mM ATP[e]	2 μl
T4 DNA ligase	1–5 units
H_2O	to 20 μl final volume.

20. For each deletion time point mixture, transform suitable *E. coli* host cells and plate out. Select several colonies for further analysis from each time point.

21. Prepare mini-prep DNA (see ref. 5) from each selected colony, linearize this with appropriate restriction enzyme(s) and examine the size of the fragments by agarose gel electrophoresis. Be aware that, although there is a general trend to finding progressively longer deletions at longer time points, individual deletions of any length may occur in reactions at any time point.

[a] Protocol kindly supplied by Dr J. Donovan, St Lukes Institute for Cancer Research, c/o Department of Pharmacology, University College Dublin, Ireland.
[b] Many suppliers of exonuclease III provide details of the activity of this enzyme in terms of the number of nucleotides one can expect to be removed per minute at specific temperatures. Note that such data are rough guides only and that the actual rate of digestion depends on the base composition of the DNA. An extreme example of this is *Dictyostelium discoideum* promoter sequences (2–3% G+C) where the rate of base removal is only about half of that expected according to the suppliers' guidelines.
[c] Mung bean nuclease is very unstable at low concentrations so carry out the dilution immediately prior to use (in 1 × mung bean nuclease dilution buffer). Note also that mung bean nuclease shows considerable dependence on the sequence composition of the target DNA. For example, with *D. discoideum* DNA it is necessary to add 10 times the amount of enzyme recommended here and to digest for a longer time (2 h) than recommended.
[d] In some cases it may be necessary to carry out a reannealing step prior to the addition of mung bean nuclease to allow the reformation of any duplexes that may have been disrupted by the 68°C incubation. Carry this out at 37°C for 30 min.
[e] Many suppliers provide buffers for use with their ligases but not all these buffers include ATP which is an essential component in the ligation reaction and so must be added.

6.2 Linker-scanning mutagenesis

6.2.1 Using matched 5′ and 3′ deletion mutants

Linker-scanning mutants are constructed by ligating 5′ and 3′ deletion mutants in the presence of a synthetic linker in such a way that short sequences within the target DNA are replaced by the linker sequence. Successful use of this method depends on generating and characterizing a sufficient number of 5′ and 3′ deletions such that pairs differing in no more than approximately 10 bp can be matched. The amount of work involved can be formidable. In the original description of the technique (27) for example, forty-three 5′ deletions and forty-two 3′ deletions were generated and sequenced to obtain fifteen matched pairs of linker-scanning mutants.

6.2.2 Alternative (simpler) strategy

An alternative strategy for constructing linker scanning mutants has been described that does not require the matching of pairs of 5′ and 3′ deletion mutants (28) (*Protocol 23*). The target DNA is introduced into the MCR of a plasmid vector such as pUC118 or pUC119. A single random nick is then introduced into the DNA by partial depurination with formic acid followed by digestion with exonuclease III. The second strand is cleaved opposite the nick or small gap by nuclease S1. This provides random linearization and a slight shortening of the DNA. Three different time points are used for the exonuclease III and nuclease S1 reactions in order to compensate for differences in the reactivity of various DNA sequences. The linearized plasmids are then treated with T4 DNA polymerase to increase the number of blunt-ended molecules.

*Bgl*II linkers (or alternative linkers if a *Bgl*II site is present in the target DNA) are ligated to the end of the linearized plasmids. Since linker ligations can be inefficient, those molecules bearing linkers attached at both ends are selected after insertion of a fragment containing the Kan^r gene from pBL2 (28) at the new *Bgl*II site. Following transformation into *E. coli*, double selection in ampicillin and kanamycin excludes any plasmids containing the Kan^r gene fragment in the β-lactamase gene of the plasmid. Plasmid DNA is then extracted from the antibiotic-resistant clones and digested with a restriction enzyme(s) which cleave(s) in the MCR and releases the target DNA fragments. These fragments are purified by agarose gel electrophoresis and subcloned into plasmid pUC12. At this point approximately 50 000 clones are plated out and used to isolate a pool of plasmid DNA bearing the Kan^r gene within the target DNA. The purified plasmid DNA is digested with *Bgl*II and the linearized plasmids purified following agarose gel electrophoresis. This DNA is then recircularized using T4 DNA ligase, and the target DNA excised by digestion of the plasmid DNA with an appropriate restriction enzyme(s) which cleaves within the MCR.

There then follows two electrophoresis steps. The first (agarose gel electrophoresis) purifies the target DNA fragments from the plasmid, and avoids overloading the second electrophoresis step (in polyacrylamide gels). Fragments of target DNA differing in length from the wild type insert by no more than ± 5 bp are purified from the polyacrylamide gel and subcloned into plasmid pUC12. Transformants containing insert DNA are selected using a blue/white colour selection. Finally, plasmid minipreparations from randomly selected white colonies are treated with topoisomerase I and the resultant relaxed DNA resolved by agarose gel electrophoresis. Following relaxation of the DNA in this manner, agarose gel electrophoresis will resolve DNA molecules that differ in length by a single base pair. Plasmids which display the wild type plasmid topoisomerase I pattern (i.e. are of the same size as the wild-type recombinant plasmid) are selected. The approximate location of the novel *Bgl*II site is determined by digesting the DNA with *Bgl*II and a restriction enzyme that cleaves in the MCR. Appropriate clones are then sequenced.

Two steps in this procedure are crucial:

- the initial linearization (this must be as random as possible)
- the screening of the individual clones with topoisomerase I.

It is essential to eliminate false-positive clones which have insertions or deletions of multiples of 10 bp, since when relaxed these fragments will co-migrate on agarose gels with relaxed wild-type plasmid DNA. Thus it is essential to perform a stringent size fractionation of the mutated inserts.

Protocol 23. Linker-scanning mutagenesis[a]

Reagents

- Recombinant plasmid DNA. To ensure that this is largely supercoiled, use a preparation that has been recently purified by equilibrium centrifugation in CsCl–ethidium bromide gradients (*Protocol 1*)
- 2% (v/v) formic acid
- 100 mM Tris–HCl pH 7.5
- Exonuclease III buffer (66 mM Tris–HCl pH 8.0, 125 mM NaCl, 5 mM $CaCl_2$, 10 mM DTT)
- Exonuclease III (50–100 units/μl)
- 50 mM EDTA pH 8.0
- Phenol:chloroform:isoamyl alcohol (25:24:1)
- TE buffer pH 7.6 (10 mM Tris–HCl, 1 mM EDTA pH 7.6)
- Materials for agarose gel electrophoresis (see *Protocol 6*)
- Materials for CsCl–ethidium bromide equilibrium centrifugation (see *Protocol 1*)
- Nuclease S1 buffer (50 mM sodium acetate pH 5.7, 200 mM NaCl, 1 mM $ZnSO_4$, 0.5% (v/v) glycerol)
- Nuclease S1 (1500 units/μl)
- 1 M Tris–HCl pH 7.6
- 10 × T4 DNA polymerase buffer (0.2 M Tris–HCl pH 7.6, 10 mM $MgCl_2$, 1 mM DTT)
- Bacteriophage T4 DNA polymerase (8 units/μl)
- 2 mM dNTP solution
- Phosphorylated *Bgl*II linkers
- 10 × ligase buffer (0.2 M Tris–HCl pH 7.6, 50 mM $MgCl_2$, 50 mM DTT)
- 10 mM ATP
- T4 DNA ligase (2 units/μl)
- *Bgl*II
- Plasmid pBL2
- *E. coli* strain DH5α
- 200 μg/μl IPTG (isopropylthio-β-D-galactoside)
- 20 μg/μl X-gal (5-bromo-4-chloro-3-indolyl-β-D-galactoside) in dimethylformamide
- 10 × topoisomerase buffer (10 mM Tris–HCl pH 8.0, 2 M NaCl, 0.5 M EDTA)
- Topoisomerase I (BRL; 15 units/μl).

A. *Depurination of plasmid DNA*

1. Dissolve 200 μg of recombinant plasmid containing the DNA fragment to be mutagenized in 0.4 ml water. Transfer the solution to a sterile 15 ml polypropylene tube. Incubate the tube for 15 min at room temperature.
2. Add 40 μl of an aqueous solution of 2% (v/v) formic acid (pH 2.0) that has been equilibrated at 15°C.
3. Incubate the mixture for 4 min at 15°C. Then add 1.6 ml 100 mM Tris–HCl (pH 7.5) to quench the depurination reaction.
4. Recover the DNA by ethanol precipitation (see *Protocol 8*, steps 4–8).

B. *Digestion of depurinated DNA with exonuclease III*

1. Dissolve the pellet of depurinated DNA in 0.8 ml of exonuclease III buffer. Transfer the dissolved DNA to a microcentrifuge tube. As a control for the exonuclease III reaction, set up two additional tubes each containing 1 μg of the original superhelical plasmid dissolved in 20 μl exonuclease III buffer.

Protocol 23. *Continued*

2. Incubate the depurinated DNA and control DNA for 5 min at 37°C.
3. Add 800 units exonuclease III to the depurinated DNA and 5 units exonuclease III to one of the control DNAs. Continue the incubation at 37°C. After 1 min, transfer 270 μl of the depurinated DNA solution to one of three microcentrifuge tubes, each containing 30 μl ice-cold 50 mM EDTA (pH 8.0). Take further samples after 3 min and 9 min incubation at 37°C.
4. After all the samples have been transferred to the EDTA, purify all the DNAs by extraction with phenol:chloroform:isoamyl alcohol and recover the DNA by ethanol precipitation (see *Protocol 8*, steps 4–8).
5. Dissolve each of the depurinated DNAs in 100 μl TE buffer, pH 7.6, and the two control DNAs in 10 μl TE buffer, pH 7.6. Analyse a small amount of each of these samples by electrophoresis through an agarose gel (*Protocol 6*). Under the conditions of exonuclease III digestion described in this protocol, a fraction of the DNA should have been converted to relaxed circular molecules, which migrate slower through agarose gels than closed circular DNA of the same size. There should be no increase in the ratio of relaxed circular molecules to superhelical molecules in the control samples.
6. Pool the samples of depurinated DNA that contain less than 50% relaxed circular molecules.
7. Separate the mixture of nicked and gapped circles from unreacted superhelical DNA by equilibrium centrifugation in a CsCl–ethidium bromide gradient (see *Protocol 1*). Collect the upper band of DNA which contains nicked circles and linear molecules. Remove the ethidium bromide from the DNA, and recover the DNA by precipitation with ethanol (see *Protocol 8*, steps 4–8).

C. *Digestion of the nicked/gapped circular DNA with nuclease S1*

1. Redissolve the DNA (60–100 μg) (from step 7 above) in 0.5 ml of nuclease S1 buffer.
2. Incubate this solution for 5 min at 37°C. Then add 5000 units of nuclease S1 and continue the incubation at 37°C.
3. After incubation for 4 min, transfer 160 μl of the DNA solution to a microcentrifuge tube containing 15 μl 50 mM EDTA (pH 8.0) and 15 μl 1 M Tris–HCl (pH 7.6). Take further samples (160 μl) after 15 min and 60 min of incubation at 37°C.
4. Purify the linear DNAs by extraction with phenol:chloroform:isoamyl alcohol, and recover the DNA by precipitation from ethanol (see *Protocol 8*, steps 4–8).

5. Dissolve each of the DNAs in 25 μl TE buffer pH 7.6. Analyse a small amount of each by electrophoresis through an agarose gel (see *Protocol 6*). Pool the samples that contain a significant proportion of linear DNA.

D. *Treatment of linear DNA with T4 DNA polymerase to repair ends*

1. Adjust the volume of the pooled DNAs (approx. 25 μg) to 80 μl with TE buffer pH 7.6.
2. Add 10 μl of 10 × T4 DNA polymerase buffer and 10 μl 2 mM dNTP solution.
3. Add 5 units of bacteriophage T4 DNA polymerase. Incubate at 12 °C for 15 min (see *Protocol 9*, footnote *b*). During this period, protruding 3′ termini are efficiently removed.
4. Purify the DNA by extraction with phenol:chloroform:isoamyl alcohol.
5. Add 100 μl 5 M ammonium acetate and 400 μl ethanol and store the tube in an ice bath for 10 min. Recover the precipitated DNA by centrifugation.
6. Dissolve the DNA in 25 μl TE buffer pH 7.6.

E. *Ligation of phosphorylated synthetic* Bgl*II linkers*

1. Set up the following reaction mixture in a microcentrifuge tube:

linearized DNA (from step 6 above)	10 μg
phosphorylated *Bgl*II linkers	1000 pmol
10 × ligase buffer	20 μl
T4 DNA ligase	2 Weiss units
10 mM ATP	20 μl
TE buffer pH 7.6	to 200 μl final volume.

2. Incubate the mixture for at least 4 h at 16 °C.
3. Heat the mixture at 68 °C for 10 min to inactivate the T4 DNA ligase. Then add 500 units of *Bgl*II and incubate the reaction mixture for at least 6 h at 37 °C.
4. Purify the DNA by extraction with phenol:chloroform:isoamyl alcohol, and precipitation with ammonium acetate and ethanol (see above).
5. Dissolve the DNA in 20 μl TE buffer pH 7.6.

F. *Preparation of a DNA fragment carrying the* Kan^r *gene*

1. Prepare a fragment of DNA carrying the *Kan*^{r} gene. A convenient source of this fragment is plasmid pBL2 (see ref. 28). Digest 100 μg of pBL2 DNA with *Bgl*II and purify the 1.4 kb fragment containing the *Kan*^{r} gene by electrophoresis through an 0.8% agarose gel (*Protocol 6*).

Protocol 23. *Continued*

G. *Ligation of the linearized plasmid DNA to the fragment carrying the* Kan^r *gene*

1. Ligate the linearized plasmid DNA to the fragment carrying the *Kan*^{r} gene (see *Protocol 13*).
2. Use this ligation mixture to transform competent *E. coli* strain DH1 or DH5α.
3. Plate out the transformation suspension on each of four 150 mm Petri dishes containing LB agar and ampicillin and kanamycin (100 μg/ml of each) and incubate for 18–24 h at 37°C.
4. At the end of the incubation, add to each plate 5 ml LB broth and harvest the colonies on each plate using a bent glass rod. Pool the bacterial suspension from each of the four plates.
5. Recover the bacterial cells by centrifugation at 5000 *g* for 10 min at 4°C. Remove the supernatant and resuspend the bacterial pellet in 5 ml LB broth containing ampicillin and kanamycin (each at 100 μg/ml).
6. Use 2.5 ml of this bacterial suspension [the remainder can be stored at −70°C following addition of 12.5% (v/v) glycerol] to inoculate a 500 ml culture of LB broth containing ampicillin and kanamycin (each at 100 μg/ml). Incubate this culture at 37°C with vigorous aeration until saturation is achieved.
7. Isolate the pool of plasmid DNA by the alkali lysis method and purify the plasmid by CsCl–ethidium bromide gradient centrifugation (see *Protocol 1*).
8. Recover the target DNA fragments containing the inserted *Kan*^{r} gene by digesting approximately 30 μg of the pooled plasmid DNAs with the appropriate restriction enzyme(s).
9. Purify the target fragments carrying the *Kan*^{r} gene by electrophoresis through a 0.7% agarose gel (see *Protocol 6*).
10. Ligate (see *Protocol 13*) the purified target fragments to an appropriately prepared plasmid vector (e.g. pUC118 or pUC119) [i.e. digested with appropriate restriction endonucleases and CIP-treated (see *Protocol 11*)]. Transform *E. coli* strain DH5α using this recombinant plasmid. Plate the transformation mixture on each of four 150 mm Petri dishes containing LB agar supplemented with ampicillin and kanamycin (100 μg/ml of each) and incubate the plates for 18–24 h at 37°C.
11. Pool the bacterial suspension from each of the plates (see steps 4–7 above).

H. *Recovery of the target DNA fragment*

1. Digest approximately 100 μg of plasmid carrying the *Kan*^{r} gene (step 4 above) with *Bgl*II.

2. Separate the products of the digestion by electrophoresis through a 0.7% agarose gel. Excise the band that corresponds to the plasmid DNA, and extract the DNA (see *Protocol 6*).
3. Recircularize the plasmid DNA using T4 DNA ligase (*Protocol 13*).
4. Recover the recircularized plasmid DNA by precipitation with ethanol and ammonium acetate (see above).
5. Resuspend the DNA in TE buffer, pH 7.6.

I. *Excision of the target DNA*

1. Digest the recircularized plasmid (approx. 50 μg) with an appropriate restriction enzyme(s) which cleaves at sites flanking the target DNA.
2. Size fractionate the digested plasmid DNA on an 0.8% agarose gel and excise a band containing the target DNA. Extract the DNA (see *Protocol 6*), then apply the DNA to a non-denaturing preparative 5% polyacrylamide gel (see *Protocol 7*). Excise a band of DNA that differs in length from the target DNA by no more than ±5 bp. Extract and purify the DNA (see *Protocol 7*).
3. Ligate the DNA recovered from the polyacrylamide gel to an appropriately prepared plasmid vector (e.g. pUC118 or pUC119) [i.e., digested with an appropriate restriction endonuclease(s) and CIP-treated (see *Protocol 11*)].
4. Use this ligation mixture to transform competent *E. coli* strain DH5α. Plate the transformants out on LB agar plates (150 mm, 20 ml agar) containing ampicillin (100 μg/ml), 4 μl IPTG solution, and 40 μl X-gal solution, and incubate the plates for 18–24 h at 37°C.
5. Pick 36 independent transformants (white colonies) and perform plasmid minipreparations from 2 ml cultures (for protocol see ref. 5). Analyse the DNAs by digestion with the appropriate restriction endonuclease(s) and agarose gel electrophoresis to obtain an estimate of the proportion of clones that carry the target DNA.

J. *Analysis of clones that contain a* Bgl*II linker in the target region by topoisomerase I digestion and agarose gel electrophoresis*[b]

1. Digest miniprep DNA with topoisomerase I[c] in a reaction mixture containing:

miniprep DNA	2 μl (approx. 300 ng)
10 × topoisomerase I buffer	1 μl
topoisomerase I	1 μl
H_2O	to 10 μl final volume.

2. Incubate for 12–18 h at 37°C.
3. Add 12 μl TE buffer pH 7.6 to each sample followed by 4 μl of gel loading buffer, and resolve the DNA topoisomers on a 1.4% agarose gel (*Protocol 6*).

Protocol 23. *Continued*

After electrophoresis, stain the gel for 30 min in water containing 1 μg/ml ethidium bromide.

K. *DNA sequencing*

1. All clones containing suitable linkers should be analysed by DNA sequencing. This is done by end-labelling the miniprep DNA using DNA polymerase I (Klenow fragment) and [α-^{32}P]dNTP (see *Protocol 21*) and sequencing by the Maxam and Gilbert chemical method (32). Selected clones are then amplified and plasmid DNA purified from a large-scale preparation (see *Protocol 1*).

[a] From ref. 28.
[b] The object here is to relax the DNA with topoisomerase I. The relaxed DNA is then analysed by agarose gel electrophoresis. In this way subtle differences in size between DNAs bearing the *Bgl*II linker and the wild-type recombinant plasmid DNA (which is analysed simultaneously for comparison) are apparent. Clones are selected that contain a *Bgl*II linker and are identical in size to the wild type plasmid.
[c] The amount of topoisomerase required must be determined empirically in pilot reactions using control superhelical plasmid.

6.3 Systematic block replacement mutagenesis

This technique (29) involves the replacement of sequences within the target DNA with chemically synthesized 20–25 bp oligonucleotides with 8–12 bp overlapping ends. This strategy was used to map mutationally sensitive sequences within the rat I insulin gene control region (30). In total, 32 block mutations were used to cover the 345 bp upstream of the insulin gene. Of these, 6 block replacements resulted in a 3-fold or greater reduction in CAT activity when transiently expressed in a pancreatic β-cell line. The main drawback of this approach is that the cost entailed in generating the required oligonucleotides make this strategy prohibitively expensive for all but the larger enterprises.

6.4 Oligonucleotide site-directed mutagenesis

Once putative transcriptional regulatory sequences have been identified by any of the above mutagenesis procedures, it is necessary to mutate specific nucleotide residues in order to finely map the sequences involved. This is achieved by oligonucleotide site-directed mutagenesis. In this method, a mismatched oligonucleotide is annealed to a region of single-stranded DNA containing the putative regulatory sequence and elongated by DNA polymerase to generate a mutant second strand containing the modification corresponding to that introduced in the oligonucleotide.

Usually, single-stranded DNA is prepared by subcloning the DNA fragment of interest into a single-stranded phage vector such as M13. However, the Bluescript-based series of CAT vectors (pBCO, pBCTKp, and pBCSVp; see *Figure 4*) contain the intergenic region of a filamentous phage, and will secrete single-stranded f1-packaged phage following transformation into *E. coli* strains harbouring an F1 episome. A method for preparing single-stranded DNA from such vectors is described in *Protocol 24*.

Protocol 24. Recovery of single-stranded phagemid DNA

Reagents

- LB broth [1% (w/v) bacto-tryptone, 0.5% (w/v) bacto-yeast extract, 1% (w/v) NaCl]
- Superbroth (see *Protocol 1*)
- Helper phage R408 (Stratagene)
- 3.5 M ammonium acetate, pH 7.5, 20% (w/v) polyethylene glycol (PEG)
- TE buffer pH 8.0 (10 mM Tris–HCl, 1 mM EDTA pH 8.0)
- Phenol:chloroform:isoamyl alcohol (25:24:1)
- 7.5 M ammonium acetate, pH 7.5.

Method

1. Plate out the *E. coli* host cells containing the phagemid-derived recombinant of interest to give single colonies.
2. Pick a suitable colony and use it to inoculate 5 ml LB broth containing 50 μg/ml ampicillin and 12.5 μg/ml tetracycline.
3. Incubate overnight with vigorous aeration. Inoculate 300 μl of this fresh overnight culture into 3 ml of Superbroth in a 50 ml conical tube. Incubate with shaking at 37°C for 2 h.
4. Add helper phage R408 (Stratagene) at a multiplicity of infection of about 1:1 [7.5 μl of fresh phage stock at a titre of 10^{11} plaque forming units/ml (p.f.u./ml)] and incubate for a further 8 h.
5. Transfer 1.5 ml of this culture to a microcentrifuge tube, centrifuge for 2 min, and then carefully transfer 1.2 ml of supernatant to a fresh microcentrifuge tube.
6. Add 300 μl 3.5 M ammonium acetate pH 7.5, 20% (w/v) PEG to the supernatant. Vortex briefly and leave at room temperature for 15 min.
7. Centrifuge for 20 min in a microcentrifuge and then remove the supernatant by aspiration.
8. Recentrifuge the tube for 1 min and remove all remaining traces of supernatant by aspiration.
9. Resuspend the PEG pellet in 300 μl TE buffer pH 8.0 and extract it with phenol–chloroform until little or no interface is visible between the aqueous and organic phases (usually about three extractions needed).
10. Precipitate the single-stranded DNA by adding 0.2 ml 7.5 M ammonium acetate pH 7.5 and 0.8 ml ethanol, followed by incubation for 15 min in a dry ice–ethanol bath.

Protocol 24. *Continued*

11. Centrifuge at 4°C for 20 min in a microcentrifuge. Wash the pellet with 70% (v/v) ethanol, dry in a vacuum desiccator for a few min, and resuspend in 10 μl TE buffer pH 7.6.

There are many systems available for subsequent site-directed mutagenesis, which differ principally in the methods used to select for mutant strands. We routinely use a mutagenesis strategy (*Protocol 25*) which depends upon the ability of some restriction endonucleases to recognize their cognate sites in hemimethylated DNA and to cleave only the non-methylated strand (31). Mutagenic primer is annealed to the template and second-strand synthesis is performed using T7 DNA polymerase and T4 DNA ligase in the presence of 5-methyl-deoxycytosine triphosphate. Thus the second (mutant) strand contains methylcytosine residues whilst the parental (wild-type) strand does not. Incubation with the methylation-sensitive restriction endonuclease *Msp*I yields a duplex which is nicked at several sites in the parental strand, and further incubation with exonuclease III leads to the digestion of this nicked strand. The mutant strand can then be used to transform an appropriate non-restrictive *E. coli* host such as SDM cells. Although one can carry out the procedure using individually purchased reagents, we routinely use the USB T7 *in vitro* mutagenesis kit. The mutation frequency is usually greater than 50% allowing mutations to be detected directly by nucleotide sequencing (32, 33).

Protocol 25. Site-directed mutagenesis

Reagents

- Synthetic oligonucleotide containing the desired base change (mutant oligonucleotide; 100 pmol)
- 10 mM ATP
- Polynucleotide kinase buffer (see *Protocol 12*)
- T4 polynucleotide kinase (1 unit/μl; NBL)
- Template DNA (see *Protocol 23*)
- 5 × annealing buffer (see *Protocol 12*)
- T7 DNA polymerase
- T4 DNA ligase
- *Msp*I
- *Hha*I
- Exonuclease III
- Competent SDM *E. coli* cells (for protocol see ref. 5).

Method

1. The first step is to phosphorylate the mutant oligonucleotides. Set up the following reaction in a microcentrifuge tube:

mutant oligonucleotide	100 pmol
10 mM ATP	2 μl
10 × polynucleotide kinase buffer	1 μl
T4 polynucleotide kinase	5 units
H_2O	to 10 μl final volume.

2. Incubate at 37°C for 30 min.
3. Incubate at 70°C for 10 min to inactivate the enzyme.
4. For the mutagenesis reaction, set up the following annealing mixture in a microcentrifuge tube:

template DNA	4 μg
phosphorylated mutant oligonucleotide mixture (from step 3)	10 pmol
5 × annealing buffer	2 μl
H_2O	to 10 μl final volume.

5. Incubate at 65°C for 5 min then allow to cool slowly to room temperature.
6. For the second strand synthesis, set up the following mixture in a microcentrifuge tube:

T7 DNA polymerase	2.5 units
T4 DNA ligase	5.0 units
annealed template DNA-primer mixture (from step 4)	2 μl
H_2O	to 10 μl final volume.

7. Incubate at 37°C for 60 min.
8. Incubate at 70°C for 10 min to inactivate the enzymes.
9. Add 5 units of *Msp*I and 5 units of *Hha*I and incubate at 37°C for 45 min.
10. Add 50 units of exonuclease III and incubate at 37°C for a further 45 min.
11. Incubate at 70°C for 10 min to inactivate the enzymes.
12. Use 1 μl of the reaction mix to transform competent SDM *E. coli*.
13. Pick 12 colonies, prepare single-stranded phagemid DNA (see *Protocol 23*) and sequence this using the dideoxy method (33).

Acknowledgements

ARC was supported by a Fellowship from the British Diabetic Association.

References

1. Miller, A. D. (1990). *Human Gene Therapy,* **1,** 5–14.
2. Graham, F. L. and van der Eb, A. J. (1973). *Virology,* **52,** 456.
3. Chen, C. and Okayama, H. (1987). *Mol. Cell. Biol.,* **7,** 2745.
4. McCutchan, J. H. and Pagano, J. S. (1968). *J. Natl Cancer Inst.,* **41,** 351.
5. Sambrook, J., Fritsch, E. F., and Maniatis, T. (1989). *Molecular cloning, a laboratory manual* (2nd edn). Cold Spring Harbor Laboratory Press, New York.

6. Potter, H., Weir, L., and Leder, P. (1984). *Proc. Natl Acad. Sci. USA,* **81,** 7161.
7. Mackett, M., Smith, G. L., and Moss, B. (1985). In *DNA cloning: a practical approach*, Vol. 2 (ed. D. M. Glover), pp. 191–211. IRL Press, Oxford.
8. Graham, F. L. (1990). *Trends Biotechnol.,* **8,** 85.
9. Brown, A. M. C. and Scott, M. R. D. (1987). In *DNA cloning: a practical approach*, Vol. 3 (ed. D. M. Glover), p. 189. IRL Press, Oxford.
10. Tooze, J. (1980). *Molecular biology of tumour viruses* (2nd edn). Cold Spring Harbor Laboratory, New York.
11. Weiringa, B., Meyer, F., Reiser, J., and Weissmamm, C. (1983). *Nature,* **301,** 38.
12. Brinster, R. L., Allen, J. M., Behringer, R. R., Gelinas, R. E., and Palmiter, R. D. (1988). *Proc. Natl Acad. Sci. USA,* **85,** 836.
13. Proudfoot, N. J. and Whitelaw, E. (1988). In *Transcription and splicing: a practical approach* (ed. B. D. Hames and D. M. Glover), p. 97. IRL Press, Oxford.
14. O'Hare, K., Benoist, C., and Breathnach, R. (1981). *Proc. Natl Acad. Sci. USA,* **78,** 1527.
15. Kaufman, R. J., Wasley, L. C., Furie, B. C., Furie, B., and Shoemaker, C. B. (1986). *J. Biol. Chem.,* **261,** 9622.
16. Kosack, M. (1989). *J. Cell Biol.,* **108,** 229.
17. Gluzman, Y. (1981). *Cell,* **23,** 175.
18. Kaufman, R. J., Murtha, P., and Davies, M. V. (1987). *EMBO J.,* **6,** 187.
19. Shakur, Y., Shennan, K. I. J., Taylor, N. A., and Docherty, K. (1989). *J. Mol. Endocrinol.,* **3,** 155–62.
20. Arrufo, A. and Seed, B. (1987). *Proc. Natl Acad. Sci. USA,* **84,** 8573.
21. Kreigler, M. (1990). *Gene transfer and expression: a laboratory manual.* Stockton Press, New York.
22. Maniatis, T., Goodbourn, S., and Fischer, J. A. (1987). *Science,* **236,** 1237–45.
23. Gorman, C., Moffat, L., and Howard, B. (1982). *Mol. Cell. Biol.,* **2,** 1044.
24. Clark, A. R., Boam, D. S. W., and Docherty, K. (1989). *Nucleic Acids Res.,* **17,** 10130.
25. Boam, D. S. W., Clark, A. R., Shennan, K. I. J., and Docherty, K. (1990). *Peptide hormone secretion: a practical approach* (ed. J. C. Hutton and K. Siddle), p. 271. Oxford University Press, Oxford.
26. Henikoff, S. (1984). *Gene,* **28,** 351.
27. McKnight, S. L. and Kingsbury, R. (1982). *Science,* **217,** 316.
28. Luckow, B., Renkawitz, R., and Schutz, G. (1987). *Nucleic Acids Res.,* **15,** 417.
29. Zenk, M., Grundstrom, T., Metthes, H., Wintzerith, M., Schatz, C., Wildeman, A., and Chambon, P. (1986). *EMBO J.,* **5,** 387.
30. Karlsson, O., Edlund, T., Moss, J. B., Rutter, W. J., and Walker, M. D. (1987). *Proc. Natl Acad. Sci. USA,* **84,** 8819.
31. Vandeyar, M., Weiner, M., Hutton, C., and Batt, C. (1988). *Gene,* **65,** 129.
32. Maxam, A. and Gilbert, W. (1980). In *Methods in enzymology* (ed. L. Grossman and K. Moldave), Vol. 65(1), p. 499. Academic Press, New York.
33. Sanger, F., Nicklen, S., and Coulson, A. R. (1977). *Proc. Natl Acad. Sci. USA,* **74,** 5463.
34. Kawai, S. and Nishizawa, M. (1984). *Mol. Cell. Biol.,* **4,** 1172.
35. Wong, T. K., Nicolau, C., and Hofschneider, P. H. (1980). *Science,* **215,** 166.
36. Schaffner, W. (1980). *Proc. Natl Acad. Sci. USA,* **77,** 2163.
37. Felgner, D. L., Gadek, T. R., Holm, M., Roman, R., Chan, H. W., Wenz, M.,

Northrop, J. P., Ringold, G. M., and Danielson, M. (1987). *Proc. Natl Acad. Sci. USA,* **84,** 7413.

38. Furasawa, M., Nishimura, T., Yamaizumi, M., and Okada, Y. (1974). *Nature,* **249,** 449.
39. Diacumakos, E. G., Holland, S., and Pecora, P. (1970). *Proc. Natl Acad. Sci. USA,* **65,** 911.
40. Capecchi, M. R. (1980). *Cell,* **22,** 479.
41. Kurata, S., Tsukakoshi, M., Kasuya, T., and Ikawa, Y. (1986). *Exp. Cell Res.,* **162,** 372.
42. Klein, T. M., Fromm, M., Weissinger, A., Tomes, D., Schaff, S., Sletten, M., and Sanford, J. C. (1988). *Proc. Natl Acad. Sci. USA,* **85,** 4305.
43. Johnston, S. A. (1990). *Nature,* **346,** 776.
44. Campo, M. S. (1985). In *DNA cloning: a practical approach*, Vol. 2 (ed. Glover, D. M.), p. 213. IRL Press, Oxford.
45. Margolskee, R. F., Kavatha, P., and Berg, P. (1988). *Mol. Cell. Biol.,* **8,** 2837.
46. Cone, R. D. and Mulligan, R. C. (1984). *Proc. Natl Acad. Sci. USA,* **81,** 6349.
47. Shih, M., Arsenakis, M., Tiollais, P., and Roizman, B. (1984). *Proc. Natl Acad. Sci. USA,* **81,** 5867.
48. Lengyel, P. (1986). *Annu. Rev. Biochem.,* **51,** 251.
49. Pelham, H. R. B. (1985). *Trends Genet.,* **1,** 31.
50. Searle, P. F., Stuart, G. W., and Palmiter, R. D. (1985). *Mol. Cell. Biol.,* **5,** 1480.
51. Israel, D. I. and Kaufman, R. J. (1989). *Nucleic Acids Res.,* **17,** 4589.
52. Alt, F. W., Kellems, R. E., Bertino, J. R., and Schimke, R. T. (1978). *J. Biol. Chem.,* **253,** 1357.
53. Wahl, G. M., Padgett, R. A., and Stark, G. R. (1979). *J. Biol. Chem.,* **254,** 8679.
54. Yeung, C. Y., Ingolia, D. E., Bobonis, C., Dunbar, B. S., Riser, M. E., Siciliano, J. J., and Kellems, R. E. (1983). *J. Biol. Chem.,* **258,** 8338.
55. Riordan, J. R., Deuchars, K., Kartner, N., Alon, N., Trent, J., and Ling, V. (1985). *Nature,* **316,** 817.
56. McConlogue, L. and Coffino, P. (1983). *J. Biol. Chem.,* **258,** 12083.
57. Andrulis, I. L., Duff, C., Evans-Blackler, S., Worton, R., and Siminovitch, L. (1983). *Mol. Cell. Biol.,* **3,** 391.
58. Kaufman, R. J. (1990). In *Methods in enzymology* (ed. D. V. Goeddel), Vol. 185, pp. 537–66. Academic Press, New York.
59. Bebbington, C. R. and Hentschel, C. G. (1987). In *DNA cloning: a practical approach*, Vol. 3 (ed. D. M. Glover), p. 163. IRL Press, Oxford.
60. De Wet, J. R., Wood, K. V., DeLuca, M., Helinski, D. R., and Subramani, S. (1987). *Mol. Cell. Biol.,* **7,** 725–37.

3

In vitro transcription with nuclear extracts from differentiated tissues

FELIPE SIERRA, JIAN-MIN TIAN, and UELI SCHIBLER

1. Introduction

Much recent research in molecular biology is focusing on the molecular interactions that control transcription. Transcription of protein-coding genes is performed by RNA polymerase II (pol II). This enzyme, however, is not capable of recognizing bona fide promoters by itself. To do so, it requires protein–protein interactions with a number of ancillary factors (1). Some of these transcription factors, the so-called basal factors, are needed for the initiation of transcription at most promoters (2). In contrast, some promoter-specific regulatory proteins stimulate transcription of only a subset of genes. These latter transcription factors include, among others, tissue-specific regulatory proteins, supposed to play an important role during cellular differentiation (3). These molecules have the capability of recognizing and interacting with sequence-specific regions in the promoter or enhancer regions of genes (the so-called *cis* elements). Complex networks of these protein–DNA, as well as protein–protein, interactions lead to the final regulated level of expression of specific genes.

Protein-coding genes can be loosely divided into two categories, housekeeping genes and cell-type-specific genes. Housekeeping genes are expressed in all cell types, all the time, and at similar levels. On the other hand, during development and in differentiated tissues, some genes are expressed only in a subset of cells at any given time, and this level of expression can, in many cases, be modified by external stimuli, such as hormones. This 'regulated' gene expression can operate as a function of several different parameters, such as time (developmentally controlled genes), physical cell location (e.g. early genes of *Drosophila*), tissue specificity, hormonal, or other humoral factors, etc.

At the molecular level, the relative level of expression of any particular gene is thought to depend on two major separable events.

(a) A pre-requisite for expression is accessibility of the gene to the transcriptional machinery (4). This is achieved through a process known as chromatin opening, which appears to be specific for a given cell type and depends on sequences generally located some distance from the gene itself (5). The cellular machinery responsible for this process has thus far eluded detailed molecular analysis.

(b) Once chromatin becomes accessible, this allows the binding of specific transcription factors to both the promoter and the enhancer regions (6, 7). These factors can have either a positive (activating) or a negative (repressing) effect on transcription.

Detailed biochemical studies of these processes requires technical manipulations of both partners of the interactions described above. At least some of these interactions can conveniently be studied by *in vitro* transcription of cloned genes in cell-free extracts. In this chapter, we concentrate on techniques that allow the preparation of nuclear extracts containing the relevant *trans*-acting factors, as well as the use of these extracts to assess transcriptional potential *in vitro*. Several methods have been described in the literature for the preparation of transcription-competent nuclear extracts (1, 8, 9). Many of these techniques were developed for cells in tissue culture, but they are not suitable for the preparation of active extracts from solid tissues derived from living animals. Since rapidly growing cells in culture are generally much less differentiated than their counterparts in the living animal, extracts prepared from such cells may not be adequate to study tissue-specific *in vitro* transcription. Therefore we have recently developed a method that does allow the preparation of transcriptionally competent nuclear extracts from a number of highly differentiated, solid tissues (10, 11). These techniques are reviewed in the following sections.

2. General practical considerations

As with most biochemical techniques, the preparation of high quality nuclear extracts depends mainly on careful attention to detail at each step of the protocol. The most important variables to be careful about are time and temperature. Keeping the time period for preparation as short as possible may be crucial for obtaining good quality extracts. This means anticipating and preparing the equipment required for each step ahead of time (pre-cool the ultracentrifuge before killing the animals, pre-warm the spectrophotometer before lysing the nuclei, etc.). Of course, it is also crucial to keep the sample well-cooled throughout the whole protocol. Vulnerable components in the sample should be as protected as possible from the action of nucleases and proteases, and, as is discussed later, it is usually advisable to take extra precautions against other hydrolytic enzymes such as phosphatases.

3. Preparation of transcriptionally active nuclear extracts from solid rat tissues

3.1 Equipment required

Most, but not all of the equipment necessary for the preparation of nuclear extracts and their utilization is standard equipment found in molecular biology or biochemistry laboratories. *Table 1* lists the important pieces of equipment required.

Table 1. Equipment required

Extract preparation

- Motor-driven Potter-Elvehjem homogenizer
- Refrigerated ultracentifuge
- Ultracentrifuge rotors: Beckman SW27 and Ti60 (or Ti50 if samples are small), or equivalent rotors for non-Beckman ultracentrifuges
- Tubes for these rotors. Thin-walled nitrocellulose or polyallomer tubes are acceptable for nuclei preparation. Thick-walled polyallomer tubes are preferable for the following step (extract preparation)
- Vacuum pump fitted with a trap
- Spectrophotometer able to measure in the ultraviolet range (230–260 nm)
- Refrigerated microcentrifuge (e.g. Eppendorf)
- Liquid nitrogen storage facility (for long-term storage of extracts)

In vitro transcription

- Radioactivity containment facility
- High-power electrophoresis power supply and apparatus
- Gel drier
- X-ray developing facility

Other equipment

The following smaller pieces of equipment are also required:

- Surgical equipment
- Glass Dounce homogenizer [fitted with a loose-fitting ('A') pestle]
- Dialysis bags
- Automatic micropipettors (e.g. Gilson Pipetman)
- Mechanical aspirator
- Shaker, vortex mixer, and stirrer
- Water baths
- Gel plates, spacers, and combs

3.2 Preparation of nuclei from rat tissues

This is the critical step in the preparation of transcriptionally active extracts. It is absolutely essential that the nuclei prepared are both pure and intact. The use of freshly isolated tissue is imperative, since it has been found that extracts prepared from frozen samples are transcriptionally inactive. In our hands, only nuclei obtained by high-speed centrifugation of tissue homogenates prepared in highly concentrated sucrose solutions are adequate for the isolation of transcriptionally active extracts. In 2.0 M sucrose solutions, only nuclei are sufficiently dense and large enough to sediment under the centrifugal conditions utilized. Unbroken cells and most of the subcellular organelles float to the top of the centrifuge tube and form a pellicle. Probably the only cellular structures other than nuclei that are more dense than 2.0 M sucrose are ribosomes. However, their sedimentation coefficient is insufficient to cause them to pellet under the centrifugation conditions employed.

The time of preparation should be kept to a minimum. Naturally, it varies somewhat depending on the tissues to be used, the number of animals and the difficulty of dissection. In general, 3–4 h are required for this part of the protocol, and no interruption is possible. The nuclei are usually fairly clean after the first centrifugation, provided that the required ratio of buffer volume to tissue mass has been maintained and the walls of the tubes are thoroughly rinsed. Sometimes, however, it is advisable to do a second centrifugation, since, as already mentioned, the purity of the nuclei is essential for obtaining good quality extracts. 'Clean' nuclei typically have a whitish, semi-transparent appearance, and any hint of reddish coloration implies contamination. A second purification step in these cases is not a waste of time, since better results will almost certainly be obtained. As will be discussed in the troubleshooting section (Section 6), the most common cause of 'dirty' nuclei is excessive enthusiasm: too much tissue being processed at the same time in not enough homogenization buffer.

The procedure for the preparation of nuclei is described in *Protocol 1*. Note that the homogenization buffer is used both for homogenization and for preparation of the cushions. However, no milk should be added to the cushions, so as to avoid contamination of the nuclei by milk proteins. This solution should be stored at −20°C at least for several hours and used very cold (not warmer than −10°C) so as to increase its viscosity. Dithiothreitol (DTT) and phenylmethylsulphonyl fluoride (PMSF) are added just before use due to their instability in aqueous solution. In addition, when preparing extracts from tissues rich in hydrolytic enzymes, the inclusion of extra protease and/or phosphatase inhibitors to the buffers might be desirable (12).

If phosphatases are perceived as a potential problem (i.e. kinase-activated transcription factors), these can be inhibited by a wide spectrum of specific substances (see for example ref. 18). We have tested several of these, and found that a few of them are rather inhibitory of *in vitro* transcription reactions (see Section 5.4). Therefore, it is better to use these substances only at the early stages of extract preparation, but not during dialysis. One exception to this is NaF, which is a potent general inhibitor of phosphatases and, at a concentration of 0.5 mM in the final extract, does not inhibit transcription. We have used 10 mM NaF during nuclei and extract preparation, and reduced the concentration to 0.5 mM during dialysis (see *Figure 1*).

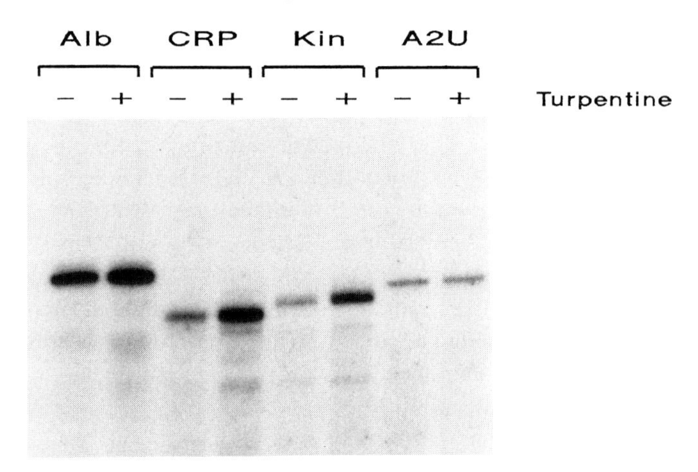

Figure 1. The effect of phosphatase inhibitors on *in vitro* transcription. Rat liver nuclear extracts were prepared essentially as described in *Protocols 1* and *2*, except that 10 mM NaF was included in both the homogenization and the nuclear lysis buffer. The NaF concentration was reduced to 0.5 mM during dialysis. The transcripts were analysed by polyacrylamide gel electrophoresis on a 4% acrylamide gel, as described in *Protocol 3*. Extracts were prepared in parallel from control animals and from rats injected subcutaneously with turpentine 18 h prior to killing. This treatment is known to induce transcription from a variety of genes, collectively called acute phase reactants. We measured *in vitro* transcription from these extracts, using templates containing promoter sequences derived from genes that do not respond to turpentine treatment [albumin (Alb) and α2U-globulin (A2U)], and templates containing promoter sequences from acute phase genes [human C-reactive protein (CRP) and rat T-kininogen, (Kin)]. The increase in *in vitro* transcription rate from acute phase promoters when extracts from turpentine-treated rats are used is not observed when NaF is omitted during extract preparation.

Protocol 1. Preparation of nuclei

Reagents

- 20% low fat milk stock solution
  Dissolve 10 g of powdered low fat milk in 50 ml H_2O. Centrifuge at 10400 *g* for 20 min. Discard the pellet and add 1/20th of this supernatant to the homogenization buffer, but not to the sucrose cushions. This solution should be made freshly each time
- Homogenization buffer
  [10 mM Hepes pH 7.6, 15 mM KCl, 0.15 mM spermine, 0.5 mM spermidine, 1 mM EDTA, 2.4 M sucrose (RNase-free), 1% low fat milk]. Just before use, add the following reagents to the stated final concentrations; 0.5 mM DTT, 0.5 mM PMSF (see below), 1% protease inhibitor peptide mixture (see below)
- PMSF stock solution
  Dissolve 174 mg PMSF in 10 ml isopropanol (final concentration 0.1 M). Store at −20°C. Before use, bring the solution to room temperature, so as to dissolve the crystals that form during storage
- Protease inhibitor peptide mixture (100 ×)
  Mix 5 ml 100% Trasylol (trademark for aprotinin), 2 ml pepstatin (250 μg/ml), and 50 μl leupeptin (10 mg/ml). Store this solution in 1 ml aliquots at −20°C
- Sucrose cushions
  Identical to the homogenization buffer, except for the absence of low fat milk.

Protocol 1. *Continued*

Method

1. Before starting, turn on the ultracentrifuge and set it so that the temperature equilibrates to −2°C. Pre-cool the centrifuge rotors to −2°C also. The following protocol assumes that an SW27 Beckman rotor is used. Equivalent rotors supplied by other manufacturers can be used, but they must have swing-out buckets.
2. Measure out the amount of homogenization buffer required for the homogenization and first centrifugation step. This depends on the volume of tissue to be processed. The following are approximate volumes for 10 g wet weight of different tissues:
   60 ml for rat liver
   6 ml for rat pancreas
   10 ml for rat spleen
   3 ml for rat parotid
   10 ml for rat thymus.
3. Add the DTT and the protease inhibitor peptide mixture (but not the PMSF or the milk) to this buffer. Due to the viscosity of the solution, this requires vigorous shaking.
4. Add 10 ml of this buffer to each SW27 rotor tube to form the cushions. Keep these on ice at all times.
5. To the rest of the homogenization buffer, add 0.2 vol. 20% low fat milk. Mix well.
6. Place the Potter–Elvehjem homogenizer in a large plastic beaker with ice-water, and add a few millilitres of homogenization buffer to the bottom of the homogenizer.
7. While the animals are being anaesthetized, add the required volume of PMSF stock solution to the homogenization buffer (see *Reagents*) and mix well as before. Also, add 50 μl PMSF stock solution to each of the cushions in the SW27 tubes.
8. Dissect the tissue and keep it in a cold beaker on ice. Mince it thoroughly with scissors.
9. Homogenize the tissue in batches by three or four strokes in the Potter–Elvehjem homogenizer. For this, resuspend the tisue in homogenization buffer. It is important that the tissue does not constitute more than 10–15% of the total volume, since dilution of the sucrose will decrease the viscosity of the homogenate, potentially resulting in nuclear leaching.
10. After each batch of tissue has been homogenized, layer it carefully on top of the 10 ml cushions in the SW27 tubes. Continue the homogenization and loading of the tubes until all the tissue has been processed.
11. Centrifuge at 24000 r.p.m. (75 000 g) for 60 min in the pre-cooled centrifuge in an SW27 rotor (it is important to use a swing-out rotor!).

12. While the sample is being centrifuged, check a small aliquot of homogenate directly (no staining required) under a light microscope. Most cells must be broken, but most nuclei should be intact.
13. At the end of the centrifugation, several fractions will be present in the tube:
    - Whole cells and unbroken pieces of tissue will float and form a pellicle.
    - Nuclei with adhering cytoplasm and broken nuclei stay at the interface.
    - Clean nuclei will pellet.

    The nuclear pellet should be transparent or slightly white, but not reddish as this indicates contamination with red blood cells or other sub-cellular debris. Should this be the case, a second centrifugation step is imperative (see step 17).
14. With a spatula, remove the solid pellicle that is floating in the tube, then carefully aspirate off the remaining liquid, wipe the sides of the tube with a tissue, and set the tube on ice in an ice bucket, almost horizontal but with the tube end slanted slightly downwards so that excess buffer runs towards the outside, and not towards the nuclei.
15. Fill a large (~50 ml) syringe with distilled water and attach to it a long (10 cm) small gauge needle with the tip bent about 90°. Holding each tube upside down, thoroughly rinse the walls by squirting water with the syringe. Carefully wipe off the excess water with a tissue.
16. If the nuclei are clean, proceed directly to prepare the extract (see *Protocol 2*). If the aim is a large scale protein isolation, the nuclei can be frozen by putting the tube directly in dry ice, and stored at −70°C for several weeks.
17. If a second centrifugation is deemed necessary, resuspend the nuclei in complete homogenization buffer containing 10% glycerol. Resuspend the nuclei by bubbling and scraping the side of the tube with a plastic pipette (*not* glass) then carefully aspirate up and down several times, making sure that foaming is avoided. Transfer to a Potter–Elvehjem homogenizer and homogenize again, only once or twice this time and with the motor set at a lower speed, then proceed as above (steps 11–16).
18. Check the nuclei again under the microscope (see step 12).

3.3 Preparation of nuclear extracts

Briefly, this part of the procedure (see *Protocol 2*) consists of preparing an ammonium sulphate fraction from lysed nuclei. The nuclei are resuspended in lysis buffer and then lysed by the addition of a low concentration (0.4 M) of $(NH_4)_2SO_4$. After centrifugation to eliminate chromatin, soluble proteins are precipitated by increasing the salt concentration. Then, excess salt is removed by extensive dialysis, followed by a brief centrifugation. In carrying out *Protocol 2*, the following precautions are advisable:

(a) Avoid locally high $(NH_4)_2SO_4$ concentrations during nuclear lysis, by rapidly mixing the contents of the tube immediately after addition of the salt.

(b) After the first centrifugation, remove the supernatant immediately, so as to avoid the chromatin re-swelling.

(c) Finally, at the last step before dialysis, carefully but thoroughly resuspend the protein pellet. After addition of the dialysis buffer, the pellet clears very fast but this does not mean that the proteins have completely dissolved. Rather, it is advisable to persevere and keep resuspending for 1–2 h.

Protocol 2. Preparation of nuclear extracts

Reagents

- Nuclear lysis buffer
  10 mM Hepes pH 7.6, 0.1 M KCl, 0.1 mM EDTA, 10% glycerol, 3 mM $MgCl_2$.
  Just before use, add the following reagents to the concentrations given: 1 mM DTT, 0.1 mM PMSF, 1% Trasylol.
  If extracts are made from tissues rich in hydrolytic activities, also add the protease inhibitor mixture (see *Protocol 1*, Materials) and/or phosphatase inhibitors (see Section 3.2). This buffer can be stored either frozen or at 4°C for a few weeks
- Nuclear dialysis buffer
  25 mM Hepes pH 7.6, 0.1 mM EDTA, 40 mM KCl, 10% glycerol. Just before use, add 1 mM DTT. Store this buffer at −20°C
- 4.0 M ammonium sulphate.
  Dissolve 52.85 g $(NH_4)_2SO_4$ in 100 ml water. Adjust to pH 7.9 with NaOH. Store this ammonium sulphate solution at either 4°C or −20°C. Any crystals must be redissolved by warming before use.

Method

1. Resuspend the nuclei (from *Protocol 1*) in nuclear lysis buffer (containing DTT, PMSF, and Trasylol) by bubbling and scraping with a *plastic* (not glass) 10 ml pipette as described in *Protocol 1*, step 17. The volume of lysis buffer used depends on the tissue:

   5 ml for rat liver
   1 ml for rat pancreas
   5 ml for rat spleen
   100 μl for rat parotid
   5 ml for rat thymus.

2. Homogenize the nuclei by several strokes with the hand-held Dounce glass homogenizer fitted with an A pestle.

3. Measure the A_{260} of a 1:50 dilution of the nuclei in 0.5% sodium dodecyl sulphate (SDS). In order to obtain a homogeneous lysate at this step, add the sample dropwise into the SDS solution, while vortexing vigorously.

From the A_{260}, calculate the concentration of DNA (1 mg/ml DNA has an A_{260} of 20.0).

4. Dilute the sample with nuclear lysis buffer so as to have a DNA concentration of 0.5 mg/ml (0.25 mg/ml for tissues with a low protein to DNA ratio, such as spleen and thymus).
5. Add 0.1 vol. 4.0 M $(NH_4)_2SO_4$, and immediately mix gently by inversion several times.
6. Leave in ice-water for 30 min with occasional mixing.
7. Centrifuge for 60 min at 90 000 *g* in a Ti60 rotor or at 100 000 *g* in a Ti50 rotor. At this step, it is important to remove and process the tubes as soon as the centrifuge stops, since otherwise the chromatin will start re-swelling.
8. Carefully pipette the supernatant into a new Ti60 or Ti50 tube, being careful to avoid any contamination with pelleted chromatin.
9. Add 0.3 g solid $(NH_4)_2SO_4$ per ml of supernatant and mix rapidly to avoid local high concentrations of salt.
10. After all the salt has dissolved, leave the sample in ice-water, shaking gently in a mechanical agitator for 20–60 min.
11. Centrifuge for 20 min at 90 000 *g* in a Ti60 rotor (or 100 000 *g* in a Ti50 rotor).
12. Aspirate off as much of the supernatant as possible and discard it. Mark the position of the pellet on the tube wall.
13. Resuspend the pellet in nuclear dialysis buffer by carefully pipetting up and down (avoid any foaming). It is *very important* that the final protein concentration of the extract should not be too low (see Section 6.2). Thus the volume of dialysis buffer to use must be calculated. This is done on the basis of DNA measurement of the nuclear lysate (step 3). Assuming that nuclei contain a 1:1 ratio of DNA: non-histone protein, the pellet should be resuspended in a volume of dialysis buffer calculated to give a protein concentration of approximately 10 mg/ml.[a] While the 1:1 ratio is true for most tissues (such as liver or brain), other tissues, like spleen or thymus, contain less soluble nuclear proteins per mg of DNA.
14. Continue resuspending every 15 min or so for approximately 1–2 h, in order to ensure complete recovery of the pelleted proteins. Avoid foaming since this denatures proteins.
15. Dialyse twice for 2 h each time in the cold against 100 vol. dialysis buffer. A white precipitate should form.
16. Remove the dialysate from the dialysis bag and centrifuge it for 2 min in a cold microcentrifuge.
17. Keep the clear supernatant. Freeze it immediately in aliquots (usually 100 μl) in liquid N_2.

Protocol 2. *Continued*

18. Use a small aliquot of the final extract to determine the A_{230} and A_{260}. Then calculate the protein concentration according to the formula:

$$187 \times A_{230} - 81.7 \times A_{260} = \mu g \text{ protein/ml}$$

A good extract should give a concentration of protein between 6 and 10 mg/ml, with a ratio of A_{230}/A_{260} of 4.0 to 6.0.

[a] In fact, we routinely resuspend the pellet in a lower concentration (two-thirds to three-quarters) that suggested by this calculation, since it is rare that excess protein causes problems, whereas extracts that are too diluted are difficult to retrieve (see Section 6.2.1).

3.4 *In vitro* transcription

The quality of the extract can be measured by a variety of different functional tests, such as *in vitro* transcription, footprinting, gel retardation analysis, etc. Of these, *in vitro* transcription is the most demanding and complex, since it requires a complex interaction between the template, several transcription factors, and RNA polymerase.

While we recognize that many variables can affect the results of *in vitro* transcription reactions, we routinely test each extract only for optimal protein concentration, using as a template a mixture of plasmids containing an ubiquitous promoter (adenovirus major late promoter) and another plasmid containing the tissue-specific promoter we are interested in studying. The easiest way to monitor the synthesis of transcripts is by using the G-free cassette developed by Sawadogo and Roeder (13), although other methods can also be used. Briefly, the G-free cassette assay relies on the use of a synthetic DNA template of 394 bp (shorter versions also exist) which does not contain G residues in the non-coding strand. When a promoter is cloned in front of this synthetic piece of DNA, the *in vitro* transcription reaction can be performed in the absence of GTP, and in the presence of a GTP analogue (*O*-methyl-GTP) which does not support chain elongation. Under these conditions, all radioactivity incorporated into macromolecules is derived from transcription of this G-free stretch, since spurious initiation sites elsewhere can not be elongated past the first G in the sequence. This way, transcripts can be visualized directly on a gel, without further manipulations. Using this system, a good extract should give a clear autoradiographic signal after 1–2 h of exposure to Kodak XAR-5 film, when using a good quality intensifying screen.

We have found that the quality of the DNA used as a template is not particularly crucial when performing *in vitro* transcription reactions. We have used relatively fast methods to isolate plasmids, and do not further clean the plasmid DNA by CsCl centrifugation.

We describe here two variants of the *in vitro* transcription protocol:

(a) If the G-free cassette is used (see *Protocol 3*), then the transcripts should be made radioactive for direct visualization by gel electrophoresis and autoradiography. This standard protocol uses radioactive UTP and the reaction mixture contains *O*-methyl-GTP instead of GTP.

(b) If, on the other hand, more classical detection techniques such as nuclease S1 analysis are used, then the transcripts should be made without radioactive nucleotide precursors (see *Protocol 4*).

Protocol 3. *In vitro* transcription with G-free cassette vectors

Reagents

- 5 × G-free cassette transcription buffer
  Mix:

1.0 M Hepes pH 7.6	100 μl
87% glycerol	350 μl
3.0 M KCl	85 μl
1.0 M $MgCl_2$	60 μl
25 mM CTP	250 μl
25 mM ATP	250 μl
10 mM UTP	35 μl
10 mM *O*-methyl-GTP	200 μl
H_2O	670 μl.

  Store this solution at −20°C. It can be thawed and re-frozen many times
- Nuclear dialysis buffer (see *Protocol 2*)
- Template DNA (0.8–1.0 μg/μl, circular or supercoiled)
- RNasin (Promega, 40 U/μl), or other RNase inhibitor
- [α-^{32}P]UTP (10 μCi/μl, >400 Ci/mmol).
- Transcription stop mixture (0.25 M NaCl, 1% SDS, 20 mM Tris–HCl pH 7.5, 5 mM EDTA). Store this solution at room temperature
- tRNA (10 mg/ml)
  Further purify commercial tRNA by a single phenol extraction, followed by ethanol precipitation. Resuspend in water and confirm its concentration by measuring the A_{260} (1 mg/ml tRNA has an A_{260} of 25.0)
- Proteinase K (10 mg/ml in H_2O)
- Formamide loading dye (see Chapter 6, *Protocol 13*)
- 4% polyacrylamide/urea gel (see Chapter 6, *Protocol 12*).

Method

The total volume of each transcription reaction is 20 μl, but this is prepared in several stages as follows.

1. Prepare the following transcription mixture (per sample):

5 × G-free cassette transcription buffer	4 μl
[α-^{32}P]UTP	0.7 μl
RNasin (or other RNase inhibitor)	1 μl
H_2O	1.3 μl.

   Keep on ice until required in step 5.
2. Set up a series of three reactions for each extract to be tested, so that the percentage of extract in the reaction can be varied as 30%, 45%, and 60% (6, 9, and 12 μl of extract, respectively). Make the volume up to 12 μl in each case by the addition of nuclear dialysis buffer.
3. To each tube, add 1 μl template DNA (this can be a mixture of two or

Protocol 3. *Continued*

more different templates, but the total amount of DNA should be kept constant at 0.8–1.0 μg per reaction).

4. Incubate on ice for 10 min.
5. Add 7 μl of the transcription mixture prepared in step 1 above.
6. Incubate at 30°C for 45 min.
7. Add a mixture of 274 μl transcription stop buffer, 2 μl 10 mg/ml tRNA, 4 μl 10 mg/ml proteinase K.
8. Incubate at 37°C for 30 min.
9. Extract once with phenol/chloroform and centrifuge for 2 min in a cold microcentrifuge to separate the organic and aqueous phases.
10. Remove the aqueous (upper) phase into a fresh microcentrifuge tube and precipitate by the addition of 750 μl ethanol (no further addition of salt required).
11. Leave in solid CO_2 for 15–20 min, then centrifuge for 10 min in a cold microcentrifuge at top speed.
12. Resuspend the pellet in 5 μl formamide loading dye.
13. Electrophorese 3 μl on the 4% polyacrylamide/urea gel until the bromophenol blue reaches the bottom of the gel.
14. Fix the gel, dry it and expose to X-ray film for 1–2 h.

Protocol 4. *In vitro* transcription with templates other than the G-free cassette vectors

Reagents

- 5 × alternative transcription buffer

  Mix:

1.0 M Hepes pH 7.6	100 μl
87% glycerol	350 μl
3.0 M KCl	85 μl
1.0 M $MgCl_2$	60 μl
25 mM GTP	250 μl
25 mM ATP	250 μl
25 mM UTP	250 μl
25 mM CTP	250 μl
H_2O	405 μl.

  Store this solution frozen at −20°C; it can be thawed and refrozen many times
- Nuclear dialysis buffer (see *Protocol 2*)
- Template DNA (0.8–1.0 μg/μl, circular or supercoiled)
- RNasin (Promega, 40 U/μl), or other RNase inhibitor
- Transcription stop mixture (see *Protocol 3*)
- tRNA (10 mg/ml) (see *Protocol 3*)
- Proteinase K (10 mg/ml in H_2O)
- DNase digestion buffer (20 mM Hepes, pH 7.6, 50 mM NaCl, 10 mM DTT, 5 mM $CaCl_2$, 5 mM $MgCl_2$)
- RNase-free DNase I (750 μg/μl, made RNase-free by treatment with iodoacetate). Note, however, that good quality RNase-free DNases that do not require pre-treatments are also commercially available
- 2 × DNase stop buffer (20 mM Tris–HCl pH 8.0, 50 mM EDTA, 1% SDS, 0.3 M NaCl)
- Reagents, gels, etc for nuclease S1 analysis or other analyses of the *in vitro* transcripts.

Method

When transcripts are to be analysed by nuclease S1 mapping, RNase mapping or primer extension, we routinely double the size of the transcription reaction to 40 μl. Therefore the following procedure applies under these conditions.

1. Prepare the transcription mixture (per sample) as follows:

5 × alternative transcription buffer	8 μl
RNasin (or other RNase inhibitor)	2 μl
H_2O	4 μl.

Keep on ice until required.

2–11. Set up a series of three reactions for each nuclear extract to be tested as described in *Protocol 3*, steps 2–11, but doubling the reagent volumes (and template DNA) since the transcription reaction volume used is 40 μl not 20 μl.

12. After the reaction is finished, the template DNA should be removed, since otherwise it will also hybridize with the probes used for the analysis. To do this, resuspend the pellet obtained after ethanol precipitation in 10 μl H_2O, and then add 90 μl DNase digestion buffer containing 3–5 μg RNase-free DNase I and 1 μl RNasin per sample (freshly added).

13. Incubate at 37°C for 15 min.

14. Add 100 μl of 2 × DNase stop buffer.

15. Heat the samples at 65°C for 5 min, extract twice with phenol/chloroform and precipitate with 500 μl ethanol.

16. Wash the RNA pellet in 70% ethanol and dissolve it in H_2O. Proceed with the nuclease S1 mapping or other type of analysis under standard conditions.

---

4. Analysis of *in vitro* transcripts

4.1 Using the G-free cassette

In general, using the G-free cassette is the easiest way of analysing the results of *in vitro* transcription reactions (13). In this case, the presence of correctly initiated transcripts is determined simply by direct autoradiography of the reaction products.

When cloning the promoter of interest in front of the G-free cassette, it is crucial that no G residues are present between the cap site and the beginning of the cassette, since even a single G residue has been found to eliminate *in vitro* transcription completely. If in the gene of interest the cap site or immediate downstream nucleotides happen to be G, or if there are heterogeneous start sites, two alternatives exist: either the sequence located downstream of the TATA box, and including the problematic cap site, can be eliminated, or a better cap site (with its own TATA box, so as to keep the distances correct) can be donated by a more appropriate gene (for example, the albumin promoter, which contains no G until nucleotide +22). In addition to the template of interest, it is usual to include a standard template as a control, such as adenovirus major late promoter or an unmodified version of the gene of interest, containing a shorter version of the G-free cassette (see, for example, ref. 10). This way, intra-sample variations can be monitored and accounted for.

The main parameters affecting *in vitro* transcription that have to be taken into account are the strength of the signal itself, the signal-to-noise ratio, the optimal protein concentration and, above all, the reproducibility of the *in vivo* control (tissue specificity, inducibility, responsiveness to promoter mutations, etc.). In general, when using strong promoters (e.g. adenovirus major late promoter, albumin promoter, α2U-globulin promoter), the signals should be strong enough to be clearly visible after 2–3 h of exposure, while the background should be low enough to cause no problem in a 24–48 h exposure. Naturally, this is not necessarily the case when working with weaker promoters. For this reason, it is advisable to use a plasmid yielding a strong signal as a control.

The protein concentration necessary for optimal *in vitro* transcription has been found to vary somewhat among extracts, and does not necessarily correlate with the total protein concentration in the extract. This phenomenon most likely reflects different degrees of contamination with either DNA (which affects the determination of protein concentration by absorbing in the UV), histones or even cytoplasmic proteins. Accordingly, some extracts work best when used at 6 μl per reaction (or less), while others require up to 12 μl per reaction for maximal activity.

Of course, the most important aspect of *in vitro* transcription assays is the potential to reproduce *in vivo* control mechanisms. Tissue specificity is usually well preserved when extracts are prepared as described in this chapter, although the magnitude of the difference between expressing and non-expressing tissues is less pronounced *in vitro* than *in vivo* (10, 14–16). Other aspects of gene control, such as humoral regulation (hormones, cytokines, etc.) are more difficult to reproduce *in vitro*. This might reflect the fact that hormones and other humoral factors often control transcription by activating or inactivating transcription factors via post-translational modifications (17). These modifications could be lost during extract preparation, unless special precautions are taken (for example, phosphatase inhibitors). Moreover, small molecular weight ligands and cofactors required for the optimal activity of transcription factors may be lost during the dialysis steps involved in preparing nuclear extracts.

4.2 Other approaches

In general, the time saved and the quality of the results gained by using the G-free cassette largely compensate for the initial effort of cloning the promoter of interest in front of the G-free cassette. However, this might not always be possible or appropriate to the particular needs of the project in question. In this case, the *in vitro* transcripts can be analysed easily by other standard methods that detect correctly initiated transcripts, such as nuclease S1 mapping, RNase mapping, or primer extension. It is beyond the scope of this article to describe these methods in detail, since excellent working protocols are available elsewhere (Chapter 1 in this book and ref. 18). Therefore, we limit our present discussion to considerations relevant to the application of these techniques to the analysis of *in vitro* transcripts.

Nuclease S1 mapping, RNase mapping, and primer extension have all been used successfully to monitor *in vitro* transcription using extracts prepared as described in this chapter. One important consideration to keep in mind is that the nuclear extracts contain small but detectable amounts of RNA, probably hnRNA. Any of this RNA that hybridizes to the probe could produce misleading results. For this reason, the probe used for the analysis should be chosen carefully so that it contains at least some vector sequences, present in the template for *in vitro* transcription, but not in the corresponding *in vivo* transcripts. This allows differentiation between these different types of hybrids (*in vivo* versus *in vitro*) by the length of the protected probe in each case.

Two methods that we feel should be avoided are Northern blotting and run-off assays.

(a) *Northern blotting*. Although we have never tried Northern blotting of *in vitro* transcripts, and cannot therefore say whether or not enough material for detection by this method is produced *in vitro*, we consider this method unacceptable because contaminating transcripts produced *in vivo* would be indistinguishable from bona fide *in vitro* ones.

(b) *Run-off assays*. These assays are unreliable when using extracts prepared as described here, since the extracts contain a wide variety of nucleic acid modification enzyme activities. Thus, it has been observed that the templates suffer many conformational modifications during the incubation so that, at the end of the reaction, circular templates become apparently concatenated, and linear templates become ligated into larger structures (11). Attempts to overcome this problem by dephosphorylating the termini of the template DNA or by creating blunt ends have been only partially successful. The presence of these activities results in very long RNA molecules that do not correspond to the expected sizes of transcripts.

5. Factors affecting the efficiency of *in vitro* transcription

Many variables are known to affect the efficiency and/or specificity of *in vitro* transcription reactions. Among the most important of these are the amount and quality of the proteins present in the extract and the concentration and, to a lesser extent, the quality of the DNA used as a template. Of course, other variables, such as temperature, time of reaction, and ionic strength are also important, but these variables are easier to control and are therefore used at a standard value for most purposes (10). These 'standard' conditions, as described in the protocols given in this chapter, were chosen based on knowledge of the requirements for RNA polymerase II transcription, but there is some evidence suggesting that particular promoters work better under slightly different conditions, presumably due to stabilization of particular transcription factors.

5.1 Protein concentration

As mentioned above, the protein concentration should be checked each time a new transcription extract is prepared. The *total* protein concentration can be determined by spectrophotometry, but spectrophotometry cannot reveal the origin or quality of the proteins present. For example, if some cytoplasmic contamination is left during the initial homogenization step, this will be retained during the rest of the protocol, and so some of the proteins measured will actually be of cytoplasmic, rather than nuclear, origin. Of course these proteins are inactive in transcription assays but in some cases might be inhibitory. Indeed, even some nuclear proteins, such as histones and post-translationally modified or partially degraded transcription factors, could act as either specific or non-specific repressors of *in vitro* transcription (19). For these reasons, it is critical to carefully titrate the protein concentration by a functional assay such as *in vitro* transcription, footprinting or gel mobility shift assay (see Chapter 6). If too much protein is present in the reaction, the repressors will compete for binding to the template, resulting in a lower level of *in vitro* transcription than if less protein is used. Conversely, if the concentration of proteins becomes too low, the formation of initiation complexes may no longer be possible.

5.2 DNA concentration

As a standard, we use 0.8–1.0 μg DNA per 20 μl transcription reaction, as it has been found that higher DNA concentrations fail to increase the specific signals obtained. We have calculated, however, that only 1–2% of the DNA molecules present in the reaction are actually transcribed. Consistent with this, we have shown that as long as the total DNA concentration is kept at approximately 1 μg/reaction, transcripts are easily detectable when as little as 6 ng of specific template is used (e.g. albumin promoter DNA). However, this is not the case if the total DNA concentration is decreased below 0.5 μg/reaction (10). The most likely explanation for these effects lies again in the presence of non-specific inhibitors of transcription in nuclear extracts. In practice, these observations indicate that, as long as the total amount of DNA is kept constant, different amounts of two or three templates can be mixed in the same reaction, so as to provide internal standards as already described. In principle at least, this should also allow the study of potential interactions and competitions among different templates.

5.3 Time course

During a standard *in vitro* transcription reaction, which includes 10 min of pre-incubation on ice before the addition of NTPs, the accumulation of transcripts occurs linearly after a short lag, and continues for a period of approximately 45 min. After this period, little or no further synthesis is detected. Experiments in which re-initiation has been inhibited by the addition of heparin at different times indicate that transcript initiation occurs at least during the first 30 min of the incubation period (11).

5.4 Hydrolytic enzymes

The main hydrolytic enzymes to be aware of when preparing nuclear extracts are proteases, even in cases such as extracts derived from pancreas or salivary glands where nucleic acid-degrading enzymes are also a major problem. In situations where the gene of interest is not constitutively expressed in the tissue under study, the possibility has to be considered that control of its expression might depend on post-translational modifications of specific transcription factors. In these instances, the addition of phosphatase and/or kinase inhibitors is advisable, but the exact choice of inhibitor depends on the particular enzymes involved among the many kinases and phosphatases present in the cell.

The protocol described in this chapter (*Protocol 1*) uses aprotinin (Trasylol), leupeptin, pepstatin, and PMSF as protease inhibitors. In addition, the low fat milk present in the homogenization buffer has been shown to minimize protein degradation very efficiently. Indeed, extracts prepared from tissues such as kidney and lung are consistently active in transcription assays only when low fat milk has been included in the homogenization buffer. For most experiments we have carried out, these precautions have proved to be sufficient. However, when tissues are particularly rich in proteases, the use of a more extensive range of protease inhibitors may be advisable or even necessary (see ref. 12).

During the preparation of the nuclear extracts (*Protocol 1*), Mg^{2+} is replaced by spermine and spermidine (all are cations and are needed to stabilize the chromatin) and divalent cations coming from the tissue itself are chelated by the addition of EDTA. These precautions are taken partially to inhibit tissue DNases. DNA degradation during the initial steps of nuclei isolation would lead to the release of short DNA fragments into the supernatant during the ammonium sulphate lysis of the nuclei (see *Protocol 2*). This in turn might inhibit *in vitro* transcription. No additional efforts other than these have been made to inhibit DNase activities. RNase activity in tissues is inhibited both by the precautions taken to minimize DNases and by the inclusion of RNasin or other RNase inhibitors during *in vitro* transcription. Not surprisingly, however, we have found that extracts prepared from tissues that are rich in DNases and RNases (pancreas, salivary glands) can contain sufficient amounts of them to degrade DNA templates and RNA transcripts, or even oligonucleotides used in gel mobility shift experiments. In these cases, further precautions would be obviously desirable. Even liver nuclear extracts usually contain small amounts of DNase. However, in the presence of ATP, the nicks it can induce are, in general, efficiently sealed during *in vitro* transcription.

As previously mentioned, phosphatases and/or kinases may be more difficult to control, due to the large variety of these enzymes. We have tested the effect of various measures to inhibit phosphatases using *in vitro* transcription reactions performed with nuclear extracts prepared by the standard protocol:

(a) As expected, *in vitro* dephosphorylation of the nuclear extracts by preincubation with calf intestinal phosphatase (CIP) completely abolishes *in vitro* transcription.
(b) Surprisingly, the addition of EGTA to the reaction mixture has no effect on the reaction.
(c) The use of non-specific inhibitors of phosphatase activity, such as pyrophosphate (PP_i) or high concentrations (10 mM) of NaF result in inhibition of *in vitro* transcription.
(d) Other inhibitors, such as molybdate, sphingosine, or Zn^{2+} have no effect on *in vitro* transcription.

We have found that NaF, PP_i and molybdate all produce transcriptionally active extracts when used during preparation of the nuclei (*Protocol 1*) and nuclear lysis (*Protocol 2*, step 1). However, only NaF (at low concentration, 1 mM) and PP_i can be included in the dialysis step (*Protocol 2,* step 13), since molybdate inhibits *in vitro* transcription (see above). When prepared in the presence of these inhibitors, extracts appear to retain the post-translational phosphorylation of the interleukin 6 (IL-6) responsive factor, as determined by the specific activation of transcription from acute phase genes in turpentine-treated animals (*Figure 1*). It should be emphasized, however, that these experiments are preliminary and have not always been reproducible. Most likely, the use of phosphatase inhibitors requires a much more detailed study.

5.5 Tissue specificity and *cis*-acting elements

Extracts prepared as described in this chapter have been shown to reproduce tissue specificity *in vitro* (10). However, it should be emphasized that even with the best extracts, the amplitude of cell-type-specific transcription is only around 100-fold *in vitro*, as compared to over 10^5-fold *in vivo*. Thus far, biochemical approaches to reproduce chromatin-mediated control of gene expression are still in their infancy (20, 21).

Tissue specificity of a nuclear extract (for example, liver) can be assessed by measuring its efficiency in transcribing *in vitro* a tissue specific promoter (e.g. from the albumin gene). Also, if tissue specificity is correctly maintained, a promoter active in another tissue (e.g. from the parotid-specific amylase gene) should not be transcribed by the same extracts. Again, this has been shown to be the case, but not with the high degree of specificity observed *in vivo*. Liver extracts utilize the albumin promoter much more efficiently than the amylase promoter, but the amylase promoter is nevertheless utilized to a certain extent by the liver extracts.

When comparing extracts from different tissues, it is necessary to standardize the extracts with respect to a ubiquitous promoter, assumed to be expressed with the same efficiency in all tissues. Many viral promoters are available that seem to have no tissue specificity. We have used the adenovirus major late promoter extensively for this purpose (see *Figure 2*). It is also theoretically possible to use promoters derived from housekeeping genes, although we

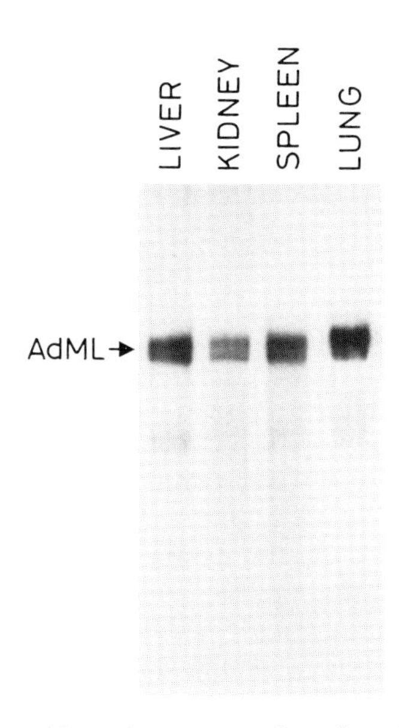

Figure 2. *In vitro* transcription with nuclear extracts from four different rat tissues. Transcriptionally competent nuclear extracts were prepared from rat liver, kidney, spleen, and lung as described in the text and analysed using G-free cassette vectors. All of these extracts efficiently utilize the adenovirus major late promoter (AdML) as shown in the figure. A doublet band was obtained in this experiment because the nuclear extracts still contained traces of GTP, rendering termination at the first G downstream of the G-free cassette inefficient. This problem can be solved by increasing the concentration of the chain terminator 3-*O*-methyl-GTP in the *in vitro* transcription cocktail (see Section 6.3.3).

have not tested this. However, it is likely that, in most case, these promoters will be weaker in activity compared to viral ones.

The effect of *cis*-acting elements on *in vitro* transcription can be evaluated by a variety of different approaches, most of which do not differ greatly from those routinely used in conventional promoter studies involving transfection into expressing and non-expressing cells. These include deletion analysis and site-directed mutagenesis (see Chapter 2). The availability of extracts does, however, help in the design of these experiments, since they can be used to perform DNA footprinting studies (22). The results of these experiments can then be used to focus the deletions and/or mutagenesis to regions of the promoter capable of interacting with nuclear proteins. The relative importance of any particular *cis* element can be further delineated by constructing artificial promoters containing several repeats of the sequence of interest, either in its wild-type or mutated forms, and testing the resulting constructs in *in vitro* transcription reactions (see *Figure 3*). Alternatively, the availability of the corresponding oligonucleotides allows the performance of experiments such as those presented in *Figure 4*. In this case, a competition between an element within the promoter in question and an excess of an oligonucleotide containing a related sequence results in a drastic decrease in *in vitro* transcription. This type of experiment provides further independent evidence for the role of a particular *cis*-acting element on *in vitro* transcription.

Many more approaches to identify and analyse the role of *cis*-acting elements, either within promoters or enhancers, have been described but it is beyond the scope of this chapter to discuss all of them in detail.

6. Trouble-shooting

6.1 Preparation of nuclei

As shown in this section, the critical step in nuclei preparation is the homogenization step itself. If the tissue is homogenized too extensively, or not enough, the result will always be that there are too few nuclei and/or they are too dirty. Unfortunately, the ideal conditions will vary according to the equipment available, so that precise conditions of speed, volume, time, etc., cannot be provided and must be determined empirically in each laboratory. We would nevertheless like to emphasize that good extracts are by no means difficult to obtain, provided that certain precautions are respected. Therefore, paying attention to the problems discussed in this section may improve the quality of your results.

6.1.1 The yield of nuclei is low

If fewer nuclei than expected are obtained, this is usually due to insufficient or excessive homogenization.

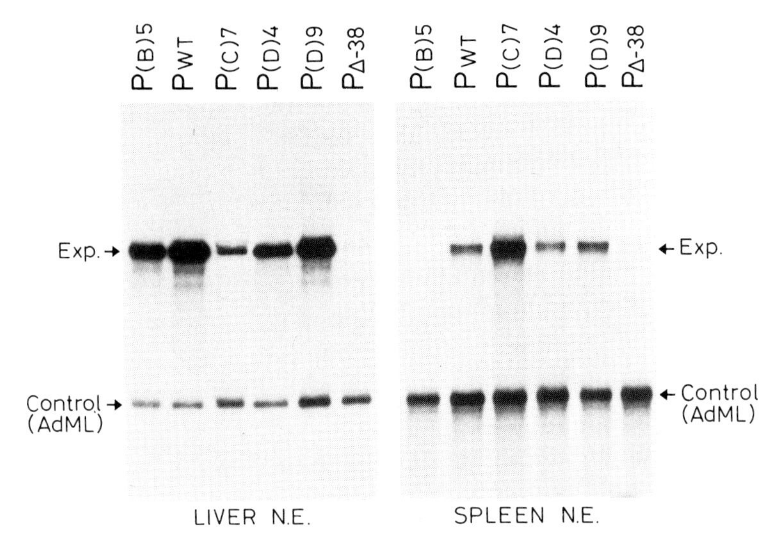

Figure 3. *In vitro* transcription from natural and synthetic promoters. G-free cassette fusion genes were constructed containing various oligomerized albumin promoter elements (B, C, D) upstream of the TATA box (for details see ref. 16). These reporter genes [P(B)5 to P(D)9] and a reporter gene bearing the albumin 'wild-type' normal promoter (Pwt) were incubated for *in vitro* transcription with liver or spleen nuclear extracts to generate transcripts of about 400 nucleotides (nt) (Exp.). A template carrying a shorter G-free cassette (190 nt) fused to the adenovirus major late promoter (AdML) was included as an internal control, since this promoter is similarly active in both extracts. The autoradiographs of *in vitro* transcripts obtained with liver and spleen nuclear extracts (N.E.) were exposed for 2 h and 8 h, respectively. The specific transcription potential of a given promoter can be expressed as the ratio of its *in vitro* transcript to the *in vitro* transcripts produced by the AdML promoter. As seen from this figure, for the templates P(B)5, Pwt, P(D)4, and P(D)9, this ratio is much higher in liver as compared to spleen nuclear extracts. This is in keeping with the cell type distribution of the cognate factors HNF-1 and LAP/C/EBP for promoter elements B and D, respectively (16), all of which are considerably enriched in hepatocytes. Relative to the activity of the AdML promoter, the figure shows that the promoter P(C)7, containing seven binding sites for the ubiquitous factor NF-Y, is as active in spleen as in liver nuclear extracts. A control promoter devoid of upstream regulatory elements, [PΔ-38), is poorly active in both spleen and liver nuclear extracts.

Insufficient homogenization

This is likely to be the problem if there are pieces of tissue floating in the tube after centrifugation (*Protocol 1*, step 13). It is normal that a few pieces of tissue escape homogenization but insufficient homogenization results in a large amount of tissue left floating with little or no material located at the interface. In this situation, check the following points:

(a) The clearance between the homogenizer and pestle might be too wide.

(b) The tissue might not have been properly minced before homogenization,

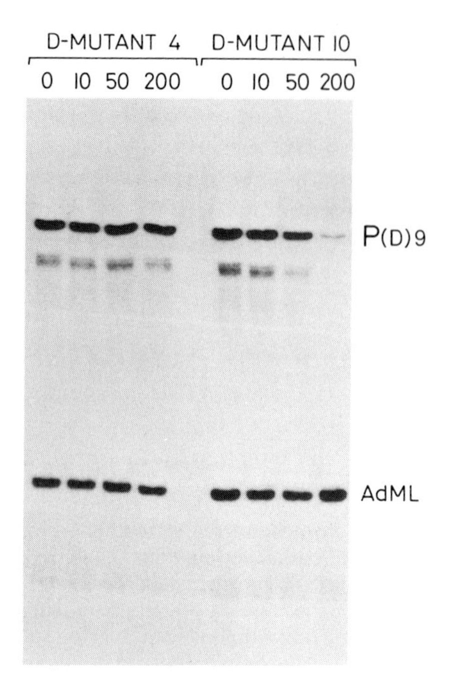

Figure 4. The effect of competitor binding sites on *in vitro* transcription. A G-free cassette vector carrying a synthetic promoter, P(D)9, with nine tandemly-repeated albumin promoter 'wild-type' D-elements was incubated with liver nuclear extract in the absence or presence of double-stranded mutant D site oligonucleotides 4 or 10. The resulting *in vitro* transcripts are shown after gel electrophoresis. The mutant binding sites 4 and 10 have a low and high affinity, respectively, for the cognate factors C/EBP and LAP. As expected, addition of the low affinity binding site oligonucleotide (mutant 4) does not interfere with D-element-specific *in vitro* transcription from P(D)9. However, 200 ng of high affinity binding site (mutant 10) very effectively reduces *in vitro* transcription from the P(D)9 promoter. Neither mutant site 4 nor mutant site 10 significantly affect transcription from the AdML control template.

so that large pieces are present. Proper attention to mincing well is particularly important when working with hard tissues, such as kidney.

(c) If too much buffer is put into the homogenizer, some tissue might be able to float into the larger portion of the apparatus, thus escaping the pestle.

(d) The speed of the homogenizer or the number of homogenizations might not be sufficient. Increase the speed of the motor and/or try longer homogenization. Adjust these carefully, though, since too much homogenization might lead to breakage of the nuclei.

Over-extensive homogenization

Over-extensive homogenization will result in low yields because of partial lysis of the nuclei. The lysed nuclei, with their extended chromatin, will stay at the interface, which will then become clogged and prevent the passage of the nuclei that are still intact. A large interface is an unequivocal signal of this problem. Reduce the extent of homogenization. In tissues that contain large amounts of connective tissue, high shearing forces might also induce nuclear lysis. In this case, the most likely solution is extensive mincing before homogenization. Of course, mechanical forces, induced for example by a pestle with too small a clearance, will also result in nuclear lysis. Try different homogenizers if possible.

Other possible causes are as follows.

(a) If too much material is processed at the same time, enough nuclei will lyse that they clog the interface, leading to a lower recovery of good nuclei in the pellet. As already discussed, too much material causes many problems, and should be avoided at all costs.

(b) If the tissue of interest contains cells with a small volume of cytoplasm, these will fail to lyse, or will leave large cytoplasmic 'tags'. This will also prevent the nuclei from entering the dense sucrose cushion. Further homogenization is likely to be better than lowering the density of the cushion.

6.1.2 Nuclei have a dirty appearance

As previously mentioned, the nuclear pellet should be whitish, with no traces of pink or brown colour. If the nuclei are coloured, clean them again through a second sucrose centrifugation (*Protocol 1*, step 17).

(a) The most common cause of dirty nuclei is the homogenization of too much tissue in too small a volume of homogenization buffer. Try reducing this ratio, even if it takes longer to homogenize all the tissue. In fact, using less material often leads to higher yields.

(b) Other sources of contamination are cytoplasmic residues adhering to the nuclei, derived from inefficient homogenization. Check under Section 6.1.1 (a) for possible remedies.

(c) If the walls of the tube are not properly rinsed with water, material from the walls will end up contaminating the nuclei.

6.2 Preparation of nuclear extract

6.2.1 The initial nuclear lysate is too dilute (low A_{260})

The concentration of the initial nuclear lysate must be checked by measuring its A_{260} before proceeding to the fractionation stages (*Protocol 2*, steps 1–4). If the overall yield of nuclei in *Protocol 1* was low, then of course the A_{260}

will also be low. Refer to the previous section for possible remedies. It is, however, possible to have a low A_{260} even when a good yield of nuclei was obtained. The most common causes are:

(a) The nuclei may have not been well resuspended. Vigorous bubbling and pipetting up and down with a plastic pipette are necessary to dislodge the nuclear pellet from the bottom of the tube, and attention should be paid to avoid their sticking to the pipette. Never use a glass pipette to handle clean nuclei. Finally, resuspend the nuclei with a hand-held pestle (all-glass Dounce homogenizer), so as to ensure homogeneity of the suspension. Here there is no choice except to use glass but, unlike the use of glass pipettes where nuclei stick inside and cannot be recovered, the recovery of nuclei is not problematic when using a glass homogenizer.

(b) Even if the nuclei are resuspended, it is possible that this suspension is not homogeneous. Do not let the nuclei suspension sit before taking the aliquot for A_{260} measurement, since nuclei will sediment even without centrifugation. Also, it is important that the nuclei are well lysed before reading the A_{260}. For this, we routinely use 0.5% SDS, and the nuclei are added to this solution while vigorously vortexing (*Protocol 2*, step 3). It is also possible to do this in 0.2 M NaOH, although some proteins may precipitate under these conditions, therefore producing light scattering, which might interfere with the A_{260} reading.

6.2.2 The chromatin pellet is too large

(a) If the lysis of the nuclei is done too roughly, some of the DNA will be broken simply by mechanical shearing. This will lead to a swollen chromatin pellet (see *Protocol 2*, step 8), which results in trapping more aqueous supernatant with the concomitant loss of soluble proteins. To avoid this problem, be sure that mixing after addition of the ammonium sulphate (see *Protocol 2*, steps 9 and 10) is done quickly, but as gently as possible. Do not shake and absolutely *do not* vortex.

(b) If the tissue is rich in DNases, increase the EDTA concentration of the homogenization and nuclear lysis buffers, so as to make sure these enzymes are not active.

(c) More trivially, chromatin will start re-swelling as soon as centrifugal force is no longer applied. Take the tubes out of the rotor as soon as the centrifuge stops (see *Protocol 2*, step 7), and proceed rapidly to the next step.

6.2.3 Low protein concentration in the final nuclear extract

In general, it is assumed that the nucleus contains roughly 1 mg of extractable non-histone proteins per mg of DNA. The volume of the final extract is based on this assumption and uses the A_{260} measurement carried out on the original nuclei preparation as a basis for deciding the volume of dialysis buffer to be used for resuspending the $(NH_4)_2SO_4$ pellet (see *Protocol 2*, step 13). This assumption is not necessarily correct for all tissues (for example, spleen nuclei contain a higher DNA/protein ratio). In these cases, it is advisable to use somewhat less dialysis buffer than the A_{260} calculation would indicate, to ensure that the nuclear extract is not too dilute. In fact, since it is rare to have so much protein as to cause problems, we routinely resuspend the extracts in about two-thirds or three-quarters of the calculated volume. Rarely have we obtained extracts with protein concentrations higher than 12 mg/ml, and therefore we do not believe that protein precipitation should cause a problem. In contrast, extracts that are too dilute are difficult to salvage. We present here some potential reasons for a low yield of protein:

(a) Some proteins might have leached from the nuclei. This could happen if the homogenization buffer was not sufficiently cold and hence not viscous enough. If too much tissue is processed at the same time, the viscosity of the resulting solution is similarly decreased, thereby leading to potential leaching of nuclear proteins.

(b) If the chromatin swells (see *Protocol 2*, step 7), many proteins will be trapped in this gelatinous structure, and protein yields will suffer.

(c) Finally, as already mentioned, resuspension of the final protein preparation can be deceptive, since the pellet becomes transparent quite rapidly, but extra time (about 1–2 h of pipetting up and down) is essential to achieve complete resuspension. Nevertheless, avoid foaming at this step since this denatures proteins.

6.2.4 The nuclear extract is contaminated with DNA (low A_{230}/A_{260} ratio)

In general, the A_{230}/A_{260} ratio of the final extract (see *Protocol 2*, step 18) is approximately 4 or 5. If a lower ratio is obtained, the chances are that the extract is contaminated with nucleic acids (mainly DNA), and often (but not always), such extracts perform poorly in functional assays. The most likely causes of DNA contamination are:

(a) Handling during the lysis step (see *Protocol 2*, step 3) might have been too rough, leading to DNA breakage by shearing.

(b) It is also possible that large amounts of DNase are present in the tissue. Try increasing the EDTA, spermine, and spermidine concentrations during homogenization. Inclusion of EGTA as well as EDTA during both homogenization and lysis might also be beneficial.

6.3 *In vitro* transcription

6.3.1 Only low amounts of *in vitro* transcripts are detected

Unfortunately, since *in vitro* transcription is the functional assay for testing the extracts, it follows that low signals in this assay can be due to a whole variety of diffferent causes. Low signals are, in general, caused by extracts that have not been prepared properly. Dilute extracts or contamination with either unwanted proteins (cytoplasmic proteins, histones) or nucleic acids are possible causes. These problems have been described in the previous sections. The following are alternative possible reasons, largely independent of the quality of the preparation undertaken:

(a) The extract might contain problematic levels of proteases, DNAses or RNases. In this case, it is probably better to shorten the transcription incubation time (see *Protocols 3* and *4*, step 8). Also try to inhibit these enzyme activities by adding DTT to the transcription reaction (to stabilize RNasin or other endogenous RNase inhibitors), increasing the DTT concentration during the DNase step (see *Protocol 4*, step 12), or by adding further inhibitors (higher concentration of RNasin, protease inhibitors, etc.).

(b) If the dialysis has not been extensive enough (see *Protocol 2*, step 15), it is possible that low incorporation of [α-^{32}P]UTP is caused by a relatively large endogenous UTP pool in the extract.

(c) While the quality of the template DNA is not too crucial for *in vitro* transcription, some residual contaminants from its preparation, such as ethidium bromide, do have a strong inhibitory effect. Avoid the use of CsCl-purified DNA as a template. Similarly, some promoters will be strongly inhibited if the template DNA is contaminated with RNA.

(d) The radioactive UTP should not have decayed by more than one half life (about 14 days), since radiolysis products are strongly inhibitory to *in vitro* transcription.

6.3.2 Some templates are transcribed, but others are not

In this case, the problem is most likely to be related to the structural features of the promoter in question:

(a) Bacterial methylases may have caused modifications that interfere with transcription factor binding sites. Use *dam*/*dcm* bacterial host strains for preparing the plasmid.

(b) Sequences downstream of the cap site might be important for promoter activity and hence lacking in the template being tested.

(c) The cap site might have been incorrectly determined, so that a G residue is left between the real cap site and the G-free cassette (see Section 4.1).

6.3.3 High background incorporation when using the G-free cassette

(a) If the background bands are higher than the expected transcript band, this is most likely due to use of insufficient *O*-methyl-GTP to stop polymerization (see *Protocol 3*). The use of *O*-methyl-GTP is necessary even though the transcription buffer lacks GTP, since some GTP is always carried over in the nuclear extract.

(b) If the RNA transcript appears degraded, add more RNase inhibitors or more DTT, and add tRNA as soon as the reaction is finished (see *Protocol 3*, step 7). Often, however, low molecular weight transcripts are the result of incomplete elongation, due to the concentration of UTP being below the K_M of RNA polymerase. In this case, after the transcription reaction is finished (see *Protocol 3*, step 8), do a 15 min chase in the presence of 5 mM UTP.

(c) Finally, if radioactive material is observed that fails to enter the gel, then the most likely reason is incomplete deproteinization of the sample (see *Protocol 3*, steps 7–9). Add more Proteinase K and/or extract the transcription products twice with phenol/chloroform before gel electrophoresis.

6.3.4 Problems encountered when using templates other than the G-free cassette (see section 4.2)

(a) If correctly initiated transcripts are present even in the absence of template, this probably reflects hybridization of the probe with endogenous hnRNA present in the extract. Make sure that the probe includes some vector sequences present in the template, so that the length of the protected fragment will allow the distinction between endogenous RNA and *in vitro* transcripts.

(b) If the full-length probe is protected in nuclease S1 mapping analysis, there is either contamination with the opposite strand or initiation has occurred at pseudo-promoter sites within the vector or other parts of the template DNA. In some cases, full-length protection has been observed, whose intensity parallels the intensity of correctly initiated transcripts. This appears to be caused by the polymerase having started at the correct site (hence the parallelism with correctly initiated transcripts), but then failing to stop. This gives rise to round-and-round transcription of the whole plasmid. This effect does not in itself represent a problem, and therefore no measures should be taken to remedy it.

Acknowledgements

The research leading to the development of the techniques described in this article has been supported by the Swiss National Science Foundation and by the State of Geneva.

References

1. Weil, P. A., Luse, D. S., Segall, J., and Roeder, R. G. (1979). *Cell,* **18,** 469–84.
2. Van Dyke, M. W., Roeder, R. G., and Sawadogo, M. (1988). *Science,* **241,** 1335–8.
3. Falvey, E. and Schibler, U. (1991). *FASEB J.,* **5,** 309–14.
4. Charnay, P., Treisman, R., Mellon, P., Chao, M., Axel, R., and Maniatis, T. (1984). *Cell,* **38,** 251–63.

5. Weintraub, H. and Groudine, M. (1976). *Science,* **93,** 848–58.
6. Wuarin, J., Mueller, C., and Schibler, U. (1990). *J. Mol. Biol.,* **214,** 865–74.
7. Johnson, P. F. and McKnight, S. L. (1989). *Annu. Rev. Biochem.,* **58,** 799–839.
8. Manley, J. L., Fire, A., Cano, A., Sharp, P. A., and Gefter, M. L. (1980). *Proc. Natl Acad. Sci. USA,* **77,** 3855–9.
9. Parker, C. S. and Topol, J. (1984). *Cell,* **37,** 273–83.
10. Gorski, K., Carneiro, M., and Schibler, U. (1986). *Cell,* **47,** 767–76.
11. Sierra, F. (1990). *A laboratory guide to in vitro transcription. BioMethods,* Vol. 2. Birkhäuser Verlag, Basel.
12. Cockell, M., Stevenson, B., Strubin, M., Hagenbücle, O., and Wellauer, P. K. (1989). *Mol. Cell. Biol.,* **9,** 2464–76.
13. Sawadogo, M. and Roeder, R. G. (1985). *Proc. Natl Acad. Sci. USA,* **82,** 4394–8.
14. Lichtsteiner, S., Wuarin, J., and Schibler, U. (1987). *Cell,* **51,** 963–73.
15. Sierra, F., Tamone, F., Mueller, C. R., and Schibler, U. (1989). *Mol. Biol. Med.,* **7,** 131–46.
16. Maire, P., Wuarin, J., and Schibler, U. (1989). *Science,* **244,** 343–6.
17. Sorger, P. K. and Pelham, H. R. B. (1988). *Cell,* **54,** 855–64.
18. Maniatis, T., Fritsch, E. F., and Sambrook, J. (1982). *Molecular cloning: a laboratory manual.* Cold Spring Harbor Laboratory, New York.
19. Croston, G. L., Kerrigan, L. A., Lira, L. M., Marshak, D. R., and Kadonaga, J. T. (1991). *Science,* **251,** 643–9.
20. Gasser, S. M. and Laemmli, U. K. (1987). *Trends Genet.,* **3,** 16–21.
21. Maniatis, T., Goodbourn, S., and Fischer, J. A. (1987). *Science,* **236,** 1237–45.
22. Galas, D. and Schmitz, A. (1978). *Nucleic Acids Res.,* **5,** 3157–70.

4

Transcriptional analysis using transgenic animals

NIALL DILLON and FRANK GROSVELD

1. Introduction

The ability to introduce a foreign gene into an organism, thus generating a genetically altered (transgenic) individual, is one of the major advances in genetics in recent years. The technology has now become widely used and has allowed a large number of fundamental biological questions to be addressed which could not have been tackled by other means. Because of their role as model organisms for genetic studies, mice and *Drosophila* have been the most widely used organisms for transgenic studies. This chapter will deal mainly with the practical aspects of using transgenic mice for the detailed analysis of transcription.

1.1 Transgenic mice

The use of transgenic mice for the analysis of transcription offers the following advantages compared with cell culture systems.

(a) Provided that the gene integrates into the germline of the founder animal, it will be present in every cell in transgenic animals from the F1 generation. There are many cell types which cannot be maintained easily in culture or transfected using existing techniques.

(b) The technique does not make use of selectable markers. This means that there is no selection for integration into 'active' chromatin domains and also avoids the need for extra genes to be introduced which might perturb the gene under investigation.

(c) Transgenic technology allows the behaviour of a gene to be studied throughout the programme of development in the live animal. This provides the means for precisely defining the regulatory elements which give rise to the pattern of expression of a gene during development. Of particular importance is the fact that the gene is integrated into the chromatin when it is presented to the cellular environment in which it is expressed. This contrasts with the situation in cell transfection systems where the gene is introduced as a naked template into the cell that will express it.

The main disadvantage of transgenic mouse technology is that it requires expensive equipment for the oocyte injection and a well organized animal facility to provide and maintain the mice. The generation of transgenic mice is also laborious and it requires several months of practice for an operator to become skilled at the procedures.

The study of transcription involves the analysis of a significant number of mutated constructs, and this in turn necessitates the generation and analysis of large numbers of transgenic animals. This chapter will deal with the specific features of a transgenic operation designed to fulfil this function. Although an outline will be given of the techniques for the generation of transgenic animals, the chapter is not intended to act as a laboratory manual of transgenic technology. Those requiring a detailed description of transgenic techniques are referred to the excellent manual of Hogan *et al.* (1).

2. Generation of transgenic mice

Procedures for generating transgenic mice are based on the fact that during a period of approximately four hours immediately prior to the first cleavage, the male and female pronuclei of the fertilized mouse egg can be clearly visualized using appropriate microscope optics, allowing them to be injected with a DNA solution using a fine glass needle. Injected eggs can then be reimplanted into the oviducts of females which have been made 'pseudopregnant' by mating with a vasectomized male (embryos that are implanted into females that have not mated do not develop). A proportion of the injected eggs develop to term and, of these, 10–20% usually have the injected DNA integrated in some or all of their cells. Those that carry the DNA in the germline will pass it to their progeny allowing the establishment of transgenic lines.

2.1 Microinjection equipment

The minimum equipment required for recovery and microinjection of mouse oocytes and their subsequent transfer into pseudopregnant females is listed in *Table 1*. We have also listed the manufacturers of the equipment used by us. Other sources for these items also exist (see *Table 1*, footnote *a*). However, the following features of microinjection equipment are particularly important whatever the source.

(a) *Injection microscope*. If at all possible, this should have Nomarski optics as these allow greatly improved visualization of the pronuclei compared with other types of optics. An inverted microscope is also the most convenient type for setting up the injecting chamber and holding pipette and needle.

(b) *Vibration-free table*. Although it may occasionally be possible to manage with an ordinary table, the vibration encountered in many large multi-

Table 1. List of equipment for micro-injection [a]

Equipment	Supplier
Diaphot TMD microscope with Nomarski optics SMZ-2B binocular microscope	Nikon Ltd.
Two sets of micromanipulators	Leitz Instruments Ltd
Schott-KL-1500 cold light source	Schott
Vibration-free table	Ealing Electro-optics
Kopf needle puller [b] model 750	David Kopf Instruments
DeFonbrunne microforge	Alcatel CIT

[a] Other commercial suppliers of suitable equipment are Carl Zeiss (injection microscope), Wild Leitz (dissecting microscope), Campden Instruments (needle puller) and Narishige Scientific Instrument Laboratory (micromanipulators). Their addresses can be found in *Appendix 3*.
[b] Settings for Kopf needle puller: heat 1:15.5; heat 2:0; Sol: 4; Delay: 0; Sol: 0.05. The filaments should be adjusted to give a pull time of 4.9–5.3 sec.

storey buildings will make it difficult to visualize the pronucleus and injecting needle, severely reducing injecting efficiency. This problem can be overcome by mounting the injection microscope and micromanipulators on a vibration-free table which works by maintaining a cushion of compressed air between the table top and the legs.

(c) *Needle puller*. The shape of the injecting needle is critically important for obtaining a high survival rate of injected eggs. The needle puller and the specific settings given in *Table 1* generate needles which have a very long taper and which, in our hands, give very high survival rates for injected eggs (up to 70%). Before purchasing a needle puller, it is important to ensure that it will provide needles which are suitable for oocyte injection.

(d) *Training*. Training of operators in the procedure for re-implantation of the injected oocytes by oviduct transfer can be made much easier if at least one binocular microscope has a double head allowing two persons to watch the procedure simultaneously. Teaching of the microinjection procedure is also facilitated by a video camera which can be attached to the injection microscope allowing the injection to be viewed on a television screen.

2.2 Animals

A reliable supply of mice is an absolute prerequisite for any transgenic mouse operation. In view of the numbers required, it is generally most economical to breed mice on site. This may be carried out by a specialized breeding facility or it may be necessary for the transgenic operator to breed them. A single operator can expect to inject on 2–3 days a week and will need 10–15 female mice aged 3–4 weeks to superovulate for each day of injection. Each operator will also need approximately six pseudopregnant females for each day of injection. In order to generate these, it will be necessary to have about 50 females aged 6–10 weeks available at any given time. Generation of this large number of females gives rise to a large excess of males. A stock of these should be kept available at all times to replace stud and vasectomized males which have become too old. It is not usually necessary to use inbred strains of mice for the analysis of transcription. Most of our analyses are carried out using eggs obtained by mating male and female F1 hybrids of C57B16 and CBA inbred strains. The 'hybrid vigour' exhibited by these F1 animals means that their eggs survive injection better and their progeny are more resistant to disease than mice from inbred strains.

2.3 Outline of transgenic procedures

The following description of transgenic procedures is intended solely to give an outline of the methodology used for generating transgenic mice. Those who require a detailed laboratory manual of transgenic methods are referred to ref. (1). However it should also be pointed out that the complexity of transgenic technology makes it strongly advisable that any written account be supplemented by direct tuition from a skilled operator. Such tuition is likely to make the difference between being able to generate small numbers of transgenic animals and rapidly acquiring the ability to make them on the large scale required for transcriptional analysis.

The complete sequence of operations for the recovery, microinjection and reimplantation of mouse eggs is shown schematically in *Figure 1*.

2.3.1 Recovery of oocytes

The oocytes used for microinjection are obtained by superovulating females (3–4 weeks old) with two sequential hormone injections. Pregnant mare's serum (PMS) is used as a substitute for follicle stimulating hormone and is injected 94–96 h before harvesting the eggs. This is followed by an injection of human chorionic gonadotrophin (HCG) 19–20 h before the time of egg harvesting. After the second hormone injection, the females are placed with individually caged stud males and left overnight. Males and females are maintained on a 12 h light/12 dark cycle which is timed so that the middle of the dark cycle occurs approximately 10 h before harvesting of the eggs. The females are checked for vaginal plugs the following morning and those that have a visible plug are killed by cervical dislocation, the oviducts are dissected out and the oocytes are recovered into Hepes-buffered medium (M2 medium; see *Table 2*). The oocytes are incubated with hyaluronidase to remove the cumulus cells and are then washed extensively by transferring through three changes of M2 medium. Finally the oocytes are transferred to CO_2-buffered

Day 1

17.00–1900: Inject 10–15 females (3–4 week old) with 5 units PMS to induce superovulation

Day 3

14.30: Inject the PMS-treated females with 5 units HCG. Place these females with stud males

Place 6–9 week old recipient females with vasectomised males

Day 4

10.30: Check the superovulated females for vaginal plugs (indicating that successful mating has occurred).
Cage out those females which have been successfully mated (now pseudopregnant).
Sacrifice the superovulated females and collect their eggs.

14.00–18.00: Micro-Inject the eggs with DNA.

18.00: Implant the surviving eggs into pseudopregnant females (the surviving eggs can be readily discriminated from those that have lysed by their appearance).

Figure 1. The complete sequence of operations for recovery, microinjection and re-implantation of mouse eggs. The times shown are for mice which are being maintained on a cycle of 07.00 to 19.00 light and 19.00 to 07.00 darkness. They should be adjusted appropriately for different lighting schedules; PMS = pregnant mare's serum; HCG = human chorionic gonadotrophin.

medium (M16 medium; *Table 2*) and kept at 37°C in 5% CO_2/air. The recovery and wash steps are most easily carried out in disposable 30 × 10 mm tissue culture Petri dishes.

The following points should be noted.

(a) Media should be made up using analar grade components which are kept separate from the normal laboratory stocks.
(b) It is advisable to make up media using disposable plastic ware (e.g. tissue culture flasks) and plastic pipettes and to avoid using glass which may contain traces of detergent.
(c) If possible, it is also advisable to buy commercially available (and relatively inexpensive) pyrogen-free water for making up media, rather than trusting to the highly variable quality of water from laboratory stills and deionizers.
(d) All media should be tested for their ability to support the development of more than 90% of a batch of embryos through to the blastocyst stage.

Table 2. Composition of M16 and M2 media.[a]

M16 Medium

Component	**Concentration**
NaCl	94.66 mM
KCl	4.78 mM
$CaCl_2$	1.71 mM
KH_2PO_4	1.19 mM
$MgSO_4$	1.19 mM
$NaHCO_3$	25.00 mM
Sodium lactate	23.28 mM
Sodium pyruvate	0.33 mM
Glucose	5.56 mM
Bovine serum albumin (BSA)	4 g/litre
Penicillin G (potassium salt)	100 000 U/litre
Streptomycin sulphate	50 mg/litre
Phenol Red	10 mg/litre

M2 Medium

M2 medium is identical to M16 except that the concentration of $NaHCO_3$ is 4.15 mM and the medium contains 20.85 mM Hepes adjusted to pH 7.4 using 0.2 M NaOH.

[a] Both M2 and M16 media are usually prepared from 10 × stock solutions which are stored frozen (see ref. 1 for details).

2.3.2 Injection

Oocytes can usually be injected during a period extending from 3–7 h after harvesting. During this period, most of them will have nuclei which can be clearly visualized and injected. At the end of this period, the nuclei will start to break down prior to the first cleavage and can no longer be injected. The injection is generally carried out on a lightly siliconized glass depression slide containing a drop of M2 medium covered by liquid paraffin (*Figure 2*). Before use, the depression slide should be washed with a Teepol-based detergent and then rinsed for a considerable period (15–20 min) under running water to ensure removal of all traces of detergent. It is then rinsed, first with distilled water and then alcohol, and allowed to air dry before setting up the drop for injection. A drop of M2 medium covered by liquid paraffin is then set up on the slide ready for injection.

The eggs are introduced into the drop from above using a handling pipette. The handling pipette consists of a narrow section (150–200 μm) into which the eggs can be drawn by suction and a wider section which is attached to a

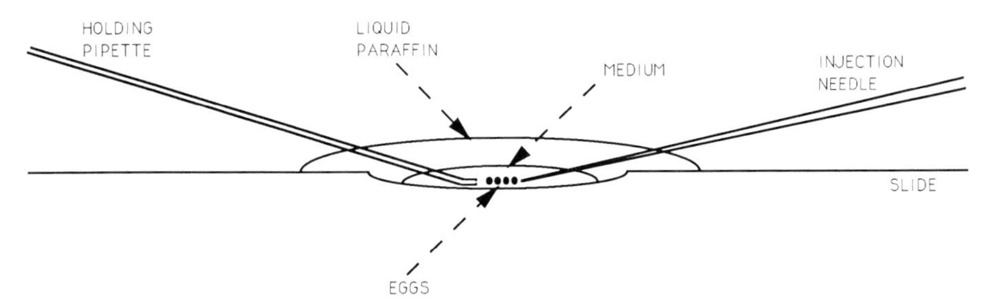

Figure 2. Arrangement of injecting needle and egg-holding pipette for microinjection on a depression slide.

length of plastic tubing and a mouthpiece. It is made by pulling a length of capillary tubing or a Pasteur pipette in a Bunsen flame (1).

The injection needle is filled by introducing the open end into the sample and allowing it to fill by capillary action. The end of the needle is then broken in the chamber by touching it against the holding pipette. This is necessary to give an opening large enough for successful injection. The oocytes are injected by first using the suction of the holding pipette to grip them and then using the micromanipulator to insert the injection needle into either the male or female pronucleus (it does not matter which). During injection, pressure is maintained on the DNA solution in the needle by means of a syringe connected to the needle holder and as soon as the needle penetrates the pronucleus it swells visibly as a result of the DNA solution entering it. Once this swelling has been seen to occur, the needle is removed as rapidly as possible to avoid sticking to the contents of the pronucleus. The oocyte is then moved to one side in the chamber and the process is repeated until all of the eggs have been injected. With a good needle it is often possible to inject as many as 50 eggs before it becomes necessary to change the needle, although 20–30 is a more usual number.

2.3.3 Reimplantation of injected eggs

After injection, the eggs are placed in M16 drop cultures (1) and returned to the incubator. The transfer into recipient female mice (0.5 days pseudopregnant) can be carried out either at the one-cell stage on the day of injection or the eggs can be incubated overnight and transferred at the two-cell stage (again into 0.5 day pseudopregnant females). We find that transfer at the one-cell stage gives better results and has the advantage that it leaves the next day free.

Pseudopregnant females are obtained by placing 6–9-week-old F1 females with vasectomized males (usually two per male) on the day before they are needed. It is advisable to set up at least five times the number of females needed and to take them from at least three cages, since females caged together tend to have their oestrus cycle synchronized. The number of matings obtained can be greatly increased by selecting those females that are in oestrus. This is done by inspection of the vaginal area for swelling and moistness (1), a technique which can be learned quite easily.

The egg transfer is carried out by anaesthetizing the mouse and making a small incision in the body wall through which the ovary and oviduct can be drawn. The oviduct is then held in place by a clip placed on the fat-pad which is attached to the ovary and the eggs are introduced into the infundibulum using a glass transfer pipette (for details see ref. 1). The actual introduction of the eggs into the oviduct is carried out using a binocular dissecting microscope. Approximately 15 one-cell eggs are introduced into each oviduct. Alternatively, 20 eggs can be transferred into one oviduct only. This has the advantage of being less traumatic for the foster mother and gives very good pregnancy rates. After the surgery, recovery of the foster mothers is assisted by warming them under an infrared lamp (care should be taken not to overheat them). They are then housed two or three to a cage. C57B1/CBA F1 foster mothers generally give birth on the night of day 18 after transfer. Foster mothers which are caged together will raise their young as a single large joint litter. Since the size of individual litters from oviduct transfers varies greatly, caging the mothers together increases the survival rate of very small litters of one or two pups which might not otherwise be reared.

3. Setting up a transgenic operation for the analysis of gene function

3.1 Scale

The analysis of gene function generally requires the abililty to test the behaviour of a significant number of mutated constructs. When this analysis is carried out in transgenic mice, the possibility of effects arising from the site of integration (position effects) means that a number of transgenic animals must be generated for each construct (a minimum of four per construct is the number which we generally aim for and many more may be required for a very position-sensitive gene). Although there is considerable variation from day to day in the number of eggs obtained from superovulated females and in their rate of survival after injection, a skilled operator can expect to inject 200–300 eggs on a good day and, of these, 150–200 should survive and be available for transfer. These would be expected to give 20–30 pups and 3–6 transgenic animals. An operator who is experienced in the technology would therefore expect to spend no more than two good days (as defined above) injecting a single construct. However, it should also remembered that microinjection is an extremely fatiguing procedure which requires intense concentration over a period of several hours. Operator fatigue can significantly reduce the efficiency of injection and even experienced operators generally find that they cannot microinject effectively for more than three days per week over extended periods. For this reason, and to maintain the interest and motivation of the staff involved, it is highly desirable that those who perform the micro-injection should also be directly involved in the analysis of the data obtained.

3.2 General considerations in the organization of a transgenic facility

There are a variety of different ways in which a transgenic facility can be organized. These range from the fully self-contained operation in which all aspects of the work (including breeding the females for superovulation) are carried out by laboratory-based staff, through to a service facility where the generation of transgenic animals is completely separated from the laboratories which provide the constructs and perform the analysis. The latter type of organization has the major attraction that individual laboratories are freed from the need to set up complex and expensive technology. The major disadvantage of service facilities is the detachment of the microinjectors from the analysis of the data and the accompanying loss of motivation. A further disadvantage is the fact that a number of laboratories are likely to be competing for a limited injecting capacity. A potentially attractive middle way between these two types of organization (and the one which we favour) is one where the superovulated and pseudopregnant females are provided by a central facility while the microinjection and transfers are performed by laboratory-based staff.

4. Isolation of DNA for microinjection

Prokaryotic vector sequences have been found to interfere with the expression of many transgenes. For this reason microinjection is usually preceded by purification of the gene away from the vector sequences. In order to do this it is necessary to design the construct containing the fragment for microinjection so that the insert is flanked by unique restriction sites.

4.1 Isolation of small (<25 kb) DNA fragments

After excision from the construct by restriction enzyme digestion, DNA inserts of less than 25 kb are most easily purified by preparative gel electrophoresis (e.g. see Chapter 2, *Protocol 6*). For the purification procedures, latex gloves should be worn which have been washed extensively under running water after putting them on, to remove any talcum powder from the outside. The choice of method for elution of the DNA fragment from the gel slice is of some importance since contamination of the purified fragment with impurities from the agarose can significantly reduce the survival of injected eggs. We find that the best elution method for microinjection is the glass powder method (2) (see *Protocol 1*). This involves disruption of the gel slice by incubation in a saturated solution of sodium iodide followed by binding of the DNA to a suspension of powdered glass and subsequent elution under low salt conditions. Although glass powder can be prepared in the laboratory (*Protocol 2*), a commercially available preparation (Geneclean from Bio-101) is vastly superior and gives much better yields.

We follow the glass powder isolation with a further purification step in which the DNA is passed through an Elutip column and filter (Schleicher and Schuell). After this, the DNA is ethanol precipitated and resuspended in microinjection buffer (10 mM Tris–HCl pH 7.4, 0.1 mM EDTA) (*Protocol 1*). It is advisable to make this buffer using pyrogen-free water and avoid contact with glass. It should then be filtered through a Millipore membrane to remove any particles which might otherwise block the injection needle.

Protocol 1. Glass powder isolation of small DNA inserts (<25 kb) from agarose gel slices

Materials

- Recombinant vector containing the DNA insert to be injected
- Appropriate restriction enzymes
- Agarose gel (normal agarose); 0.5–0.8% agarose depending on the size of DNA fragment to be isolated (see *Appendix 2*)
- Tris-acetate EDTA (TAE) electrophoresis buffer (see Chapter 6, *Protocol 5* for preparation)
- Ethidium bromide (5 μg/ml)
- 6 M sodium iodide, 0.5% sodium sulphite
- 50% ethanol, 0.1 M NaCl, 1 mM EDTA, 10 mM Tris–HCl pH 7.0
- TE buffer (10 mM Tris–HCl, 1 mM EDTA pH 8.0)
- Microinjection buffer (10 mM Tris–HCl pH 7.4, 0.1 mM EDTA).
  Prepare this buffer using pyrogen-free water, avoid contact with glass and filter it through a Millipore filter (pore size 0.2 μm)
- Glass powder slurry; either prepared in the laboratory (*Protocol 2*) or, preferably, a commercially available preparation (Geneclean from Bio-101).

Method

Note: wear disposable latex gloves throughout the purification procedures. After putting them on, wash the outside of the gloves extensively with running water to remove any talcum powder.

1. Excise the DNA insert to be injected from the recombinant vector by digestion with an appropriate restriction endonuclease using the supplier's recommended buffers and conditions.
2. Electrophorese 20–50 μg of the restriction digested DNA on a 0.5–0.8% agarose gel (depending on the insert size) using TAE electrophoresis buffer (not Tris-borate type buffers which can interfere with DNA recovery).
3. Stain the gel with ethidium bromide (5 μg/ml) for 20 min, visualize the bands on a long wave UV transilluminator and cut out the appropriate band.

4. Weigh the gel slice in a 20 ml plastic tube and add 2–3 vol. 6 M sodium iodide, 0.5% sodium sulphite. Incubate at 55°C until the gel slice is completely dissolved (5–10 min).
5. Add 1 μl glass powder slurry per μg DNA and incubate for 20 min on ice with occasional swirling.
6. Spin for 2 min at 1000 *g* in a bench centrifuge. Carefully remove the supernatant and resuspend the glass powder pellet in 1 ml of 6 M sodium iodide, 0.5% sodium bisulphite. Transfer the suspension to a 1.5 ml Eppendorf microcentrifuge tube for the remaining steps.
7. Spin the glass powder at top speed in a microcentrifuge for 10 sec, remove the supernatant and resuspend the glass powder in 1 ml 50% ethanol, 10 mM Tris–HCl pH 7.0, 0.1 M NaCl, 1 mM EDTA by pipetting up and down. Pellet by spinning for 10 sec in a microcentrifuge.
8. Repeat step 7.
9. Resuspend the glass powder in 100 μl TE buffer and incubate at 55°C for 15 min.
10. Pellet the glass powder by centrifuging for 5 min in a microcentrifuge and remove the supernatant containing the eluted DNA.
11. The DNA can be further purified by passing it through an Elutip column. This step is carried out according to the manufacturer's instructions.
12. Adjust the concentration of NaCl in the purified DNA solution to 0.3 M by adding an appropriate volume of a 5 M stock NaCl solution and precipitate the DNA by adding 2 vol. ethanol. Leave on ice for 15 min.
13. Pellet the precipitated DNA by centrifuging in a microcentrifuge for 10 min.
14. Add 1 ml ice-cold 70% ethanol, vortex and centrifuge for 2 min. Remove as much of the supernatant as possible.
15. Resuspend the DNA in 100 μl injection buffer (it is not necessary or desirable to dry the pellet before resuspending it). Check the concentration by running 1 μl on an agarose gel against an appropriate standard and then dilute the sample to the correct concentration for micro-injection.

Protocol 2. Preparation of glass powder

Materials

- Flint glass powder 325 mesh (this is available from most glass supply companies)
- Concentrated nitric acid.

Method

Take care since the preparation method involves using boiling concentrated

Protocol 2. *Continued*

nitric acid. This step must be carried out in a fume cupboard with appropriate safety measures, including the wearing of gloves and safety spectacles.

1. Suspend 250 ml glass powder in 500 ml distilled water. Allow to settle for 1 h, discard the settled particles, and centrifuge the supernatant to recover the fines.
2. Resuspend the powder in 200 ml water and carefully add an equal volume of concentrated nitric acid. Bring to the boil in a fume cupboard.
3. Allow to cool to room temperature. Recover the glass by centrifugation, resuspend in water, and wash repeatedly until the supernatant is neutral.
4. Make a 50% slurry in water and store as aliquots at −20°C.

Glass powder prepared in this way shows a considerable amount of batch variation depending on the starting material. A commercially available preparation (Geneclean from Bio-101) can be used instead, which in our hands gives uniformly good results.

4.2 Isolation of large DNA fragments

The glass powder method results in a significant degree of shearing of fragments larger than 25 kb, while other methods for eluting large fragments from gels suffer from problems of low yield and agarose contamination. For this reason we purify large inserts (25–45 kb) using sodium chloride gradients (see *Protocol 3*). Although slightly more laborious than gel purification, this method gives good yields of very clean intact DNA. Large DNA fragments cannot be filtered since this causes shearing. Instead, after dilution to the appropriate concentration for microinjection, the DNA is given two successive 30 min spins at 12 000 *g* in a bench-top centrifuge and the supernatant is carefully removed to a fresh tube. The second of these spins is carried out just before microinjection. When injecting unfiltered solutions it is generally necessary to use a needle with a slightly wider opening than for filtered solutions.

Protocol 3. Isolation of cosmid inserts using sodium chloride gradients[a]

Materials

- 100 μg cosmid DNA containing the insert of interest
- Appropriate restriction enzymes
- Phenol/chloroform (1:1, buffered with 10 mM Tris–HCl pH 8.0)
- Chloroform
- Microinjection buffer (for preparation, see *Protocol 1*)
- 0.5% agarose gel made using TBE buffer (89 mM Tris base, 89 mM boric acid, 2 mM EDTA)
- TE buffer (10 mM Tris–HCl, 1 mM EDTA pH 8.0)
- 5% NaCl (w/v), 3 mM EDTA pH 7.5
- 25% NaCl (w/v), 3 mM EDTA pH 7.5.

Method

1. Digest 100 μg cosmid DNA with the appropriate restriction enzyme. Check that the digest has gone to completion by electrophoresis of an aliquot on a 0.5% agarose gel.
2. Extract the digested DNA once with phenol/chloroform and once with chloroform, then precipitate with ethanol (see *Protocol 1*, steps 12–14) and resuspend in 200 μl TE buffer.
3. Prepare a 5% to 25% NaCl gradient in a 14 ml polycarbonate centrifuge tube suitable for centrifugation in a Beckmann SW41 swing-out rotor or its equivalent. Pour the gradient using a standard gradient maker and the same procedure as for a sucrose gradient. The component solutions are 5% NaCl and 25% NaCl each containing 3 mM EDTA. The solutions should be autoclaved and then filtered through a Millipore filter immediately before use. Wear washed latex gloves during all stages of the procedure.
4. Layer the DNA solution gently on to the top of the NaCl gradient and centrifuge at 37 000 r.p.m. for 5.0–5.5 h (the exact time depends on the size of the insert; larger inserts should be centrifuged for shorter times).
5. Harvest the gradient in 0.5 ml aliquots in microcentrifuge tubes. This is most easily achieved by passing a capillary connected to a peristaltic pump down through the gradient and collecting from the bottom.
6. Take 10 μl from each aliquot, dilute it to 30 μl with TE buffer and electrophorese on a 0.5% agarose gel in TBE buffer at 150 V; the high salt concentration will cause the DNA samples to run into the gel very slowly.
7. Pool those gradient fractions containing the required DNA fragment, dilute with an equal volume of TE buffer and precipitate with two volumes of ethanol. Leave in ice for 15 min.
8. Collect the DNA by centrifugation, wash twice by centrifugation with 70% ethanol.
9. Resuspend the DNA pellet in 100 μl injection buffer (be careful not to let the DNA pellet dry after precipitation as it will then be very difficult to redissolve). The DNA sample is now ready for dilution to the appropriate concentration for microinjection (see Section 4.3).

[a] Adapted from ref. (3).

4.3 Determination of the concentration of DNA for microinjection

The figure which is usually given for the concentration of DNA for microinjection is 1 μg/ml. In practice the optimal concentration varies between operators, presumably because different injectors inject slightly different volumes into the pronucleus. For this reason, it is advisable for each operator to determine his/her own optimal concentration empirically. Once a concentration of DNA insert has been obtained which gives good rates of transgenesis, this can be used as a concentration standard. Adjusting the concentration of a solution containing a new DNA fragment involves electrophoresing 20–30 μl of the solution on an agarose gel, together with an equal volume of the DNA standard, and then diluting the solution of the DNA fragment until the intensity of ethidium bromide staining is equal to that of the standard.

The optimal concentration for microinjection also varies with the size of the insert. Paradoxically we find that the relationship is an inverse one, with large DNA inserts giving the best rates of transgenesis at lower concentrations. This observation is surprising since a larger fragment will mean that fewer copies are injected at a given concentration and indicates that the frequency of integration is not a simple function of the number of copies injected into the pronucleus.

5. Integration of injected genes

5.1 Mosaics

During the microinjection procedure, about 200–500 copies of the transgene are usually introduced into the pronucleus. Little is known at present about the parameters which affect the subsequent integration of the injected genes into the mouse genome. In a proportion of the eggs, integration will take place at the one-cell stage giving rise to animals which are transgenic in all tissues. However, in some embryos the gene does not integrate until one or more cell divisions have taken place, resulting in mosaic animals in which the transgene is present in some tissues but not in others. No systematic study has been carried out on mosaicism, but our experience of analysing large numbers of transgenic animals suggests that two different types of mosaic are likely to occur among transgenic founders. Integration of the transgene at the two-cell stage probably results in founder animals which have the transgene in every tissue but in only a proportion of the cells in each tissue. This type of mosaic is very difficult to detect. Mosaics which have the gene in some tissues but not in others probably result from integration at later stages. These mosaics are readily detected by Southern blotting of DNA from several different tissues (see below).

5.2 Copy number

Although a small proportion (5–10%) of transgenic mice carry single copies of the transgene, the majority will contain more than one copy and these are usually arranged as head-to-tail tandem repeats. The mechanism by which this tandem repeat arrangement is generated is not known. There is at present no means for specifically generating single-copy animals, although in our experience the frequency with which such animals are obtained can be increased by reducing the concentration at which the fragment is injected to about one-half that which is normally used. Significant numbers of single-copy transgenic animals can be generated only by making many times the number of transgenic animals that are needed and selecting out those in which there have been single-copy integration events.

5.3 Identification of transgenic animals

Animals that carry the transgene are identified by analysis of DNA prepared from a small segment cut from the end of the tail. The tail biopsy is rendered less traumatic for the animal by carrying it out at around 10 days of age when the tissue is soft and cartilagenous and by use of a local anaesthetic (ethyl chloride). Approximately 0.5 cm of tail is removed using sharp scissors and DNA is prepared from it according to *Protocol 4*. DNA prepared using this protocol can be digested with most restriction enzymes. However, tail DNA is not generally as clean as that obtained from soft tissues and so it is advisable to carry out digestions in 100 μg/ml bovine serum albumin (BSA) and to use an excess of enzyme for extended periods of digestion.

The other major problem which is commonly encountered during analysis of large numbers of tail samples is contamination of the samples with plasmids in the laboratory. This leads to false positives or to all the samples giving a positive signal. Since such contamination renders the biopsy sample useless, stringent precautions should be taken to avoid it by aliquoting all solutions and keeping them completely separate from plasmid manipulations.

Protocol 4. Preparation of DNA from mouse tails[a]

Materials

- Ethyl chloride B.P.
- Tail buffer (50 mM Tris–HCl pH 8.0, 0.1 M EDTA, 0.1 M NaCl, 1% SDS)
- Proteinase K (10 mg/ml)
- Phenol/chloroform (1:1) buffered at pH 8.0 with 10 mM Tris–HCl
- Chloroform
- Isopropanol
- TE buffer (10 mM Tris–HCl, 1 mM EDTA pH 8.0)

To avoid contamination with plasmid DNA used in the laboratory, keep all solutions separate from plasmid manipulations and store them in aliquots.

Method

1. Freeze the tail by spraying with ethyl chloride and use sharp scissors to cut approximately 0.5 cm from the end of the tail into a 1.5 ml snap-cap tube (e.g. Eppendorf microcentrifuge tube) containing 0.7 ml of tail buffer.
2. To each tube, add 25 μl of 10 mg/ml Proteinase K and then incubate overnight at 55 °C (it is not necessary to chop or homogenize the tails).

Protocol 4. *Continued*

3. Add 0.7 ml phenol/chloroform to each tail digest and place on a benchtop shaker for 15 min.
4. Centrifuge each tube for 10 min in a microcentrifuge and then remove the aqueous phase to a fresh tube taking care not to take any of the interphase.
5. Repeat steps 3 and 4.
6. Add 0.7 ml chloroform and place the tube on the benchtop shaker for 5 min.
7. Centrifuge for 5 min in a microcentrifuge. Remove the aqueous phase to a fresh tube.
8. Add 0.6 vol. isopropanol to each tube, mix by inversion and immediately hook out the precipitated DNA using a sealed off Pasteur pipette.
9. Immerse the DNA at the end of the Pasteur pipette briefly in 70% ethanol then place it in a 1.5 ml microcentrifuge tube containing 50 μl TE buffer. Leave it to stand for approximately 5 min to allow the DNA to dissolve and then remove and discard the Pasteur pipette.
10. Use a 2–3 μl aliquot of the dissolved DNA to determine the concentration by measuring the A_{260}. A yield of 50–100 μg is normally obtained.

[a] If the tail DNA is to be digested with enzymes such as *Eco*RI, *Bam*HI or *Hin*dIII which are resistant to inhibition, it is possible to use a shortened protocol in which the Proteinase K digest is subjected to a single extraction with phenol/chloroform mixture (steps 3 and 4). The DNA is then precipitated and redissolved as in steps 8–10. Restriction enzyme digestion of DNA which has been prepared by this shortened protocol should be carried out in the presence of 300 μg/ml BSA.

The most reliable method for analysing mouse tail DNA samples for the presence of a transgene is by Southern blotting. A 5–10 μg aliquot of each tail DNA sample is digested in the presence of 100 μg/ml BSA with an excess of a restriction enzyme which generates an internal fragment. This is then separated on an agarose gel, blotted using standard procedures (4), and hybridized with an appropriate probe from within the transgene fragment. For mammalian transgenes it will be necessary to have a probe which does not hybridize with repeat sequences in the mouse genome.

PCR analysis and slot-blotting are alternative methods by which transgenic animals can be identified. Although polymerase chain reaction (PCR) analysis initially appeared to offer a rapid and easy alternative to Southern blotting for transgenic identification, our experience has been that the benefits are outweighed by the disadvantages. The principal disadvantage is the high rate of false positives as a result of cross-contamination of samples which means that all positives have to be verified by Southern blotting. A second disadvantage is the cost of *Taq* polymerase which makes the procedure expensive for screening large numbers of samples.

Slot-blotting (which requires special apparatus) is probably the simplest and least laborious method for screening large numbers of samples, but requires a hybridization probe which has a very low rate of background hybridization with endogenous mouse sequences. Where such a probe is available, this method is particularly useful for monitoring the transgene and hence the routine maintenance of breeding lines. However, slot-blotting does not give any information on the physical integrity of the transgene which must be checked by Southern blotting. A procedure for slot-blotting tail DNA is described in *Protocol 5*.

Protocol 5. Slot-blot analysis of tail DNA

Materials

- Slot-blot apparatus (e.g. Schleicher and Schuell)
- Nitrocellulose membrane strip (size depends on the apparatus used)
- 2.0 M ammonium acetate
- 1.0 M ammonium acetate
- 5 μg of each tail DNA sample
- 4 M NaOH
- Radiolabelled hybridization probe.

Method

1. Wash a nitrocellulose strip of the correct size in 1.0 M ammonium acetate and assemble the slot-blot apparatus according to the manufacturer's instructions.
2. Switch on the vacuum pump and fill each well with 1.0 M ammonium acetate to check that all of the wells are clear. When the wells have emptied, switch off the vacuum pump.
3. For each sample, add 5 μg DNA to 180 μl water and then add 20 μl 4 M NaOH. Allow to stand at room temperature for 15 min.
4. To each sample, add 200 μl of ice-cold 2 M ammonium acetate. Mix by vortexing and place on ice.
5. Switch on the vacuum pump and apply each sample to a separate well of the slot-blot apparatus.
6. When all of the wells have emptied, dismantle the apparatus, and bake the filter for 1 h at 80°C. The area of the filter under the wells will turn brown during the baking due to the action of the sodium hydroxide.
7. Hybridize the filter with the appropriate radiolabelled probe using the same procedure as for Southern blotting.

5.4 Determination of copy number

5.4.1 Southern blot comparison with a single-copy mouse gene

The copy number of a transgene is determined by probing a Southern blot of tail DNA with two probes, one of which is specific for the transgene while the other detects an endogenous mouse gene which acts as a standard. The control gene can be any single-copy mouse gene for which a good probe giving low background is available. Obviously a restriction enzyme must be used which gives fragments of different sizes for the control and transgene. Ideally the hybridization should be carried out using a mixture of the two probes labelled to equal specific activities. An example of a copy number blot is shown in *Figure 3*. Quantitation of the ratio of the transgene to the control gene can be carried out either indirectly by densitometric scanning of an autoradiograph of the gel, or directly using a phosphorimager. If scanning of an autoradiograph is used, it should be remembered that linearity is rapidly lost with increasing exposure of X-ray film which may result in underestimation of the copy number for high-copy animals. For this reason, it is advisable to dilute high-copy number samples to bring the transgene signal down to a level which is similar to that of the control gene in an undiluted sample.

5.4.2 Probing for end fragments

Determination of copy number using the above method, by comparison with the signal from an endogenous gene, suffers from problems with achieving

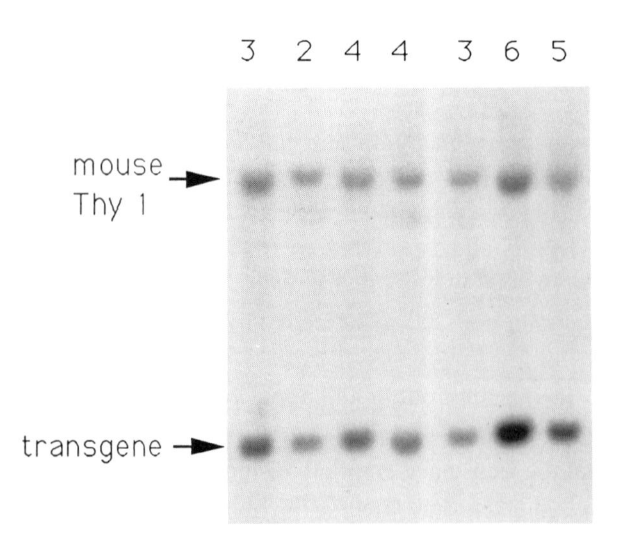

Figure 3. Estimation of the copy number of a human γ-globin transgene by Southern blotting. The signal from the transgene has been compared with that of the mouse Thy 1 gene by densitometric scanning of the autoradiograph. The number above each lane indicates the copy number estimated by this means.

equal specific radioactivities for the two probes used and problems of accurate quantification of the signals from blots. This loss of precision can make it difficult, for example, to distinguish between single-copy animals and those carrying two copies. An alternative method for obtaining information on copy number (outlined in diagrammatic form in *Figure 4*) is to use a probe which hybridizes to a region close to one end of the injected DNA fragment. The Southern blot is then prepared using tail DNA cut with a restriction enzyme which cuts internally but does not cut between the probed region and the end. The fragments which are recognized after Southern blotting will then provide information on the type of integration event which has occurred.

(a) A multi-copy head-to-tail tandem repeat will give a strong junction fragment and a single weaker end fragment.
(b) A single copy animal will give no junction fragment and one end fragment.
(c) An animal that has a tail-to-tail integration or multiple separate integrations will give two or more end fragments.

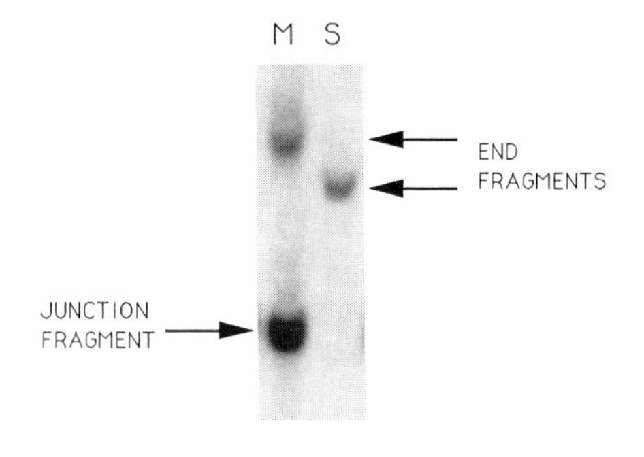

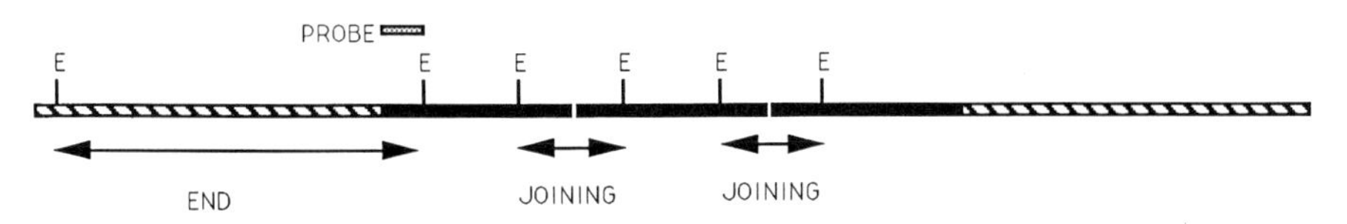

Figure 4. Determination of copy number by probing for end fragments. The strategy is shown diagrammatically with the transgene represented by a filled line and surrounding mouse genomic sequences represented by a hatched line. A restriction enzyme (E) is chosen which cuts within the injected fragment but does not cut between the probed region and the end of the fragment. DNA from a multi-copy animal (M) shows a strong junction fragment resulting from head-to-tail joining of multiple copies of the injected fragment and a weaker end fragment (to simplify the diagram, only two copies are shown). A single copy animal (S) shows a single end fragment.

A further advantage of this method is the fact that it gives an indication of the real copy number in mosaic founders since it is based on a qualitative assessment of the structure of the integrated DNA rather than a quantitative comparison with an endogenous gene. An example of this type of blot is shown in *Figure 4*.

5.5 Injection of very large DNA fragments

Transgenic mice have been generated routinely with cosmid inserts of up to 45 kb. Injection of DNA fragments of this size presents no special problems and gives rates of transgenesis which are similar to those obtained with smaller fragments. Until now, 45 kb has been the upper size limit for microinjection because this is the largest fragment which will fit into a cosmid. Although much larger fragments can be cloned into yeast artificial chromosomes (YACs), attempts to generate transgenic mice by microinjection of YACs have not so far been successful. This appears to be due mainly to the difficulty of purifying sufficient amounts of YAC DNA for microinjection as well as the very large size of most YAC inserts.

We have recently developed a novel method for generating large fragments for microinjection which involves joining two cosmid inserts together [J. Strouboulis *et al.* (1992). *Genes Dev.* (in press)]. This method involves cloning short (15 bp) homopolymeric GC oligonucleotides at the joining point (*Figure 5*). The cosmids are first cut with a restriction enzyme which cuts at one end of the homopolymeric region. This is followed by incubation with T4 polymerase in the presence of A and T only. The exonuclease activity of the T4 polymerase digests one strand (3′ to 5′) until it reaches the first A or T residue leaving a single-stranded homopolymeric tail (see *Figure 5*). When the cosmids are incubated together under appropriate conditions, the tails anneal at high efficiency (70%) giving a large fragment which is incubated with T4 DNA ligase and then purified by preparative pulse-field gel electrophoresis (5). Microinjection of a 70 kb fragment prepared by this method has given a good frequency of animals carrying the intact fragment (>10% of total pups born) extending the upper size limit for successful microinjection. There appears to be no reason why even larger fragments should not be microinjected, although there will almost certainly be an upper size limit at which shearing of the DNA during passage through the injection needle prevents the integration of intact fragments.

5.6 Co-injection

Another method which has been used in an attempt to circumvent size limits for microinjection is the injection of a mixture of two DNA fragments. It has been found that the fragments ligate together at high frequency following microinjection so that theoretically this approach offers a means for generating very large fragments *in vivo*. Unfortunately, however, multiple ligations of the two fragments occur in a largely random manner (P. Fraser, D. Whyatt, and N. Dillon, unpublished) giving rise to extremely complex structures

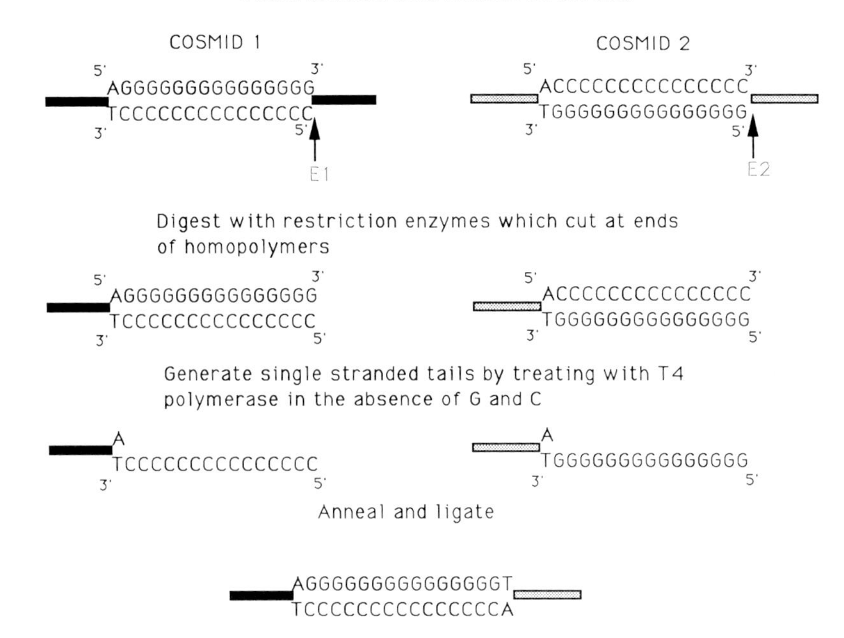

Figure 5. Strategy for linking two cosmid inserts for microinjection [J. Strouboulis *et al.* (1992). *Genes Dev.* (in press)]. Homopolymeric oligonucleotides are cloned into the cosmids at the point where the join is to be made. The cosmids are then cut at the ends of the homopolymers with unique restriction enzymes E1 and E2, and treated with T4 DNA polymerase in the presence of dATP and dTTP only. Under these conditions, the 3′ exonuclease of the T4 polymerase will digest one strand of each GC homopolymer until it reaches the first A or T residue (see reference 4). This gives rise to complementary single-stranded tails which can then be annealed to give a single large fragment.

which are different in every transgenic line and which cannot be resolved by conventional Southern blotting. This complexity makes the procedure useless for analysing complex transcriptional behaviour, although it can be useful for analysing gross effects such as the presence or absence of a strong enhancer on a co-injected fragment. Homologous recombination has also been observed to occur between overlapping fragments following microinjection, together with random end-to-end joining (ref. 6 and S. Eccles, personal communication). Co-injection of overlapping cosmids could therefore be used as a method for joining together coding segments of a large gene in order to obtain expression of a protein product in mice. In this case, the occurrence of random joining events in addition to homologous recombination would not be a problem.

6. Generation of transgenic mice using embryonic stem cells

Embryonic stem (ES) cells offer an alternative route to microinjection for generating transgenic mice. ES cells are isolated from the inner cell mass of blastocysts and can be maintained in culture. Provided they are maintained under specific conditions, they remain totipotent and, following introduction into the embryo by injection at the blastocyst stage, can contribute to all cell types including the germline. This means that ES cells can be transfected with a gene (usually by electroporation) and cells carrying the integrated gene can be selected using a selectable marker such as neomycin resistance. A clone of cells which has the gene integrated can then be cultured and the cells can be injected into blastocysts to generate chimaeric mice. In some of these chimaeras, the injected cells should contribute to the germline giving progeny which are transgenic for the transfected gene (for a detailed treatment of ES cell methodology, see ref. 7).

It will be obvious from the above account that generating transgenic mice using ES cells is much more laborious than using microinjection. For this reason, ES cells are likely to be used for transgenesis only in certain circumstances. One such circumstance is where a specific type of integration event is required; for example low copy integrations or integration into an active gene in 'gene trap' experiments. ES cells also offer a potential means for generating transgenics carrying large YAC inserts; it may prove to be easier to obtain intact integration of very large fragments by transfection rather than by microinjection. However, as yet there have been no reports of transgenic mice carrying YACs generated by this (or any other) means.

Embryonic stem cells also have the potential to be used as a 'pseudo-transgenic' system for studying gene expression during early development. The cells are normally maintained in a totipotent state by culturing them on a layer of fibroblast feeder cells. When they are removed from the feeder layer and grown in suspension, they aggregate to form embryoid bodies which have endoderm, basal lamina, mesoderm, and ectoderm and are in many cases morphologically similar to embryos of the 6–8-day egg-cylinder stage (8). After 8–10 days of culture, many embryoid bodies develop myocardium as well as blood islands which contain most of the haematopoietic colony-forming cells found in bone marrow (9). Developing embryoid bodies have been shown to express the mouse embryonic globin genes synchronously with the development of blood islands and a transfected human ε-globin gene was found to be developmentally regulated in a similar manner (10). This system offers a potential means for characterizing the sequences involved in the regulation of genes during early development without having to go through the full procedure for generating transgenic mice.

7. Breeding of transgenic animals

Transgenic lines are established by breeding the founder animals with non-transgenic animals. Pups are born after an 18–21 day gestation period (depending on the strain). Tail biopsies are carried out at 10 days to identify transgenic pups and the animals are weaned and the sexes separated at

3 weeks. The mice become sexually mature and are ready to breed at around 6 weeks. If a male is left with a female, they will usually mate again within 24 h of the birth of a litter and the female will continue to produce litters at three week intervals.

Females should be pregnant with their first litter before they are three months old, otherwise they may fail to breed. It is also advisable to have males mate before they reach three months. Once males have mated, they will generally continue to do so until they are six months old. After six months of age the breeding performance of both males and females declines and it is not advisable to rely on such animals for the maintenance of transgenic lines.

Maintaining transgenic lines is made much easier if they can be bred to homozygosity. Candidate homozygotes can be detected by careful estimation of the copy number (see Section 5.4). However, the difficulty of accurately estimating a twofold difference in signal on a Southern blot means that homozygosity should be confirmed by breeding with a non-transgenic animal. All of the progeny from such a mating should be transgenic. A proportion of transgenic lines (>10%) will fail to give homozygotes due to the insertional inactivation of essential genes.

8. Design of transgenic experiments

The large amount of work involved in generating and analysing transgenic mice makes it particularly important that experiments be designed carefully in advance so as to maximize the value of the results obtained. This section will deal with the experimental approaches which can be used to address a variety of questions using transgenic technology and the problems which affect experimental design.

The following aspects of transcriptional behaviour are among those that can be probed using transgenic techniques:

- control of the rate and cell-type specificity of transcription
- regulation of transcription during development
- identification of new developmentally-regulated genes by means of 'gene trap' experiments
- role of a particular gene product in regulating the expression of other genes and thereby controlling processes such as cell differentiation and proliferation.

8.1 Analysis of the control of rate and cell-type specificity of transcription

Transgenic mice can be used as an alternative to cell transfection systems for directly analysing the sequences involved in controlling transcription. They have the advantage of providing a more authentic *in vivo* environment for such analysis and have in some cases given markedly different results compared with those obtained in transfected cells. They also allow access to all cell types.

This type of analysis is likely to require the testing of a large number of mutant constructs so the experiment needs to be streamlined as much as possible to keep the time-scale and labour input within reasonable bounds. The following aspects should be carefully considered.

8.1.1 Transient analysis versus breeding

There are many advantages to be gained from analysing the founder animals rather than breeding through to the next generation. Breeding is laborious, requires a considerable amount of cage space, and greatly extends the time required for an experiment. In transient transgenic experiments, the founder animals are killed without being bred and the expression of the transgene in one or more tissues is analysed. This type of analysis can allow very rapid characterization of substantial numbers of mutants (11, 12). However, such an approach can only succeed if it includes rigorous procedures to exclude mosaics as these will seriously affect the results if they are included. Our procedure for excluding mosaics involves careful estimation of the copy number in three different tissues by Southern blotting and comparison of the transgene signal with that of an endogenous single-copy mouse gene. Any animal which does not have an identical ratio of transgene to endogenous gene in all three tissues is excluded from the analysis. This procedure will detect mosaics with substantial differences in the proportion of cells containing the transgene in different tissues. However it will not detect the second type of mosaic described in section 5.1, that is, one which has the transgene fairly evenly distributed among the tissues but missing from some cells in each tissue. This second type of mosaic is likely to cause a certain amount of unavoidable variation in the analysis of transgene expression. For this reason, transient transgenic analysis works best where sequences which have large effects on transcription (e.g. enhancer elements) are being analysed. For the analysis of more subtle effects which result in only small differences in expression levels, it will probably be necessary to use breeding lines.

The problem of mosaics can also be circumvented by including a control gene in the construct. Since all tissues will contain equal amounts of both control and test gene, it is possible to carry out a very precise quantification of the relative expression level of the test gene. This approach cannot be used to test the functioning of elements which act over a long distance since such elements will affect both genes. It is likely to be most useful in testing the function of promoter elements which are located close to the cap site.

8.1.2 Position effects

It has long been known that the position of integration of a transgene into the host genome can have very substantial effects on its expression. Typically the transgene shows a much lower level of expression than that which is observed when it is in its native environment and the level, while remaining constant within a line derived from a single integration event, shows wide variation between different lines. There is also little correlation between the copy number of the transgene and its level of expression.

Recent work has begun to cast some light on these effects and suggests that the term 'position effect' actually describes several different phenomena. The genes of the human β-globin locus were considered to be highly position-sensitive showing expression levels in transgenic mice which were only a few percent of their normal level and with many transgenic lines giving no expression at all. However, it was then found that a region at the 5′ end of the locus (the locus control region or LCR) gave rise to full expression of a linked β-globin gene in all transgenic lines and that this expression was directly related to the copy number of the transgene (12). Although it was initially thought that this position insensitivity might be the result of a boundary effect insulating the gene from surrounding sequences, subsequent work has indicated that it results from a dominant positive activation effect which is insensitive to the position of integration. This activation effect appears to have at least two components, a chromatin-opening function which creates a region of DNase I sensitivity and a second function which results in very large direct enhancement of transcription (reviewed in ref. 13). A locus control region (LCR) with similar properties has also been identified close to the human T-cell specific CD2 gene (14).

A second type of sequence which renders a gene position-insensitive has been discovered in the *Drosophila* HSP locus. These SCS (specific chromatin structure) sequences appear to act as true boundary elements which insulate a gene from neighbouring sequences in transgenic *Drosophila* (15). They confer position insensitivity only when placed on both sides of a gene and they do not seem to be able to enhance transcription. Sequences from the flanking region of the chicken lysozyme gene (A elements) have also been found to have similar insulating properties in transgenic mice (16).

From one viewpoint, the position sensitivity of most genes is a major inconvenience. The variation in levels of transcription between mice caused by position effects can make it very difficult to compare the levels of transcription from different constructs. However, the discovery of LCR and boundary sequences places a different perspective on the problem. It now seems likely that position sensitivity results largely from the absence of specific sequences which are required for the establishment of transcriptionally active domains. This function constitutes a further level of gene regulation, the analysis of which represents a major challenge. Transgenic systems provide the only really effective means for analysing position sensitivity. Stable cell transfection systems tend to select for integration into active domains which limits their usefulness for studying this phenomenon.

The identification of sequences which confer position insensitivity is not a trivial exercise since such elements are likely in many cases to be located a long way from the gene itself. In the case of the β-globin locus, the LCR is located 50 kb away from the β-globin gene and was discovered only by using an artificial construct which brought the gene and LCR together. Mapping of DNase I hypersensitive sites provides a method for identifying candidate sequences which can be then be cloned into constructs containing the gene. The location of such sequences would also be greatly facilitated by methods for generating transgenic mice carrying very large DNA fragments (Section 5.5).

8.1.3 Transgenes with harmful effects

Since transgenes will generally encode proteins which have specific functions in the normal animal, their expression in transgenic animals may have harmful effects. This may be caused by an excess of transgene product, or it may be the result of inappropriate tissue distribution or timing of expression during development. The human β-globin genes again provide a good example of this phenomenon. The high level, copy-dependent expression which results from linkage of the β-globin gene to the LCR gives rise to a globin-chain imbalance in the erythrocytes of muti-copy animals which causes a lethal anaemia (12). Genes which are involved in the regulation of complex developmental phenomena (e.g. homeobox genes) are also likely to have lethal effects. The mechanisms by which these effects are exerted may be of considerable interest in their own right but such effects also present a serious problem since they are likely to make it difficult to obtain live transgenic animals.

The solution to the problem of lethality of a transgene generally lies in a careful consideration of the probable mechanism of the lethal effect and the pattern of expression of the transgene. In the case of the β-globin gene for example, the problem of analysing transcriptional function was solved by carrying out a transient transgenic analysis in which the foetuses were dissected from the foster mothers at 13.5 days gestation (12). High level expression of the β-globin gene first occurs in the foetal liver at 12.5 days, so it was possible to harvest the foetuses before the anaemia became severe enough to cause the death of the foetus. A very detailed characterization of the sequences which mediate the functioning of the LCR has been carried out using this approach (reviewed in ref. 13).

For genes such as homeobox genes which are expressed very early in development, this approach may not be feasible. An alternative strategy is to alter the transgene so that it no longer encodes a functional protein product. This can be achieved by disrupting the initiator codon or by introducing a non-sense codon which causes premature termination. A disadvantage of this strategy is the fact that the reduced loading of ribosomes on the mRNA is likely to make the mRNA less stable and may reduce the mRNA level compared with its endogenous counterpart. For some genes it may be possible to avoid this problem by removing specific regions which are involved in the functioning of the protein (e.g. zinc fingers, homeodomains) while leaving the rest of the protein intact. Yet another strategy is the replacement of the coding region with a reporter gene (e.g. β-galactosidase). Judicious choice of the reporter gene can facilitate rapid analysis of transgenic animals. However, it should be borne in mind that the removal of coding and intron sequences may result in the loss of elements involved in transcriptional regulation. The stability of the reporter gene transcript is also unlikely to be the same as the native transcript, a factor which must be taken into account when comparing expression levels.

8.2 Analysis of developmental regulation of gene expression using transgenic mice

8.2.1 Important initial considerations

Transgenic analysis provides the only effective means for defining the control sequences which give rise to a particular pattern of expression during development. This will generally involve the generation of breeding lines and careful characterization of the expression pattern of the transgene at different stages of development. The central aim of any programme for mutagenizing the control regions of a developmentally regulated gene is to disrupt specific parts of the expression pattern and thereby localize their control to particular sequences. Before attempting to do this, it is very important to establish clearly the pattern of developmental expression of the complete transgene so as to establish a baseline with which mutants can be compared. Ultimately the aim will be to reproduce the exact pattern of expression of the endogenous gene so that the sequences controlling this expression can be dissected. This will often prove to be a difficult task. Even when working with mouse genes, all of the problems which have been described for general transcriptional analysis in transgenics will apply. Important control sequences may be missing from small constructs leading to low levels and aberrant patterns of expression. Position effects are also likely to give altered patterns of expression in some animals. Therefore, it is highly desirable to begin by injecting the largest construct available as this is more likely to contain all of the required control sequences. Hopefully, when methods for injecting very large fragments become well established, it will become possible to begin an experiment by injecting an entire gene domain and then specifically mutagenize sequences within the domain.

When working with heterologous (e.g. human) genes in mice, it is necessary to be even more careful in interpreting expression patterns. There are substantial differences between the physiology of humans and mice and so interpretation of developmental expression patterns will often involve a degree of extrapolation between the two systems. It is therefore very important not to make assumptions as to what is the 'correct' expression pattern of a heterologous gene in mice.

8.2.2 Dissection of mouse embryos or foetuses

Analysis of gene expression at different stages of development requires the dissection of embryos or foetuses at various time points as described in *Protocol 6* (the foetal stage is taken as post-12.5 days gestation).

Protocol 6. Obtaining dissected mouse embryonic or foetal tissue at different developmental stages

Materials

- PBSa (8 g/litre NaCl, 0.2 g/litre KCl, 1.15 g/litre Na_2HPO_4, 0.2 g/litre KH_2PO_4).

Method

1. Place non-transgenic females with a transgenic male carrying the gene of interest. Usually two females are placed with each male.
2. Inspect the females daily for the presence of a vaginal plug. Cage those that have been plugged separately and note the plugging date. Mating is assumed to take place in the middle of the dark cycle, although in practice the exact time does show some variation. The day the mating plug is observed is taken as day 0.5.
3. For dissection of the embryos on the appropriate day, kill the female by cervical dislocation. Dissect out the uterus containing the embryos and place it in a Petri dish containing PBSa.
4. Dissect out each individual embryo by cutting through the uterus above and below it and then gently pulling away the uterine wall from around the embryo using two pairs of watchmaker's forceps. Dissections of pre-12.5 day embryos are most easily carried out in a dish of PBSa. Flash freeze tissues from which RNA is to be extracted in liquid nitrogen and store them at −70°C.
5. Use the placenta to prepare DNA for identification of transgenic embryos. Prepare the placental DNA using the same protocol as for tails (see *Protocol 4*).

8.2.3 Use of reporter genes

When analysing tissue-specific expression of transgenes, murine development can effectively be divided into two periods. Before 12.5 days gestation, the embryo is extremely tiny and the developing organs are microscopic and very difficult to dissect. Although it is possible to dissect out some tissues at this stage (e.g. yolk sac), for most tissues a very high degree of skill is required and the small amount of tissue obtained can present difficulties for conventional analytical methods for detecting specific transcripts. This problem can be circumvented by linking the transgene to a reporter gene the expression of which can be detected by staining of whole embryos or tissue sections. The gene which is most commonly used for this purpose is the β-galactosidase gene. When cells containing β-galactosidase protein are treated with the chromogenic substrate, X-gal (5-bromo-4-chloro-3-indolyl-β-D-galactopyranoside), they stain an intense blue colour. A procedure for X-gal staining of whole embryos is given in *Protocol 7*. Methods for sectioning embryos for staining can be found in ref. 17.

If a suitable antibody specific for the product of a heterologous gene is available, then antibody staining of sections can also be used. Alternatively, an antibody epitope can be introduced into the coding region of a gene, generating a fusion protein which can be visualized by immunostaining. Transcripts can also be directly visualized by *in situ* hybridization. Protocols for this procedure can be found in ref. 17 and Chapter 1.

Protocol 7. Staining of whole embryos for β-galactosidase expression[a]

Materials

- Phosphate-buffered saline (PBS) (7.6 g/litre NaCl, 3.8 g/litre Na_2HPO_4, 0.42 g/litre NaH_2PO_4)
- Fixative (1% formaldehyde, 0.2% glutaraldehyde, 2 mM $MgCl_2$, 5 mM EGTA, 0.02% NP-40, made up in PBS)
- Washing solution (0.02% NP-40 in PBS)
- Stock X-gal substrate (40 mg/ml X-gal dissolved in dimethylformamide). Store dark at 4°C
- Staining solution (5 mM $K_3Fe(CN)_6$, 5 mM $K_4Fe(CN)_6.3H_2O$, 2 mM $MgCl_2$, 0.01% sodium deoxycholate, 0.02% NP-40; 1 mg/ml X-gal diluted from stock).

Method

1. Wash the mouse embryos in PBS.
2. Fix the embryos in cold (4°C) fixative for 30–90 min at 4°C.
3. Wash the embryos three times at room temperature in washing solution (30 min per wash).
4. Place the embryos in staining solution and leave to stain in the dark at room temperature, usually overnight. The staining solution can be re-used several times if it is stored at 4°C in the dark.
5. Rinse the embryos in PBS and store them at 4°C in PBS.

[a] Provided by J. Whiting.

8.2.4 Analysis of RNA from embryos

All staining procedures such as those described in section 8.2.3 have the disadvantage that they do not give much information about the level of expression of a transgene. It may therefore be necessary to analyse RNA from early embryos. Small amounts of tissue can be disrupted in 200 μl 6.0 M guanidine hydrochloride (buffered at pH 5.2 with 50 mM sodium acetate). Carrier tRNA (5 μg) is added and the lysate extracted twice with phenol/chloroform and once with chloroform followed by precipitation with two volumes of ethanol. Total nucleic acid prepared in this way can be used for primer extension, nuclease S1 protection, and RNase protection analysis.

During the foetal stage (post-12.5 days gestation) many of the organs are quite well defined and can be dissected out more easily for RNA preparation. Depending on the amount of tissue obtained, the procedure described above can be used, or for large organs such as liver and brain, the lithium chloride method may be preferred (*Protocol 8*).

Protocol 8. Lithium chloride/urea method for isolation of RNA from mouse tissues

Materials

- 3.0 M LiCl, 6.0 M urea
- 10 mM Tris–HCl pH 7.6, 1 mM EDTA, 0.5% SDS
- Phenol/chloroform (1:1) buffered with 10 mM Tris–HCl pH 7.6
- 3.0 M sodium acetate, pH 5.2.

Method

1. Homogenize the tissue in 5–10 ml 3.0 M LiCl, 6.0 M urea per gram of tissue.
2. Sonicate the sample on ice for 1 min. This step shears the DNA and prevents it from precipitating with the RNA. Leave the sample to precipitate overnight at 4°C. Pellet the RNA by spinning at 9000 *g* for 20 min.
3. Pour off the supernatant, resuspend the pellet in a half-volume of cold 3.0 M LiCl, 6.0 M urea, and centrifuge again.
4. Remove the supernatant and dissolve the pellet in 10 mM Tris–HCl pH 7.6, 1 mM EDTA, 0.5% SDS at about half the volume of the original homogenate. If the RNA is slow to dissolve, step 5 can still be initiated and the solution vortexed with the phenol/chloroform to finish dissolving the RNA.
5. Extract with an equal volume of phenol/chloroform, separate the phases by centrifugation and remove the aqueous phase. One extraction is generally sufficient.
6. Add one tenth volume of 3.0 M sodium acetate pH 5.2 and two volumes of ethanol. Allow the RNA to precipitate either on ice or at −20°C for 30 min.
7. Collect the RNA by centrifugation at 9000 *g* for 20 min, wash with 70% ethanol and resuspend in water.

8.3 Gene trap experiments

Transgenesis can be used as a method for identifying novel developmentally regulated genes using a 'gene trap' protocol. In this approach, the transgene is used as a neutral template to detect regulatory sequences close to its site of integration. When interesting patterns of regulation are detected, the transgene can then be used as a marker to clone out the endogenous DNA sequences associated with this pattern.

In one type of approach (18), the transgene was a β-galactosidase gene under the control of a thymidine kinase promoter from herpes simplex virus (HSV). In this type of 'enhancer trap' protocol, novel patterns of expression were generated by integration of the transgene close to endogenous enhancers which could then be cloned out.

A more rigorous approach has made use of 'intron trap' constructs. In intron trap protocols, a β-galactosidase gene is flanked at the 5′ end by a splice acceptor site and part of an intron (18). Integration (in the correct orientation) into an intron of an endogenous gene will generate a splice and a β-galactosidase fusion protein which can be detected by X-gal staining. In comparison with enhancer trap protocols, this approach has the advantage that it selects for integrations within genes, facilitating their subsequent cloning. Such integration events will also disrupt the gene and in many cases will give rise to a null mutation whose phenotype can be determined by breeding the line to homozygosity. An obvious problem with intron trap protocols is the fact that integration in the correct orientation in an intron from a developmentally regulated gene is likely to be a relatively rare event. This problem can be solved by transfecting the construct into embryonic stem cells and looking for developmentally regulated expression of the β-galactosidase gene in differentiating embryoid bodies (see Section 6). Cells which show developmentally regulated expression can then be used to generate transgenic mice by blastocyst injection and breeding of germline chimaeras.

A number of laboratories are now engaged in efforts to use gene trap protocols of the type described above on a large scale to isolate libraries of murine genes in which specific patterns of expression are correlated with mutant phenotypes. Such libraries of developmentally regulated genes would be analogous to those which were obtained for *Drosophila* by genetic means and which have proved to be of such value in analysing pattern formation in flies.

8.4 Transgenesis as a means for probing complex cellular and physiological processes

Transgenic procedures offer a means for examining the role of a transgene product in the regulation of cellular and physiological processes in the live animal. The role of oncogenes in tumorigenesis is the best example to date of a process which has been studied by this means (19). Another example is the confirmation of the *SRY* gene as the mammalian sex-determining gene (20).

Transgenic mice have also been widely used to probe the complex cellular interactions of mammalian immune systems.

In general, mutant phenotypes induced by expression of a transgene will be dominant. This can present problems since the animals will often not survive for long enough to breed, making it difficult to establish lines. It would be highly desirable to be able to place a transgene under the control of a fully inducible promoter which would normally be silent in every tissue but which could be activated by breeding with a transgenic animal expressing the inducer in a specific tissue. Although this is simple in theory, the difficulties of putting it into practice are formidable. Designing the 'perfect' inducible promoter presents major problems since a homologue of almost any vertebrate inducer is likely to be present in at least some mouse tissues. Yeast transcriptional inducers such as Gal-4 offer a possible alternative. Gal-4 has been shown to activate a promoter containing a Gal UAS sequence in mammalian cells (21) but it is not yet clear whether the level of induction would be sufficient to allow it to be used as a universal inducer in transgenic mice. Inducible systems will also suffer from the problem of negative position effects which will prevent efficient expression and positive position effects giving rise to inappropriate expression independent of induction. To insulate the gene from these effects it will probably be necessary to include locus boundary elements (see Section 8.1.2) in any construct.

In transgenic studies using oncogenes, an alternative to using inducible promoters is the use of a temperature-sensitive mutant oncogene. Transgenic mice have been generated which have a temperature-sensitive SV40 T-antigen gene under the control of a mouse histocompatibility H-2K^b promoter (22). The fact that the normal body temprature of mice is close to the non-permissive temperature for this mutant appeared to prevent the formation of tumours, although thymic hyperplasia was observed. When thymus cells were isolated and cultured, they were found to be conditionally immortalized at the permissive temperature while growth was arrested at the non-permissive temperature. The success of this method in preventing oncogenesis in the live animal is likely to depend on the level of expression of the temperature-sensitive oncogene being fairly low as incomplete inactivation of the oncogene may become a problem at higher levels.

The development of mouse models for human diseases is another of the exciting possibilities offered by transgenic methodology. One example of this is the generation of transgenic mice carrying a human β-globin gene containing the mutation which causes sickle-cell anaemia (23). Sickling of red cells was observed in animals which expressed high levels of sickle haemoglobin providing a potential model for testing of drugs to alleviate this condition.

8.5 Cellular ablation in transgenic mice

By placing a transgene encoding a cytotoxic gene product under the control of a cell-type specific promoter, it is possible to specifically ablate cell lineages in which the transgene is expressed. The early cell ablation protocols made use of the highly cytotoxic diphtheria toxin gene (24). Subsequent protocols have made use of the HSV-thymidine kinase (HSV-TK) gene which is itself non-toxic but which makes cells susceptible to the cytotoxic effects of the drug, gancyclovir (25). This allows transgenic lines to be established and animals can then be treated with gancyclovir at a specific stage of development and the effects of ablating specific cell types can be studied. One problem which has been encountered with this protocol is the fact that the HSV-TK gene under the control of a variety of promoters has been found to cause sterility in male mice (26). This appears to be due to testis-specific expression of the TK gene from a cryptic internal promoter. This male sterility means that lines can only be maintained by breeding female animals.

In general, the success of cell-ablation experiments in answering biological questions is likely to depend on the particular features of the system being studied and on the availability of a very tightly regulated gene to target a specific cell-type. Leaky expression of the transgene, incomplete cell killing, and the complexity of interactions between multiple cell-types may combine to make interpretation difficult. Like many other types of transgenic experiment, it will often be impossible to assess the usefulness of this approach for a particular problem without testing it first.

9. Other transgenic systems

Among mammalian transgenic systems, transgenic mice remain the method of choice for transcriptional analysis because of the facility which they can be generated, maintained, and bred. There are also a number of non-mammalian transgenic systems which have been used for the analysis of gene expression.

9.1 *Drosophila*

Since the inception of genetics as a science, *Drosophila* has been by far the most important model organism for the study of the genetics of metazoans. The utility of this organism for genetic studies has been extended even further in recent years by the development of a highly effective system for generating transgenic *Drosophila* (reviewed in ref. 27). Generation of transgenic *Drosophila* makes use of vectors based on transposable P elements. A potential transgene is cloned within an internally deleted P element which is then co-injected into an M strain embryo together with an intact P element. P elements are genetically stable in P strain embryos but undergo transposition at high frequency in M strain embryos. The full P element encodes a transposase which catalyses the transposition of the transgene and P element sequences into the genome. This procedure gives rise to mosaic flies. The very short generation time of *Drosophila* means that these mosaics can then be easily bred to give genuine transgenics.

9.2 *Xenopus laevis*

The South African clawed frog, *Xenopus laevis*, has been the prototype vertebrate for the study of early development. The ability to make transgenic *Xenopus* would greatly aid the probing of the genetic control of these processes. Unfortunately the construction of transgenic *Xenopus* faces problems which have so far proved insurmountable. Fertilized eggs from *Xenopus* are large and opaque which makes it impossible to locate and inject the pronuclei. The DNA must therefore be injected into the cytoplasm. The injected DNA replicates rapidly and permissively and persists into the swimming tadpole stage. Some of it ultimately integrates to produce animals which are highly mosaic (28). Theoretically it would be possible to produce truly transgenic animals by breeding these mosaics, but the very long generation time of *Xenopus* (2 years) means that this not a useful method for producing transgenics. However, it has been possible to analyse the function of genes (such as the actin genes) which are expressed early in development by transient analysis following cytoplasmic injection (29). The globin genes which are expressed in the swimming tadpole stage were not expressed when introduced by this method (ref. 30 and N. Dillon, unpublished).

9.3 Zebrafish

The use of Zebrafish as an alternative system for studying early development in vertebrates has recently attracted much attention. Zebrafish have many of the advantages of *Xenopus* for embryological studies but in addition have a generation time which is similar to that of the mouse (3 months). This makes the organism amenable to genetic analysis. As with *Xenopus*, DNA can be microinjected only into the cytoplasm of Zebrafish eggs giving rise to mosaic animals. However, the short generation time of Zebrafish means that these mosaics can be bred rapidly to give genuinely transgenic animals (31). So far, only a limited amount of work has been done on this system and the facility with which transgenic animals can be generated remains to be established.

9.4 *Dictyostelium discoideum*

The cellular slime mould *D. discoideum* is probably the simplest organism which can be regarded as a true transgenic system. *D. discoideum* exists in two states: as independent amoebae and as a more complex multicellular fruiting body which results from aggregation of amoebae followed by a programme of differentiation. This two-stage life cycle makes the generation of transgenic *Dictyostelium* relatively straightforward since the amoeboid cells can be grown in culture and transfected by electroporation. Most transfection protocols involve the use of vectors carrying a neomycin resistance gene which allow selection of stable integrants using the drug G418. Stable transformants can then be induced to differentiate into multicellular fruiting bodies. For an example of how these techniques can be used to study gene expression during *Dictyostelium* differentiation, see ref. 32.

References

1. Hogan, B., Costantini, F., and Lacy, E. (1986). *Manipulating the mouse embryo—a laboratory manual.* Cold Spring Harbor Press, New York.
2. Vogelstein, B. and Gillespie, D. (1979). *Proc. Natl Acad. Sci. USA,* **76,** 615–19.
3. Kaiser, K. and Murray, N. E. (1985). In *DNA cloning: a practical approach,* Vol. 1, (ed. D. M. Glover), pp. 41–2. IRL Press, Oxford.
4. Sambrook, J., Fritsch, E. F., and Maniatis, T. (1989). *Molecular cloning—a laboratory manual,* (2nd edn). Cold Spring Harbor Press, New York.
5. Schwartz, D. and Cantor, C. (1984). *Cell,* **37,** 67–75.
6. Pieper, F. R., de Wit, I., Pronk, A., Kooiman, P., Srijker, R., Krimpenfort, P., Nuyens, J., and de Boer, H. (1992). *Nucleic Acids Res.,* **20,** 1259–64.
7. Robertson, E. J. (ed.) (1987). *Teratocarcinomas and embryonic stem cells: a practical approach.* IRL Press, Oxford.
8. Doetschmann, T. C., Eistetter, H., Katz, M., Schmidt, W., and Kemler, R. (1985). *J. Embryol. Exp. Morphol.,* **87,** 27–45.
9. Schmitt, R. M., Bruyns, E., and Snodgrass, H. R. (1991). *Genes Dev.,* **5,** 728–40.
10. Lindenbaum, M. and Grosveld, F. (1990). *Genes Dev.,* **4,** 2075–85.
11. Hammer, R. E., Krumlauf, R., Camper, S. A., Brinster, R. L., and Tilghman, S. M. (1987). *Science,* **235,** 53–7.
12. Grosveld, F., Blom van Assendelft, G., Greaves, D. R., and Kollias, G. (1987). *Cell,* **51,** 975–85.
13. Dillon, N., Talbot, D., Philipsen, S., Hanscombe, O., Fraser, P., Lindenbaum, M., and Grosveld, F. (1991). In *Genome analysis,* Vol. 2, *Gene expression and its control.* Cold Spring Harbor Laboratory Press, New York.
14. Lang, G., Wotton, D., Owen, D., Owen, M., Sewell, W., Brown, H., Mason, D., Crumpton, M. J., and Kioussis, D. (1988). *EMBO. J.,* **7,** 1675–82.
15. Kellum, R. and Schedl, P. (1991). *Cell,* **64,** 941–50.
16. Stief, A., Winter, D., Stratling, W., and Sippel, A. (1989). *Nature,* **341,** 343–5.
17. Copp, A. J. and Cockroft, D. L. (eds) (1990). *Postimplantation mammalian embryos: a practical approach.* IRL Press, Oxford.
18. Gossler, A., Joyner, A., Rossant, J., and Skarnes, W. (1989). *Science,* **244,** 463–5.
19. Groner, B., Schonenberger, C., and Andres, A. (1987). *Trends Genet.,* **3,** 306–8.
20. Koopman, P., Gubbay, J., Vivian, N., Goodfellow, P., and Lovell-Badge, R. (1991). *Nature,* **351,** 117–21.
21. Webster, N., Jin, J., Green, S., Hollis, M., and Chambon, P. (1988). *Cell,* **52,** 169–78.
22. Jat, P., Noble, M., Ataliotis, P., Tanaka, Y., Yannoutsos, N., Larsen, L., and Kioussis, D. (1991). *Proc. Natl Acad. Sci. USA,* **88,** 5096–100.
23. Greaves, D. R., Fraser, P., Vidal, M., Hedges, M., Ropers, D., Luzzato, L., and Grosveld, F. (1990). *Nature,* **343,** 183–5.
24. Palmiter, R., Behringer, R., Quaife, C., Maxwell, F., Maxwell, F., Maxwell, I., and Brinster, R. (1987). *Cell,* **50,** 435–43.

25. Borrelli, E., Heyman, R., Hsi, M., and Evans, R. (1988). *Proc. Natl Acad. Sci. USA,* **85,** 7572–6.
26. Al-Shawi, R., Burke, J., Wallace, H., Jones, C., Harrison, S., Buxton, D., Maley, S., Chandley, A., and Bishop, J. O. (1991). *Mol. Cell. Biol.,* **11,** 4207–16.
27. Rubin, G. M. (1988). *Science,* **240,** 1453–9.
28. Etkin, L. D., Pearman, B., Roberts, M., and Bektesh, S. L. (1984). *Differentiation,* **36,** 194–202.
29. Wilson, C., Cross, G., and Woodland, H. (1986). *Cell,* **47,** 589–99.
30. Bendig, M. M. and Williams, J. G. (1984). *Mol. Cell. Biol.,* **4,** 567–70.
31. Stuart, G., McMurray, J., and Westerfield, M. (1988). *Development,* **103,** 403–12.
32. Ceccarelli, A., Mahbubani, H., and Williams, J. G. (1991). *Cell,* **65,** 983–9.

5

Identification and characterization of eukaryotic transcription factors

STEPHEN P. JACKSON

1. Introduction

An important and widely used mechanism for regulating gene expression in eukaryotic organisms is that of modulating the efficiency of transcriptional initiation. Molecular genetic analysis of transcriptional promoters recognized by RNA polymerase II has revealed that accurate and efficient transcription requires a variety of *cis*-acting sequence elements. These elements are recognized and bound by sequence-specific DNA-binding transcription factor proteins that then serve to either activate, or in some cases repress, transcriptional initiation by the general transcriptional apparatus (1–4). A great many sequence-specific transcription factors exist; they differ from one another in the DNA sequence(s) that they recognize and in the way that they are regulated themselves. For example, some transcription factors are restricted to particular cell types, others are expressed only at certain stages of development, while yet others are activated only in response to a particular physiological agent such as a hormone. By employing different combinations of these sequence-specific factors, highly complex and unique patterns of gene expression are achieved. This chapter describes various approaches that are used to define, purify, and characterize sequence-specific transcription factor proteins.

2. DNA binding assays for identifying transcription factors in crude cell extracts

Most studies on sequence-specific transcription factors begin by analysing which regions of a transcriptional promoter are required for the correct expression of a particular gene (see Chapters 2 and 6). Once an important promoter region has been defined in this way, the next step often is to detect an activity in crude cell-free extracts that recognizes this element; such an entity is then a candidate for regulating transcription of the gene of interest *in vivo*. Since even relatively short fragments of promoter DNA can often bind many different factors, it is important to characterize the DNA region of interest thoroughly (see Chapter 6 for details) before proceeding with the DNA binding assays described below.

2.1 Gel mobility shift assay and DNase I footprinting

Two methods commonly used to detect specific protein–DNA interactions are the gel mobility shift assay and the DNase I footprinting assay. Since each has its advantages and drawbacks, it is often beneficial to use them both in combination. A detailed description of these methods is given in Chapter 6; listed below are elaborations of the basic techniques that are designed to determine the nature of the protein component of a particular DNA-protein complex.

In the gel mobility shift assay, the binding of a protein to a radiolabelled DNA fragment reduces the mobility of the DNA in a non-denaturing polyacrylamide gel, and thus results in a complex that can be distinguished electrophoretically from the unbound probe (*Figure 1a*). This method is very simple to perform and is more sensitive than DNase I footprinting. Furthermore, the gel mobility shift assay readily provides a quantitative measure of the amount of a particular DNA binding activity. These features make this method particularly ideal for rapid, quantitative analysis of a large number of samples. Another advantage of the gel mobility shift assay is that complexes between a DNA fragment and a series of different proteins usually differ from one another in electrophoretic mobility. Thus, it is possible to distinguish between multiple proteins interacting with the same DNA recognition element.

In DNase I footprinting, the binding of a protein to an end-labelled DNA probe protects the region that it interacts with from digestion by DNase I, resulting in a region of DNase I protection. An example is shown in *Figure 1b*. Unlike the gel mobility shift assay, footprinting has the advantage that it readily reveals which region of a DNA probe is bound by the protein. Disadvantages, however, are that it is generally less sensitive than the gel mobility shift assay and it is generally very difficult to distinguish between multiple alternative protein complexes assembled on the same region of DNA.

An alternative to DNase I footprinting is to use chemical footprinting reagents such as hydroxyl radicals (see Chapter 6 for details). These reagents are much smaller than the DNase I enzyme and thus provide more detailed information as to the precise sites of contact between the protein and DNA. However, because chemical footprinting is generally more difficult to perform than DNase I footprinting (particularly when crude extracts are used), this method is usually employed only after the region of DNA interaction has first been defined by DNase I footprint analysis.

2.2 Involvement of previously characterized transcription factors

Because a great many transcription factors have been identified already, it is likely that some of the proteins binding to a chosen promoter will correspond

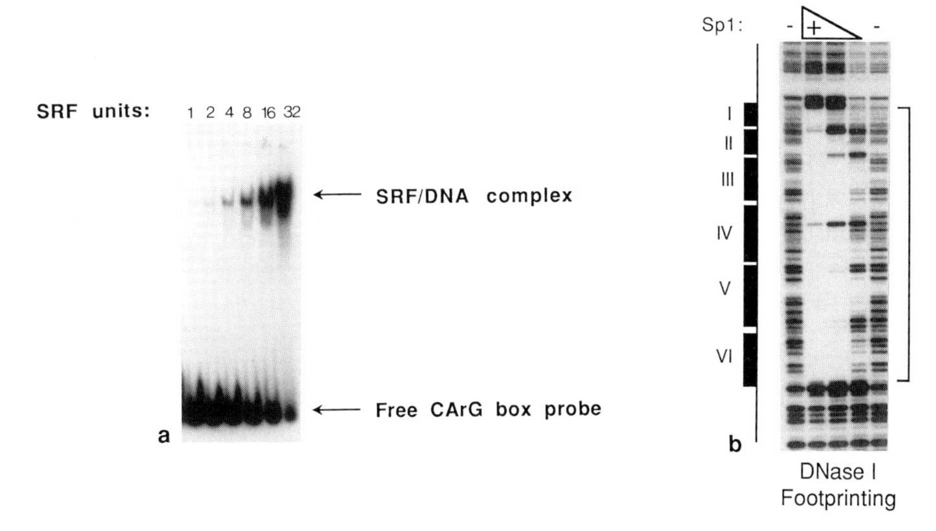

Figure 1. (a) The gel mobility shift assay. An end-labelled oligonucleotide probe containing a wild-type SRF binding site (the CArG box 1 of the *Xenopus* cardiac actin gene) was incubated with increasing amounts of the *Xenopus* SRF protein. The amount of SRF in each reaction (arbitrary units) is given above each lane. The location of the unbound probe and SRF/DNA complex are indicated. These data were provided by M. V. Taylor. (b) The DNase I footprinting assay. This experiment employed an end-labelled DNA fragment containing the six Sp1-binding GC-box elements of the SV40 intergenic control region (from pSV07; ref. 5). Assays were performed in the absence of Sp1 (lanes 1 and 5), with 20 ng of Sp1 (lane 2), with 4 ng of Sp1 (lane 3), or with 1 ng of Sp1 (lane 4). The Sp1 used in this experiment had been purified by DNA affinity chromatography. After incubation of Sp1 with the DNA probe for 10 min on ice followed by 1 min at room temperature, the DNA was digested by DNase I, then the reactions were stopped and analysed on a denaturing 6% polyacrylamide gel/6 M urea. To the left of the figure is a representation of the six GC box elements (I–VI). The region of the DNA protected by bound Sp1 is bracketed on the right. Note, not all bands in the region of the footprint disappear even when the highest levels of factor are used (lane 2). For other factors the level of protection may be substantially less than that which can be achieved by Sp1. In many cases, in addition to reducing the intensity of some bands, factor binding may cause some other bands to become darker. These DNase I hypersensitive sites frequently surround the footprinted region. These data are from Jackson and Tjian, 1989 (ref. 6).

to previously characterized factors. Therefore, before embarking on the considerable task of analysing a particular protein, it is wise to determine whether it corresponds to any known transcription factor. Although this is sometimes obvious if the promoter fragment contains sequences that conform precisely to the consensus binding site for a known protein (see *Appendix 1*), many transcription factors have considerable flexibility in the DNA sequences that they can recognize. In most cases, even where strong consensus sites exist for a previously characterized factor, the binding of the known factor to the chosen promoter fragment should be determined empirically, as described in

Section 2.2.1 below. On the other hand, if the sequence of interest has no obvious homology to the binding sites of any known factors, the next step is to investigate the polypeptide composition of the DNA binding activity that has been detected (see Section 3).

2.2.1 Use of purified transcription factors

One approach to investigate whether a known factor binds to a promoter sequence is to obtain a purified sample of the known factor and determine whether its footprinting protection pattern and gel mobility shift complex mobility correspond with the activity detected in crude cell extracts. (A number of common transcription factors such as TFIID and Sp1 are now commercially available.) Although providing evidence either for or against an identity with the known factor, this type of approach on its own is often far from conclusive. For example, slight differences in the footprint pattern could indicate either that the factor is different, or that it is the same factor but its interaction with DNA is altered in some way by association with other protein(s). Conversely, although identical footprints between the purified factor and the activity detected in crude extracts may indicate that the two activities are one and the same, they could represent different proteins that recognize the same sequence element. (There are many known examples where multiple factors bind to the same sequence; refs 7, 8.) This approach should therefore be used in conjunction with the oligonucleotide competition or antibody approaches described below (Sections 2.2.2 and 2.2.3).

2.2.2 Oligonucleotide competition studies

Another approach to determine whether a DNA binding activity corresponds to a previously characterized factor is to test whether an oligonucleotide fragment containing the consensus sequence for this factor effectively abolishes the gel mobility shift complex by competition with the promoter fragment. Ideally, these studies should be conducted with both the wild-type consensus sequence and with a variety of point mutated derivatives whose effects on binding of the known factor have been determined; only if competition precisely parallels the binding characteristics of the known factor can it be concluded with a high degree of confidence that the binding activity indeed corresponds to the known factor (*Figure 2a* and *2b*).

2.2.3 Use of antibodies specific to a previously characterized factor

One of the most conclusive ways of demonstrating that the identified DNA binding activity contains a certain transcription factor is to utilize antibodies that specifically recognize this factor in gel mobility 'super-shift' experiments. The rationale behind this approach is that if this factor is part of a gel mobility shift complex, then inclusion of antibodies against this factor in the binding reaction will result in the antibodies interacting with the factor/DNA complex and thus further retard its electrophoretic mobility (*Figure 2c*). Because of non-specific interactions that can occur between protein-DNA complexes and components in antibody preparations, it is imperative to use appropriate pre-immune controls. In many cases, it has been found that the inclusion of non-ionic detergents such as Nonidet P-40 to 0.1% in the DNA binding reaction (and, in some cases, in the gel also) reduces the degree of non-specific interactions with components of the antibody preparation (13). Occasionally, the inclusion of antibodies in the gel mobility shift assay can abolish factor binding, causing loss of the gel mobility shift complex.

3. Identifying the DNA-binding polypeptide

Once a sequence-specific DNA-binding activity has been detected in a crude cell-free extract, one can design experiments to identify the protein species responsible. Although purification of this factor to homogeneity (Section 4) or cloning of its cDNA (Section 7) provide obvious paths towards this goal, it is often a useful preliminary step to gain an indication of the relative molecular mass of the transcription factor by using one of the two methods described below that rely on the sequence-specific nature of the DNA–protein interaction. These methods can be used with crude extracts or with purified fractions.

3.1 UV crosslinking of transcription factors to DNA

In this method (14, 15), a cell-free extract containing the transcription factor is incubated with a uniformly radiolabelled DNA fragment that bears a high affinity binding site for the protein. Next, the resulting DNA–protein complex is subjected to UV light, causing covalent bonds to be formed between the DNA and the bound protein. DNase I is then added to remove from the protein all but the portion of the DNA fragment that is directly attached to it. The molecular mass of the DNA-binding protein complex (now radiolabelled) is then estimated by comparing its migration on an SDS-polyacrylamide gel with that of known size markers. The fact that a fragment of DNA is covalently attached to the protein often results in the protein electrophoresing slightly more slowly than it would if it were not linked to DNA. However, since the size of the DNA fragment is very small after DNase treatment, the effect on electrophoretic mobility is usually minimal.

For the UV crosslinking procedure, it is normal to use radiolabelled DNA that has had its thymidine residues substituted by bromodeoxyuridine (BrdU). The reason for this is twofold: firstly, BrdU-containing DNA is generally crosslinked to protein much more readily than unsubstituted DNA; secondly, when BrdU-substituted DNA is used, lower wavelength UV light can be used, which causes less damage to the protein. Although this procedure can be employed when using crude cell extracts, better results are

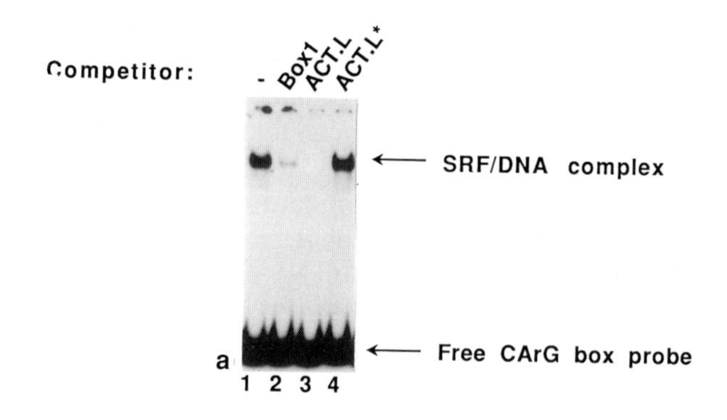
Competitor:
-
Box1
ACT.L
ACT.L*
SRF/DNA complex
Free CArG box probe
a
1 2 3 4

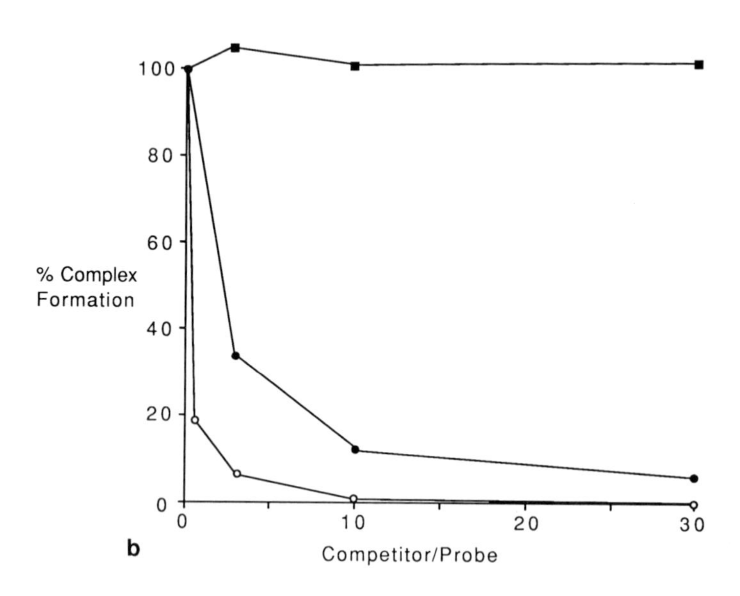
100
80
60
40
20
0
% Complex Formation
0
10
20
30
b
Competitor/Probe

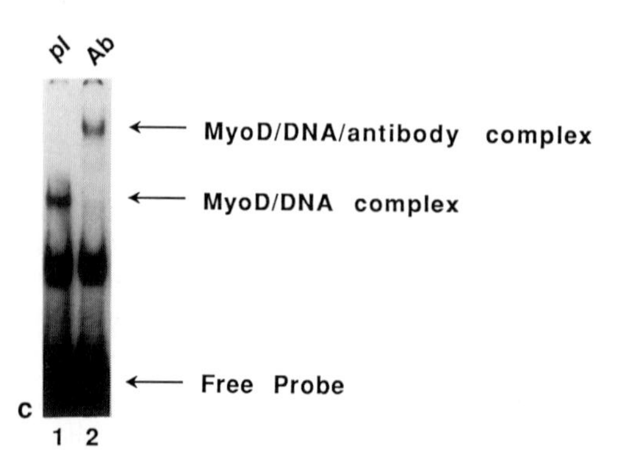
pI
Ab
MyoD/DNA/antibody complex
MyoD/DNA complex
Free Probe
c
1 2

Figure 2. Determining the identity of the transcription factor in an electrophoretic gel mobility shift complex. (a) Oligonucleotide competition studies. Binding of *in vitro* translated *Xenopus* SRF to an end-labelled DNA fragment bearing a wild-type SRF binding site (CArG box 1 of the *Xenopus* cardiac actin gene). For lanes 2 to 4, competitor oligonucleotides were included in the binding reactions at a 10-fold molar excess over the radioactive probe. Competition was by the wild-type SRF binding site (lane 2), an 'up mutant' (ACT.L) binding site that binds SRF more strongly than the wild-type sequence (lane 3), and a 'down mutant' site (ACT.L*) that does not bind SRF effectively (lane 4). These data are from Mohun *et al.*, 1991 (ref. 9) with permission but see also ref. 10. (b) Graphical representation of oligonucleotide competition titrations. Varying amounts of the three oligonucleotides described in (a) were used to compete the binding of *Xenopus* SRF (present in a *Xenopus* embryo extract) to an end-labelled DNA fragment bearing a wild-type SRF binding site (CArG box 1 of the *Xenopus* cardiac actin gene). The horizontal axis shows the amount of competitor used expressed as a molar ratio to the amount of radiolabelled probe and the vertical axis denotes the amount of complex formed as a percentage of that obtained in the absence of competing oligonucleotide. ●, Self competition (i.e. competition with unlabelled CArG box oligonucleotide); ○, competition with the ACT.L 'up mutant' oligonucleotide; ■, competition with the ACT.L* 'down mutant' oligonucleotide. These data are from Taylor *et al.*, 1989 (ref. 11) with permission. (c) Use of antibodies in 'supershift' studies. Somite whole cell extract from late neurula *Xenopus* embryos was incubated on ice for 15 min with an end-labelled probe derived from the *Xenopus* cardiac actin gene. Next, 0.1 μl of pre-immune serum (pI; lane 1) or 0.1 μl of antiserum raised against the *Xenopus* MyoD protein (Ab; lane 2) was added and incubation continued for 15 min on ice. Finally, complexes were resolved by electrophoresis in a 3.5% native polyacrylamide gel. Arrows indicate the positions of the MyoD/DNA complex (lane 1) and the MyoD/DNA/antibody 'supershifted' complex (lane 2). These data are from Taylor *et al.*, 1991 (ref. 12) with permission.

---

generally achieved when the transcription factor has been at least partially purified.

A modification of the UV-crosslinking procedure is to UV crosslink protein-DNA complexes that have been resolved by the gel mobility shift assay. This can be of particular benefit if multiple complexes form on the same DNA probe. In this procedure, gel mobility shift complexes are electrophoresed on an agarose gel made using low melting point agarose and run in TBE buffer, then subjected to UV irradiation, and finally located by autoradiography. Individual complexes are excised from the gel, and subjected to crosslinking (14, 16, 17).

3.1.1 DNA probes for UV crosslinking

In the method described below (*Protocol 1*), a DNA fragment is prepared that is radiolabelled at cytosine residues and has thymidine residues substituted by BrdU. Although probes of this type will be effective in most cases, it should be noted that the most efficient probes will be ones that contain the maximal number of labelled residues in the vicinity (within 10 residues) of the protein binding site, and that also contain BrdU residues within the DNA region contacted by the protein. If the DNA chosen does not fulfil one or other of these criteria, then it is advisable to do one or more of the following:

- use another probe that has its binding site embedded in more favourable sequences
- label the other DNA strand by using the M13 vector with the fragment cloned in the opposite orientation
- use another radiolabelled nucleotide instead of [α-^{32}P]dCTP.

Protocol 1. Preparation of radiolabelled probe for UV crosslinking

Equipment and reagents

- High voltage electrophoresis power supply and gel apparatus for DNA fragments
- Heating block or water baths that can be set to 90°C for template annealing, 16°C for DNA labelling, and 37°C for restriction enzyme digestion
- Single-stranded form of an M13 vector containing the binding sequence of interest cloned into the multiple cloning site
- M13 'universal' (−20 nt) sequencing primer (Stratagene)
- 0.1 M dithiothreitol (DTT)
- Klenow fragment of *E. coli* DNA polymerase I (5 U/μl)
- 10 × 'medium salt' restriction enzyme buffer (0.5 M NaCl, 100 mM $MgCl_2$, 10 mM DTT, 100 mM Tris–HCl pH 7.5)
- Appropriate restriction endonucleases (see below)
- [α-^{32}P]dCTP (10 mCi/ml; 3000 Ci/mmol)
- 10 × crosslinking nucleotide mixture [0.5 mM dGTP, 0.5 mM dATP, 0.5 mM BrdU (5-bromo-2′-deoxyuridine triphosphate), and 50 μM dCTP].

Method

1. To 5 μg single-stranded M13 vector containing the binding sequence of interest, add an equimolar quantity (usually 6 to 15 ng, depending on size of clone) of M13 'universal' sequencing primer. Add 10 μl of 10 × medium salt restriction enzyme buffer and then make volume up to 74 μl final with distilled water. Heat to 90°C for 5 min in a heating block or water bath, then switch off the heating unit and allow to cool slowly to room temperature (generally 1–3 h). During this time, the primer and M13 DNA anneal. When cool, briefly spin in a microcentrifuge to collect water droplets from the inside of the tube lid.
2. Add in the following order:

10 × crosslinking nucleotide mixture[a]	10 μl
0.1 M DTT	1 μl
[α-^{32}P]dCTP	10 μl.

   Mix the tube contents by gentle pipetting. Add 5 μl of Klenow fragment enzyme (5 U/μl) and gently mix the tube contents again. Incubate for 1.5 h at 16°C.
3. Inactivate the Klenow fragment enzyme by incubating at 65°C for 10 min.

Allow the reaction to cool to room temperature, then briefly spin in a microcentrifuge to collect water droplets from the inside of the tube lid.

4. Add 25 units of the appropriate restriction enzyme(s) required to excise a double-stranded radiolabelled DNA fragment of fifty to a few hundred base pairs long that contains the binding site of interest. Incubate at the appropriate temperature (usually 37°C) for 2 h to digest the DNA. Note that if the enzyme requires a salt concentration substantially different from that in the labelling reaction, the salt concentration of the tube contents must be adjusted by the addition of either distilled water or NaCl before adding the restriction enzyme.
5. Ethanol precipitate the DNA and resuspend it in distilled water. Purify the radiolabelled DNA fragment on either a low melting point agarose gel or non-denaturing polyacrylamide gel (see Chapter 6).
6. Measure the level of radioactive incorporation in the purified DNA fragment by liquid scintillation counting; specific activities of 10^8–10^9 c.p.m./μg should be achieved. Because of the high specific activity of the DNA, use the probe as soon as possible, and certainly within a week of preparation.

[a] Note that, if a radiolabelled nucleotide other than [α-^{32}P]dCTP is used, the composition of the 10 × crosslinking nucleotide mixture should be altered in a corresponding fashion.

3.1.2 UV crosslinking procedure

Cell-free extracts prepared by a variety of procedures and with varying degrees of purification can be employed successfully in UV-crosslinking studies. As a rule, if sequence-specific DNA binding by a protein can be shown by DNase I footprinting or the gel mobility shift assay, then, using a similar set of conditions, it should be possible to crosslink this protein specifically to its DNA recognition element. Obviously the conditions for UV-crosslinking should be compatible with efficient binding of the protein to DNA. One of the most important parameters is salt concentration; for most proteins this will be in the range of 50–100 mM KCl (Z′ buffer containing 50–100 mM KCl, which is used in a variety of DNA binding assays, is very effective; see *Protocol 2*). Another parameter that must be chosen carefully is the amount and type of non-specific DNA which is included in the reaction to abrogate interactions between the radiolabelled probe and non-specific DNA-binding proteins in the extract. In most instances, conditions that have been optimized for the detection of specific DNA complexes using DNase I footprinting or the gel mobility shift assay are ideal. *Protocol 2* lists a set of standard conditions for UV crosslinking that works well with a number of sequence-specific DNA-binding proteins. Since UV light is absorbed by plastics and by aqueous solutions, most effective crosslinking is achieved by shining the light directly down (using an inverted UV transilluminator) on top of reactions contained in uncapped round-bottom vials or in a microtitre plate.

Protocol 2. UV crosslinking

Equipment and reagents

- UV transilluminator (305 nm, approximately 7000 μW/cm^2) or Stratalinker (Stratagene)
- Round-bottom freezing vials (2 ml capacity e.g. Sterilin) or microtitre plate (Nunc)
- BrdU-substituted radiolabelled probe (see *Protocol 1*)
- DNase I (Worthington; >2000 U/mg)
- Micrococcal nuclease (Worthington; >6000 U/mg)
- Z′ buffer containing 50 mM KCl [25 mM Hepes-KOH, pH 7.6, 50 mM KCl, 12.5 mM $MgCl_2$, 20% glycerol, 0.1% Nonidet-P40, 10 μM $ZnSO_4$ 1 mM DTT (added freshly before use)]
- Poly(dI-dC).poly(dI-dC) DNA (Pharmacia)
- 0.1 M $CaCl_2$.

Method

1. Set up the binding reaction on ice in a round-bottomed tube or in a microtitre plate well by adding in the following order:

Z′ buffer containing 50 mM KCl	a volume such that the final reaction volume is 40 μl
BrdU-substituted radiolabelled probe	10^5 c.p.m. in, at most, 5 μl
poly dI:dC competitor DNA	10 μg in 1 μl
extract containing the DNA-binding activity of interest	volume as appropriate

2. Mix by pipetting gently, then seal each tube by placing a piece of cling film over the top and sealing it around the side with a small rubber band or with parafilm.

3. Incubate at an appropriate temperature (usually 30 °C) for 10 min to allow the protein to bind to the DNA.

4. Remove cling film from the tube(s). Place the tube(s) containing the binding reaction in a rack, then place on top of the rack an inverted UV transilluminator. The distance between the light source and solution should be 2–5 cm. *Whilst wearing a protective UV face shield*, turn on the UV source. The optimal time for irradiation varies from one protein-DNA complex to another and should initially be determined empirically by performing a series of time-points between 2 and 60 min.

5. Add to each tube the following:

0.1 M $CaCl_2$	5 μl
micrococcal nuclease	4 U
DNase I	10 μg.

Mix by pipetting gently and incubate at 37 °C for 30 min. This nuclease treatment digests away all but the 10–20 bp of DNA that is protected by the specifically-bound, crosslinked protein.

3.1.3 SDS-polyacrylamide gel electrophoresis of crosslinked proteins

The proteins crosslinked to the DNA under investigation can be resolved according to their relative molecular weights by analysing the crosslinking reaction mixture using denaturing SDS-PAGE (18) as described in *Protocols 3* and *4*. After this is complete, the gel is dried, subjected to autoradiography, and the molecular masses of specific protein complexes estimated by comparing their gel mobilities with those of protein molecular weight standards. Usually, an approximately 8% polyacrylamide separating gel is used in the first instance to resolve throughout most of the protein size-range. Should the protein of interest be smaller than 30 kDa or greater than 100 kDa, however, its size will be estimated more accurately using higher or lower percentage separating gels, respectively.

Protocol 3. Preparation of SDS-PAGE gels

Equipment and reagents

- Mini-slab gel electrophoresis apparatus, giving 0.5–1.0 mM thick mini-gels (e.g. BioRad Mini Protean II system)
- 4 × lower gel buffer (1.5 M Tris–HCl pH 8.8, 0.4% SDS)
- 4 × upper gel buffer (0.5 M Tris–HCl pH 6.8, 0.4% SDS)
- 30% acrylamide gel mixture (30% acrylamide and 0.8% *N,N'*-methylenebisacrylamide)
- 10% ammonium persulphate
- TEMED (*N,N,N',N'*-tetramethylethylenediamine)
- 2-methyl-1-propanol (alternatively, water-saturated 2-butanol)
- 10 × electrophoresis buffer stock (0.25 M Tris base, 1.92 M glycine).

Method

1. Assemble the two gel plates (8 × 10 cm) so that they are separated by 1 mm spacers.
2. Prepare the *separating* gel mixture of the correct concentration as shown below.

	Stacking gel mixture	Separating gel mixtures (final % acrylamide)				
		6	8	10	12	15
distilled water (ml)	3.0	2.75	2.42	2.09	1.75	1.25
4 × lower buffer (ml)	0	1.25	1.25	1.25	1.25	1.25
4 × upper buffer (ml)	1.25	0	0	0	0	0
30% acrylamide gel mixture (ml)	0.7	1.00	1.33	1.66	2.00	2.50

Protocol 3. *Continued*

3. To start polymerization, add 11 μl TEMED and 30 μl 10% ammonium persulphate to the *separating* gel mixture, swirl briefly to mix, then transfer to between the glass plates with a Pasteur pipette, until the meniscus is approximately 1.5–2.0 cm from the top.
4. Immediately layer 0.5 ml of 2-methyl-1-propanol on top of the acrylamide solution, and leave the gel to polymerize at room temperature at least 30 min. The polymerized separating gel can be left for a few hours at room temperature, or even weeks at 4°C if sealed in an airtight bag, before adding the stacking gel. Once the stacking gel has been poured, however, the gel should be run within 3 h.
5. Wash out the 2-methyl-1-propanol with distilled water and then remove any remaining drops of water with tissue paper.
6. Add 11 μl TEMED and 30 μl 10% ammonium persulphate to the *stacking* gel mixture (see step 2 above), mix and transfer to between the glass plates with a Pasteur pipette. Place a well-forming comb between the glass plates taking care to avoid bubbles under the comb. Leave at room temperature for at least 20 min to polymerize.
7. Carefully remove the comb, then place the gel in an electrophoresis tank containing 1 × electrophoresis buffer (made by diluting 10 × electrophoresis buffer stock to 1 × with distilled water, then adding SDS to a final concentration of 0.1%). Flush out the sample wells with electrophoresis buffer using a Pasteur pipette.

Protocol 4. SDS-PAGE of UV-crosslinked proteins

Equipment and reagents

- Power supply and apparatus for SDS-PAGE of proteins
- Vacuum gel drier
- UV-crosslinked, nuclease-treated protein samples (see *Protocol 2*)
- [^{14}C]methylated protein molecular weight standards (Amersham), or prestained non-radioactive markers (e.g. Rainbow markers; Amersham)
- Fluorographic agent (e.g. Amplify; Amersham)
- 2 × SDS sample buffer (10% glycerol, 5% 2-mercaptoethanol, 3% SDS, 62.5 mM Tris–HCl pH 6.8, 0.2% Bromophenol Blue). Store in aliquots at −20°C
- Fixing solution (25% 2-propanol, 10% acetic acid).

Method

1. Transfer the crosslinked, nuclease-treated samples into 1.5 ml microcentrifuge tubes and combine with an equal volume of 2 × SDS sample buffer. To estimate the size of any crosslinked protein resolved in the gel, also prepare an additional sample containing [^{14}C]methylated protein

molecular weight standards or prestained non-radioactive markers in SDS sample buffer.

2. Pierce a hole in the lid of each tube with a needle, then denature the protein samples by placing in a boiling water bath for 5 min.
3. Load the samples on to the SDS-polyacrylamide gel and electrophorese at an appropriate voltage (check the manufacturer's guidelines; usually 100 V) until the sample has reached the interface with the separating gel, then at 160 V thereafter until the dye reaches bottom of the gel. The total running time for a typical mini-gel is 1.5 h.
4. Remove the gel from the glass plates and fix it by incubating twice for 10 min in fixing solution at room temperature with gentle agitation.
5. If [^{14}C]methylated protein molecular weight standards have been used, incubate the gel for 10 min with Amplify fluorographic agent at room temperature with gentle agitation. This step, which results in the impregnation of the gel with the fluorographic agent, is necessary to detect ^{14}C emissions efficiently.
6. Place the gel on a piece of Whatman 3MM paper and dry it on a gel drier. Expose the dried gel to X-ray film. Suitable exposure times are usually 2–24 h.

3.2 Identification of transcription factors by South-Western blotting

In this method, a cell-free extract containing the DNA-binding protein of interest is first electrophoresed on an SDS-polyacrylamide gel to resolve protein species according to their relative molecular mass. Since the strength of the final signal obtained by this technique is proportional to the quantity of protein run on the gel, best results are obtained by running the maximum amount of extract that does not overload the gel. In the first instance, this must be determined empirically, usually by running a number of lanes containing varying amounts of extract. For 0.5–1 mm thick protein minigels (see *Protocol 4*), this is usually 30–150 μg of crude cell extract protein per gel lane. The proteins in the electrophoresed gel are then electrophoretically transferred on to a nitrocellulose membrane (Western blotting; ref. 19) and subsequently renatured on the membrane by treatment with and then withdrawal of protein denaturants such as guanidinium chloride or urea. (In some protocols, the denaturation step is omitted; ref. 20.) The nitrocellulose membrane is finally probed with a radiolabelled DNA fragment bearing the binding site for the transcription factor. After washing off non-specifically bound DNA, the relative molecular mass of the protein is determined following exposure to X-ray film. Failure of this technique may indicate either that the DNA-binding protein of interest has not been efficiently renatured, or that greater than one polypeptide species is required to reconstitute the binding activity.

3.2.1 Electrophoretic blotting of protein extracts

The method described in *Protocol 5* is designed for use with the BioRad Mini Protean II system, but is applicable to other types of electrophoresis apparatus. If another system is used, the voltages and currents chosen should correspond to those suggested by the manufacturer. In order to estimate the molecular mass of proteins detected by this method, one should also electrophorese protein size standards in the gel. Ideal for this purpose are pre-stained marker proteins (e.g. Rainbow markers; Amersham), which can be observed following transfer to the nitrocellulose membrane. Alternatively, unstained markers can be visualized by treating the nitrocellulose membrane with the dye Ponceau S (*Protocol 5*).

Protocol 5. Electrophoretic blotting of proteins

Equipment and reagents

- Apparatus and power supply for SDS-PAGE of proteins
- High power supply and apparatus for electrophoretic transfer of proteins
- Prestained protein molecular weight standards (e.g. Rainbow markers; Amersham)
- Nitrocellulose filter membranes (Schleicher & Schuell BA85, 0.45 μm) and Whatman 3MM paper
- Ponceau S solution (0.5% Ponceau S in 1% acetic acid)
- Western transfer buffer for proteins 20–200 kDa (50 mM Tris base, 380 mM glycine, 0.1% SDS, 20% methanol). Do *not* adjust the pH by the addition of acid or alkali
- An alternative Western transfer buffer for use when proteins are <80 kDa (25 mM Tris base, 190 mM glycine, 20% methanol)
- Methylene blue solution (0.1% methylene blue in 20% glycerol)
- Tris-buffered saline–Tween (TBST) buffer (10 mM Tris–HCl pH 7.9, 150 mM NaCl, 0.05% Tween 20).

Method

1. Resolve the proteins and molecular weight standards on an SDS-polyacrylamide gel (see *Protocols 3* and *4*).
2. (Optional.) When the bromophenol blue dye is approximately 1 mm from the bottom of the gel, disconnect the power supply and apply a small quantity (1–5 μl) of methylene blue solution to each lane. Reconnect the power supply and electrophorese for 1 min. During this time the methylene blue dye runs into the stacking gel, marking the positions of the lanes.
3. Disconnect the power supply, take the gel out of the apparatus, turn it over and put it on the bench so that lanes 1, 2, 3, 4, 5. . . are orientated from right to left. (This way, when the proteins have been transferred to nitrocellulose, the lane order will be increasing numerically left to right.) Remove the top gel plate.
4. Prepare an appropriate volume of Western transfer buffer. Pour two-thirds of this into the Western transfer apparatus, and the remaining one-third into a dish.

5. Cut two pieces of Whatman 3MM paper each to a size a little larger than the gel. Wet them by floating them on water and then briefly soak them in Western transfer buffer. Wearing gloves, cut a piece of nitrocellulose slightly larger than the gel and soak this in water then transfer buffer.
6. Wearing gloves, assemble a Western blot sandwich in the following order (see *Figure 3*):
   - plastic former facing the negative electrode
   - one fibre pad
   - one piece of prepared Whatman 3MM paper
   - the gel
   - the piece of prepared nitrocellulose
   - the second piece of Whatman 3MM paper
   - the second fibre pad
   - the plastic former facing the positive electrode.

   After putting each layer in place, gently smooth with gloved fingers to remove any air bubbles trapped between the layers.
7. Close the sandwich and place it in the electrophoresis tank. Transfer the proteins on to the membrane at 100 V (250 mA) for 1 h at room temperature, or at 30 V (40 mA) overnight at 4°C. The best protein transfer is usually achieved by longer transfer times.
8. When transfer is complete, disconnect the apparatus from the power supply, disassemble the sandwich, and place the Whatman sheets (with gel and nitrocellulose inside) on the bench. Remove the top Whatman sheet and cut the nitrocellulose to the size of the gel with a clean, sharp razor blade. Gently peel the nitrocellulose off the gel and wash the membrane briefly with TBST buffer.
9. If unstained marker proteins have been co-electrophoresed, stain them by incubating the blot for 5 min in Ponceau S solution immediately after step 8. Wash the membrane in water for 2 min and then mark the location of the marker proteins with indelible ink.

3.2.2 Probing protein blots with radiolabelled DNA

The DNA chosen for probing protein blots should contain a binding site for only the protein of interest. This can be achieved by using a carefully selected promoter DNA fragment or a synthetic double-stranded DNA oligonucleotide (*Protocol 6*). When using oligonucleotide probes, it should be noted that, although the minimal consensus recognition sequence for a transcription factor may only be 6–10 bp, efficient DNA binding usually requires additional DNA sequence on each side of the recognition element. It is therefore advisable to use an oligonucleotide in which the binding site is flanked by at

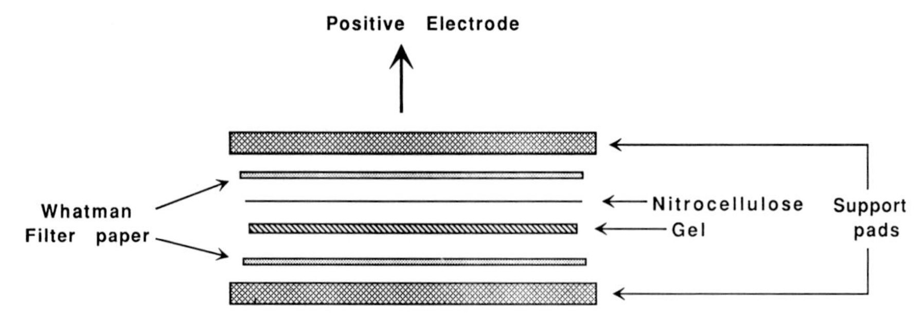

Figure 3. Assembly of sandwich for electrophoretically transferring proteins from an SDS-polyacrylamide gel to a nitrocellulose membrane.

least 10 bp of non-specific sequence on either side. For optimal sensitivity, the probes used should be of as high a specific activity as possible. Labelling of promoter fragments is best achieved by nick-translation or by random oligonucleotide priming (see Chapter 1, *Protocol 2* and refs 21, 22) whereas oligonucleotide labelling is best achieved by incubation with [γ-^{32}P]ATP and T4 polynucleotide kinase. Because of the presence of non-specific DNA binding proteins in cell extracts, an important control in these experiments is to include a control DNA probe that does not bear a binding site for the protein of interest (23).

Protocol 6. Probing protein blots with radiolabelled oligonucleotides

Equipment and reagents

- Nitrocellulose blot bearing the transferred proteins (see *Protocol 5*)
- ^{32}P-labelled double-stranded DNA probe bearing the binding site of interest
- Z′ buffer (25 mM Hepes-KOH pH 7.6, 12.5 mM $MgCl_2$, 20% glycerol, 0.1% Nonidet P40, 0.1 M KCl, 10 μM $ZnSO_4$, 1 mM DTT). Finally adjust the pH to pH 7.6 with KOH. Add the DTT from a 1 M stock solution just before use
- Blocking buffer [Z′ buffer containing 3% non-fat dried milk (e.g. Cadbury's Marvel, or Carnation)]
- Z′ buffer containing 6 M guanidinium chloride
- Binding buffer (Z′ buffer containing 0.25% non-fat dried milk)
- TBST buffer (10 mM Tris–HCl pH 7.9, 150 mM NaCl, 0.05% Tween 20).

Method

Perform all steps at 0–4°C.

1. After the electrophoretic transfer of the proteins (*Protocol 5*), place the nitrocellulose membrane in a plastic dish and add a sufficiently large volume of Z′ buffer containing 6 M guanidinium chloride to cover the gel

fully (usually 20–30 ml). Incubate with gentle rocking or shaking for 10 min.

2. Decant the solution and replace it with an equal volume of the same buffer. Incubate for 10 min with gentle rocking or shaking.
3. Decant the solution into a measuring cylinder and mix with this solution an equal volume of Z′ buffer (without guanidinium chloride). Add half of this solution to the filter and incubate for 10 min with gentle rocking or shaking.
4. Repeat step 3 five more times; the guanidinium chloride concentration will be gradually reduced to approximately 0.1 M.
5. Wash the membrane twice with Z′ buffer lacking guanidinium chloride (5 min each time), then transfer the membrane to a fresh dish.
6. Incubate the membrane for 30 min in blocking buffer to block non-specific binding sites on the membrane, then wash for 5 min with binding buffer.
7. Replace the buffer with binding buffer containing the ^{32}P-radiolabelled double-stranded DNA probe bearing the binding site of interest. To make the probe concentration as high as possible, the volume of buffer should be the minimum needed to cover the membrane (smallest volumes are achieved if the gel is sealed in a plastic bag). Incubate for 30 min. For some proteins, incubation at room temperature may increase the extent of DNA binding.
8. Carefully decant the radioactive probe solution and dispose of it safely. Wash the membrane three times with Z′ buffer for a total time of 15 min.
9. (Optional.) Wash three times with Z′ buffer containing 0.2–0.3 M KCl. This higher salt wash can reduce non-specific binding of probe to the membrane, and can dissociate complexes between the probe and weakly interacting proteins. However, the salt concentration in this wash must not be sufficient to dissociate complexes between the probe and the protein of interest.
10. Place the nitrocellulose between two pieces of plastic cling film and expose to X-ray film. If non-radioactive markers have been used, the edges of the cling film should be marked with pieces of tape labelled with radioactive or fluorescent ink. These marks allow the autoradiogram to be aligned with the protein size markers on the filter (unstained markers detected by Ponceau S, or prestained markers; see *Protocol 5*). If radioactive protein size markers have been used (see *Protocol 4*) the position of these will be apparent on the X-ray film without the need to use the radioactive or fluorescent ink procedure.

4. Purification of sequence-specific transcription factors using DNA binding site affinity chromatography

4.1 Introduction

To characterize a sequence-specific transcription factor biochemically, it is necessary to obtain it in pure form. Although this may be achieved by isolating its cDNA clone and then expressing this, the most direct approach is to purify the factor from cell-free extracts. In addition to being useful in biochemical studies, the purified transcription factor can also, for example, be employed to generate antibodies against the factor and to clone its corresponding cDNA (see Section 7). Initial attempts to purify sequence-specific transcription factors using conventional biochemical approaches were hampered by the fact that these proteins are present at very low levels in the cell. However, the development of sequence-specific DNA affinity chromatograph now makes their purification relatively straightforward (14, 24–26). This procedure takes advantage of the highly specific DNA-binding properties of transcription factor proteins; a transcription factor is selectively retained on a column consisting of an insoluble support bearing synthetic oligonucleotides that contain binding sites for this protein. The basic procedure for purifying a transcription factor is as follows (see *Figure 4*):

(i) a crude cell-free extract is generated;
(ii) the extract is fractionated to separate transcription factors from nucleases;
(iii) this material is passed over a DNA affinity resin and the unbound proteins are removed by washing;
(iv) the purified transcription factor is dissociated from the column by washing the column with elevated salt concentrations.

4.2 Preparation of cell-free extracts

Many types of cell-free extract have been used successfully to purify transcription factors. When choosing the source for the transcription factor of interest, two main considerations should be borne in mind. First and foremost is the consideration of cost; since transcription factors are not generally abundant proteins, large amounts of cells or tissue must be used to generate substantial (>μg) quantities of the factor. In general terms, a minimum of several grams of tissue or packed tissue culture cells is required. Second, the source should have already been demonstrated to contain the factor of interest. Indeed, if alternative sources exist, it is often worthwhile determining by quantitative DNA binding assays such as the gel mobility shift assay which source contains the factor in the greatest quantities and then use this.

In many instances the source of a transcription factor is a nuclear extract in

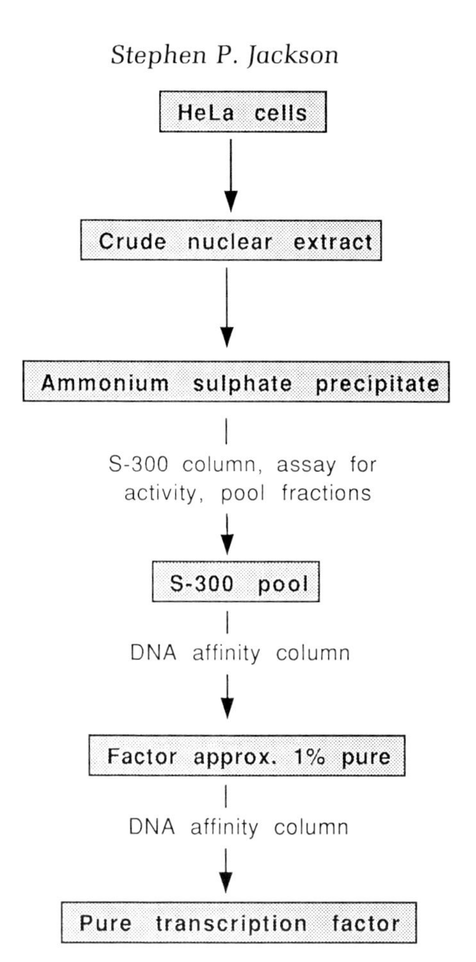

Figure 4. Standard transcription factor purification scheme. HeLa cells are harvested from a suspension culture by centrifugation, washed, and disrupted with a Dounce homogenizer. Nuclei are then collected by centrifugation and nuclear proteins are extracted with high salt buffer. Proteins are precipitated from the crude nuclear extract by the addition of ammonium sulphate, dialysed, then subjected to gel filtration on Sephacryl S-300. Column fractions are tested for specific DNA binding activity by the gel mobility shift assay and those containing the factor of interest are pooled. After the addition of non-specific competitor DNA, the extract is passed over a sequence-specific DNA affinity column. Unbound proteins are washed off and the factor of interest is eluted using a salt gradient. Usually, the factor is approximately 1% pure at this stage. For further purification, the salt concentration of the partially pure fraction is adjusted to 0.1 M. The factor is then mixed with additional non-specific competitor DNA and subjected to another round of DNA affinity chromatography. The factor preparation is generally 50–95% pure at this stage. If further purification is desired, a third DNA affinity chromatography step can be employed.

which its activity has been demonstrated by *in vitro* transcription assays. For detailed discussions on the generation of transcriptional extracts, the reader is directed to Chapter 3. *Protocols 7* and *8* describe a procedure for generating nuclear transcription extracts from human HeLa tissue culture cells grown in suspension culture. This has been the starting point in the purification of a great many transcriptional regulatory proteins (6, 24, 27). The basic strategy for preparing nuclear extract is as follows:

(i) HeLa cells are grown and harvested by centrifugation;
(ii) the cells are swollen by incubation in a hypotonic buffer, then disrupted in a Dounce homogenizer;
(iii) nuclei are collected by centrifugation;
(iv) nuclei are incubated at elevated salt concentrations. During this step, sequence-specific transcription factors dissociate from the DNA and leak out of the nuclei.

Note that *Protocol 8* uses 1 volume of high salt buffer to extract the nuclei. However, if the extract is to be used only for factor purification purposes (i.e. not as an extract for *in vitro* transcription), then the use of larger volumes (up to 4 vol.) of high salt buffer can sometimes lead to higher factor yield.

In order to generate extracts with maximal activity, it is necessary that the cells are at a density of 4–7 $\times$ 10^5/ml and are growing exponentially prior to harvesting. Since the nuclei can be stored at −70°C for considerable periods of time, it is usual to harvest the cells in batches and accumulate nuclei over several days or weeks before pooling them and going on to factor purification. For example, approximately half of a HeLa cell suspension culture can be harvested each day. The remaining half can be diluted and incubated further to produce fresh cells the following day. In this way, even relatively small 2 litre or 4 litre cultures can rapidly generate large amounts of material. The protocol given for extraction of nuclear proteins (*Protocol 8*) was devised for nuclei derived from 6 litres of HeLa culture, although it can be scaled down accordingly. Six litres of HeLa culture generally yields approximately 9 g of cells (wet weight) and results in a crude nuclear extract containing approximately 200 mg of total protein. Note that all buffers used in extract preparation from HeLa cells should contain DTT and the protease inhibitors phenylmethylsulphonyl fluoride (PMSF) and sodium metabisulphite. For other cell types, it may be necessary also to include a variety of other protease inhibitors (see Chapter 3).

Protocol 7. Preparation of HeLa cell nuclei

Equipment and reagents

- Wheaton Dounce homogenizer fitted with a B-type pestle
- HeLa cell culture grown to 4–7 $\times$ 10^5 cells/ml
- PBSM (prepared by dissolving 8.0 g NaCl, 0.2 g KCl, 0.92 g anhydrous Na_2HPO_4, 0.2 g KH_2PO_4, and 1 g $MgCl_2 \cdot 6\ H_2O$ in distilled water to a final volume of 1 litre)
- 10 $\times$ buffer A stock (100 mM Hepes-KOH pH 7.6, 15 mM $MgCl_2$, 100 mM KCl)
  Generally, the buffer is made using a 0.5 M Hepes-KOH pH 7.6 stock solution, then the final mixture is adjusted to pH 7.6 with KOH]
- 0.1 M PMSF in 2-propanol; store at −20°C
- 1 M DTT; store at −20°C

- 0.1 M sodium metabisulphite; prepare fresh before use
- 1 × buffer A: dilute 10 × buffer A stock with 9 vol. distilled water; add DTT, PMSF, and sodium metabisulphite (from stocks) to 1 mM each, just before use
- Buffer B (0.3 M Hepes-KOH pH 7.6, 1.4 M KCl, 30 mM $MgCl_2$).

Method

Perform all steps at 0–4°C

1. Harvest the HeLa cells (grown to 4–7 × 10^5/ml) by centrifuging at 4600 *g* for 15 min.
2. Decant the supernatant and gently resuspend the cells with a wide-bore pipette in 200 ml PBSM. Centrifuge as in step 1.
3. Decant the supernatant and resuspend the cells in 50 ml PBSM. Distribute into 50 ml disposable plastic test tubes and centrifuge at 2800 *g* for 10 min.
4. Decant the supernatant and estimate the packed cell volume (PCV). Add 5 times the PCV of 1 × buffer A to the cells and resuspend by gentle pipetting. Incubate on ice for 20 min.
5. Centrifuge at 2800 *g* for 10 min. Decant the supernatant and resuspend the cells in 2 PCV of 1 × buffer A.
6. Transfer the cell suspension into the Dounce homogenizer and disrupt the cells by 10 strokes with the B pestle. Since many transcription factors quickly leak out of the nuclei of disrupted cells, it is important to proceed immediately to steps 7 and 8.
7. Centrifuge at 2800 *g* for 10 min. During this step, the nuclei pellet, whereas the cytoplasmic constituents remain in the supernatant.
8. Pipette off the upper cytoplasmic phase and add 0.11 vol. buffer B. Mix and freeze in liquid nitrogen. Also freeze the nuclear pellet and store it at −70°C until required (storage can be for many months without appreciable loss of transcriptional activity of the resulting extract).

Protocol 8. Extraction of nuclear proteins

Reagents

- Solid ammonium sulphate
- Fresh or frozen HeLa cell nuclei (see *Protocol 7*)
- Stocks of 0.1 M PMSF in 2-propanol, 1 M DTT, and 0.1 M sodium metabisulphite; see *Protocol 7* reagent list for preparation
- Buffer C (20 mM Hepes pH 7.6, 25% glycerol, 0.42 M NaCl, 5 mM $MgCl_2$, 0.2 mM EDTA) [Make by using 0.5 M Hepes-KOH pH 7.6 stock solution, then adjust the pH of the final mixture to 7.6 with KOH. Add DTT, PMSF, and sodium metabisulphite (from stocks) to 1 mM each, just before use]
- Buffer D (20 mM Hepes-KOH pH 7.6, 20% glycerol, 50 mM KCl, 5 mM $MgCl_2$, 0.2 mM EDTA) [Add DTT, PMSF, and sodium metabisulphite (from stocks) to 1 mM each, just before use]
- 10 M KOH.

Protocol 8. *Continued*

Method

This protocol was devised for nuclei derived from 6 litres of HeLa cell culture, but can be scaled down for smaller culture volumes. Perform all steps at 0–4 °C.

1. Resuspend the nuclei (thawed or freshly prepared) from a total of 6 litres of HeLa cell culture in an equal volume of buffer C. Gently stir the suspension for 30 min on ice with a magnetic stirrer (do not stir so fast as to cause frothing, which denatures proteins).
2. Remove the nuclear debris (including chromatin and membrane components) by centrifugation in a fixed angle rotor at 21 000 *g* for 30 min (e.g. 15 000 r.p.m. in a Sorvall SS34 rotor). Carefully decant the supernatant into a small beaker and measure its volume.[a]
3. Grind an appropriate quantity of ammonium sulphate into a fine powder with a mortar and pestle or motorized grinder. Whilst stirring the extract on a magnetic stirrer, slowly add, over a period of 10 min, 0.33 g ammonium sulphate powder for every ml of extract. Then, slowly add 4 μl 10 M KOH for every gram of ammonium sulphate added. Stir for a further 45 min.
4. Collect the protein precipitate by centrifugation in the fixed angle rotor (e.g. Sorvall SS-34) at 21 000 *g* for 15 min. Discard the supernatant. (Protein pellets can be stored overnight at 4 °C if desired.)
5. Resuspend the protein pellet in 1 ml of buffer D for every litre of HeLa cells used to make the extract. Effective resuspension is achieved by gently pipetting up and down with a wide-bore Pasteur pipette.
6. Dialyse the protein extract four times for 1 h against 500 ml buffer D.
7. Remove insoluble material by centrifugation at 10 000 *g* in the fixed angle rotor for 10 min. The supernatant fraction is now ready for use in *in vitro* transcription reactions, or for preliminary fractionation. Extracts can be frozen in liquid nitrogen and stored at −70 °C indefinitely.

[a] In some cases, the transcription factor of interest can be purified directly by lectin affinity chromatography at this stage without the need for ammonium sulphate fractionation (see section 4.3.3).

4.3 Preliminary fractionation of crude nuclear extracts

Before proceeding to DNA affinity chromatography, it is usually desirable to pre-fractionate the extract. In addition to purifying the DNA binding protein of interest somewhat, this also separates it from nucleases that might otherwise destroy the DNA affinity resin. However, in cases where the factor of interest does not require magnesium ions to bind to DNA, one can apply the crude extract directly to the DNA affinity column. The reason for this is that most DNases do not function in the absence of magnesium ions. In this case, resuspend the ammonium sulphate pellet generated in *Protocol 8* (step 4) in a buffer that lacks magnesium and then dialyse the mixture extensively against this buffer. However, since pre-fractionation is relatively straightforward, it is normally advisable to use a preliminary column before affinity chromatography in any case. Described below in sections 4.3.1–4.3.3 are three of the most common preliminary fractionation procedures; Sephacryl S-300 gel-filtration chromatography, heparin-Sepharose affinity chromatography, and lectin affinity chromatography.

4.3.1 Sephacryl S-300 gel-filtration chromatography

Gel-filtration columns separate proteins on the basis of their size. Since proteins often exist as complexes in crude extracts, the elution profile does not always give an accurate indication of their relative molecular mass. A commonly used gel-filtration matrix is Sephacryl S-300 superfine (Pharmacia). On this resin, many transcription factors, such as Sp1, CTF, AP-1, and AP-2, elute at V_e/V_o (the ratio of elution volume to void volume) of about 1.35–1.4 (*Figure 5*; ref. 28).

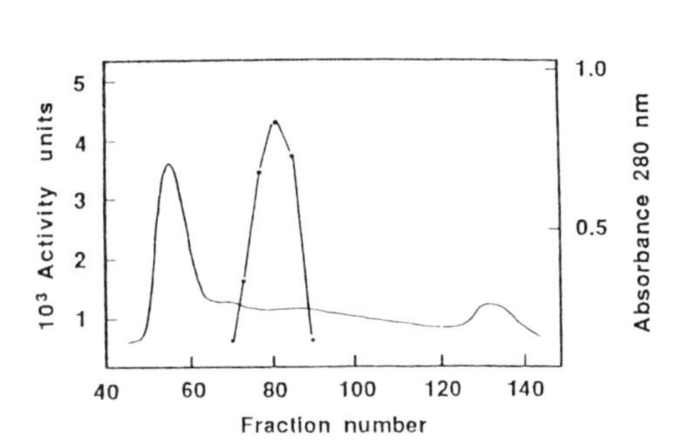

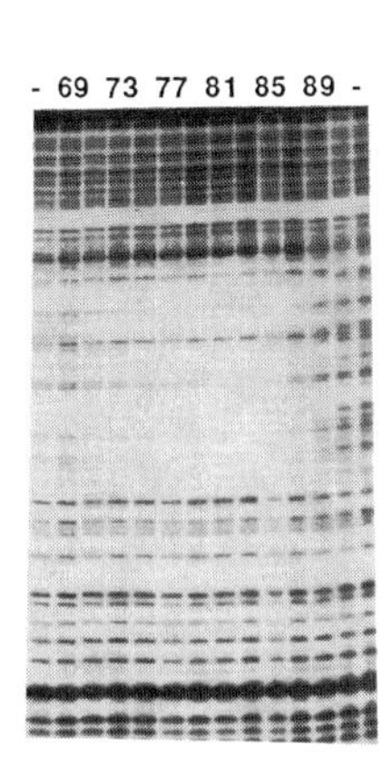

Figure 5. Sephacryl S-300 column chromatography of Sp1. Nuclear extract derived from 60 g (wet weight) of HeLa cells was precipitated by 53% saturated ammonium sulphate then resuspended in TM buffer (50 mM Tris–HCl pH 7.9, 12.5 mM $MgCl_2$, 1 mM EDTA, 1 mM DTT, 20% glycerol) to a final protein concentration of about 30 mg/ml. The soluble protein extract was then applied to a Sephacryl S-300 column (1 litre bed volume; column dimensions 51 cm × 5 cm) equilibrated with TM buffer containing 0.1 M KCl. Protein elution was monitored by A_{280} (left panel; continuous line) and Sp1 activity (activity peak indicated in left hand panel) was determined by DNase I footprint analysis of column fractions (right hand panel). The ratio of elution volume to void volume (V_e/V_o) for Sp1 was approximately 1.4. Fractions containing Sp1 were pooled (40 mg protein in 40–60 ml). These data are from Briggs *et al.* (ref. 28) with permission.

To carry out the fractionation, connect the column to a buffer reservoir via a peristaltic pump and collect the output using a fraction collector with a drop-counter. The size of the fractions collected should be 0.5–1% of the column volume. Ideally, the column should be connected to a flow-through UV-monitor (280 nm) with a chart recorder that has an event marker and is connected to the fraction collector so that fractions can be identified on the printout. Alternatively, the UV absorbance of fractions can be determined manually using a spectrophotometer after the column has been run.

There are a number of general rules to be borne in mind when preparing and running the column.

- The column should be large enough that the sample applied is at most 3% of the bed volume.
- To aid separation, the column should be long and relatively thin (for example, a column approximately 2 cm in diameter should be used for 200 ml of resin).
- 'Fines' should be removed from the Sepharose before use according to the manufacturer's instructions.
- The column should be packed in one step by gravity flow.
- Column flow rates should not be greater than 0.1 column volume per hour.
- All procedures must be performed at 0–4°C.

Protocol 9. Sephacryl S-300 gel-filtration

Equipment and reagents

- Glass column of the appropriate dimensions (see section 4.3.1 for discussion)
- Peristaltic pump and fraction collector fitted with (optional) flow-through UV monitor linked to a chart recorder
- Sephacryl S-300 superfine (Pharmacia)
- Crude nuclear extract (see *Protocol 8*)
- Gel filtration column buffer e.g. Buffer D (see *Protocol 8* reagents for preparation).[a]

Method

Perform all steps at 0–4°C.

1. Equilibrate the column by passing through three column volumes of buffer D.[a]
2. Disconnect the upper buffer reservoir and allow the buffer level to fall until it just reaches the top of the column. Apply the nuclear extract carefully so as to avoid disturbing the surface of the Sephacryl bed. Allow the extract to enter the column by gravity flow. When the meniscus reaches the gel surface, add 3 ml buffer D and allow this to enter the

column. Then add buffer D to fill the column and reconnect the upper buffer reservoir.

3. Collect fractions (overnight runs are often convenient).
4. Remove small aliquots of each fraction and assay for DNA-binding activity by the gel mobility shift assay (or by DNase I footprinting). Freeze the remainder of each fraction in liquid nitrogen and store at −70°C.
5. When those fractions containing the protein of interest have been identified, pool them and subject the pooled preparation to DNA affinity chromatography (see Section 4.4). If the elution profile of the desired protein from the column has already been established and no assay is required, the appropriate fractions can be pooled and applied to affinity columns directly, thus avoiding the freeze/thaw cycle.

[a] A variety of buffers can be used. The protocol given here uses buffer D but other buffers are equally suitable e.g. TM buffer (50 mM Tris–HCl pH 7.6, 12.5 mM $MgCl_2$, 1 mM EDTA, 20% glycerol) containing 0.1 M KCl.

4.3.2 Heparin-agarose affinity chromatography

Heparin is a sulphated mucopolysaccharide with a high negative charge that binds a wide variety of DNA-binding proteins. Heparin can be purchased from a number of manufacturers already attached to insoluble supports such as agarose or Sepharose. Alternatively, this matrix can be prepared in the laboratory (29, 30). Both heparin-agarose and heparin-Sepharose matrices can be used as described in *Protocol 10*. In a normal fractionation procedure, crude extracts are passed through a heparin affinity column under fairly low salt concentrations. After washing off unbound material, bound proteins are eluted with buffers containing elevated salt concentrations. Although this is often done stepwise, the use of linear salt gradients may result in higher factor purity since it gives greater resolution of the eluted proteins. After heparin affinity chromatography, the eluate fractions are assayed for the DNA binding activity of interest, usually by the gel mobility shift assay. Once identified, the appropriate fractions are pooled and subjected to DNA affinity chromatography after having their salt concentration adjusted to 0.05–0.1 M by the addition of buffer lacking KCl. The binding capacity of most commercially available heparin affinity matrices (approx. 10 mg heparin/ml hydrated matrix) is about 10 mg of protein per ml of hydrated gel. In general terms, one should use at least 1 ml of matrix for every 10 mg of total protein to be fractionated. Most transcription factors elute between 0.2–0.4 M KCl from heparin columns (28, 31) but this should be determined empirically for each factor under study.

Protocol 10. Heparin-agarose affinity chromatography

Equipment and reagents

- Column of appropriate dimensions
- Peristaltic pump, fraction collector and (optional) flow-through UV monitor with chart recorder
- Heparin-agarose (e.g. Sigma) or heparin-Sepharose (Pharmacia)
- Buffer D: see *Protocol 8* reagents
- Crude nuclear extract (see *Protocol 8*)
- Buffer D containing higher salt concentrations: composition as for buffer D, except with higher concentrations of KCl appropriate to the elution regime chosen (see step 4 below).

Method

Perform all steps at 0–4°C.

1. Pour an appropriately sized heparin-agarose column. A 15 ml column is usually sufficient to fractionate extract derived from 6 litres of HeLa cell culture. Connect the column to a peristaltic pump and equilibrate with buffer D containing 0.1 M KCl.
2. Apply the nuclear extract at a flow rate of approximately 1.0–1.5 column volumes per hour.
3. Wash the column with buffer D until the A_{280} decreases to the same value as that of the input buffer (this usually requires 2–3 column volumes).
4. Elute the bound proteins with either a 0.1–1.0 M KCl buffer gradient or in a series of steps (e.g. 0.2, 0.4, 0.7, and 1.0 M KCl steps). For step gradients, ensure that the previous elution is complete before applying the next step.
5. Identify which fraction(s) contain the protein of interest by the gel mobility shift assay.
6. Pool the appropriate fractions and adjust the KCl concentration to 0.1 M by dialysis or by mixing with a buffer lacking KCl.
7. Subject the material to DNA affinity chromatography (see Section 4.4).

4.3.3 Purification of transcription factors by lectin affinity chromatography

A number of transcription factors are glycoproteins bearing *N*-acetylglucosamine (GlcNAc) moieties (32). This feature confers upon some of these factors an ability to bind tightly to the lectin, wheat germ agglutinin (WGA). This avid binding has been used to devise purification schemes that employ WGA affinity chromatography (6, 31). If a transcription factor is glycosylated sufficiently to allow WGA binding, then a lectin affinity chromatography-based purification scheme (see *Protocol 11*) will probably be the purification method of choice. This is because it usually yields larger quantities of transcription factor than conventional purification schemes, and it is also more rapid and much easier to perform.

The method for generating the cell-free extract for lectin affinity chromatography can be the same as that employed for the other two pre-fractionation procedures given above (for extract preparation, see Section 4.2). However, since some glycosylated factors bind to WGA even at high salt concentrations, it is often possible to load the high salt nuclear extract (*Protocol 8*, step 2) directly on to the WGA column (6). Thus, the ammonium sulphate precipitation step can be omitted. For some factors, however, significant binding to the lectin column occurs only under lower salt conditions. In these cases, the extract should be precipitated with ammonium sulphate and dialysed as usual (*Protocol 8*). The general strategy for using lectin affinity chromatography is as follows:

(i) The extract containing the factor is loaded onto the WGA column and unbound (non-glycosylated) proteins are washed off.

(ii) The bound glycosylated proteins are then eluted using a buffer containing the competitor sugar, GlcNAc. This can result in an approximately 100-fold purification of the transcription factor of interest. In certain cases where the transcription factor binds extremely tightly to the lectin resin and is not eluted efficiently by GlcNAc, more efficient elution may be achieved by eluting with 10 mM triacetylchitotriose.

(iii) Next, the eluate is subjected to a single round of DNA affinity chromatography to yield pure product. This compares to the two or three successive rounds of DNA affinity chromatography that are usually required if either of the other two pre-fractionation methods are used (see Section 4.4 below). *Figure 6* shows a typical purification of a transcription factor on a WGA column. It is worth noting that, since components of the general transcriptional apparatus are not retained on WGA-agarose, the WGA column flow-through can be used as a source of general factors depleted of glycosylated sequence-specific factors (6).

Protocol 11. WGA affinity chromatography

Equipment and reagents

- Small glass or plastic columns (e.g. BioRad Econo-columns)
- Crude nuclear extract (see *Protocol 8*)
- WGA-agarose (>5 mg lectin/ml gel). Available from many manufacturers, e.g. Vector labs, Sigma, and Pharmacia
- *N*-Acetyl-D-glucosamine (Sigma)
- Buffer D (see *Protocol 8* reagents)
- Z′ buffer containing 0.1 M KCl (25 mM Hepes-KOH pH 7.6, 0.1 M KCl, 12.5 mM $MgCl_2$, 10 μM $ZnSO_4$, 20% glycerol, 0.1% Nonidet-P40, 1 mM DTT). [Adjust the pH to pH 7.6 with KOH. Add the DTT freshly before use]
- Elution buffer (Z′ buffer containing 0.3 M *N*-acetylglucosamine).

Protocol 11. *Continued*

Method

Perform all steps at 0–4°C.

1. Equilibrate a 10 ml plastic column containing 1 ml of WGA-agarose by washing with 100 ml of buffer in which the protein extract is dissolved (usually buffer D). This wash is essential to remove the competing sugar GlcNAc which is present in the storage buffer for WGA-agarose.
2. Apply the crude nuclear extract (derived from 6 litres of HeLa cell culture) at a flow rate of approximately 15 ml/h. Collect the flow-through and use it, if desired, as a source for general transcription factors and/or sequence-specific factors that do not bind to the WGA-agarose.
3. Wash the column with 4 ml of the buffer used in step 1, then four times with 4 ml of Z′ buffer containing 0.1 M KCl.
4. Elute GlcNAc-bearing proteins with two batches of 5 ml elution buffer. Collect all 10 ml of the eluate.
5. Add non-specific competitor DNA[a] and further purify the factor by DNA affinity chromatography (see Section 4.4.3). Since the factor is already partially purified, it needs to be passed through a DNA affinity column only once. Therefore apply it to 1 ml of the sequence-specific DNA affinity matrix and elute as described in *Protocol 14*, step 7.
6. Regenerate the WGA column by washing it with 20 ml of elution buffer containing 0.02% NaN_3. Turn off the column, add a further 5 ml of this buffer, and store the washed column at 4°C. Columns can be reused at least four times over a period of 6 months without substantial loss of activity.

[a] For Sp1, 50 μl of 10 A_{260} units/ml of poly(dI-dC)·poly (dI-dC) is optimal.

4.4 DNA binding site affinity chromatography

In this procedure, an extract containing the DNA-binding protein of interest is passed over a matrix bearing high affinity binding sites for this factor. Before applying the extract to the column, it must first be mixed with non-specific competitor DNA for which the factor of interest has very low affinity. Upon chromatography, the factor binds to the DNA affinity matrix rather than the competitor DNA in solution. However, other proteins bound to the competitor DNA flow through the column. After washing off unbound material, the factor of interest is then eluted by increasing the salt concentration of the column buffer. Many transcription factors can be purified approximately 1000-fold by two sequential DNA affinity chromatography steps, with an overall recovery of approximately 30% (see *Figure 7*).

Because of its importance to the success of this technique, the sequence of the DNA binding site must be chosen carefully to represent the strongest

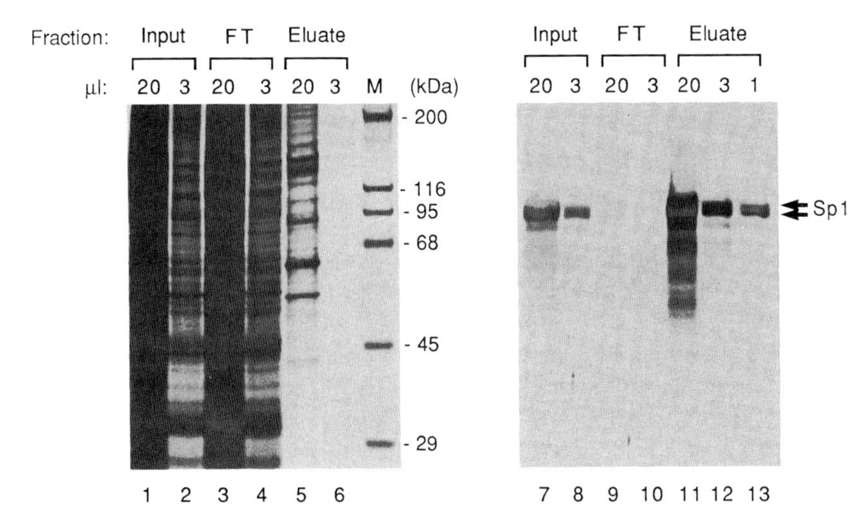

Figure 6. Fractionation of Sp1 by WGA affinity chromatography. Duplicate samples of fractions generated by WGA affinity chromatography of a crude HeLa cell nuclear extract were analysed on an SDS/8% polyacrylamide gel. The gel was then cut in half; one half (lanes 1–6) was stained with silver, whereas for the other half (lanes 7–13) proteins were transferred on to a nitrocellulose filter and Sp1 protein was detected by probing with an anti-Sp1 monoclonal antibody. Lanes 1 and 7 contain 20 μl (approximately 180 μg of total protein) of the initial nuclear extract (input) wheras lanes 2 and 8 contain 3 μl of the input fraction. Lanes 3 and 9 contain 20 μl (approximately 180 μg of total protein) of the material that flowed through the column (FT fraction) while lanes 4 and 10 contain 3 μl of the FT. Lanes 5 and 11 contain 20 μl (approximately 2.8 μg of total protein) of the material eluted from the WGA affinity matrix by 0.3 M GlcNAc (eluate); lanes 6 and 12 contain 3 μl of the eluate; and lane 13 contains 1 μl of the eluate. Lane M, protein standards (sizes in kDa). The locations of the Sp1 polypeptides are indicated by arrows. From lanes 7–13, it is estimated that ~95% of Sp1 in the input fraction is bound to, and then eluted from, the WGA-agarose. The total protein applied to the resin was ~270 mg and the total protein eluted from the WGA resin was ~1.4 mg. This figure is reproduced from Jackson and Tjian, 1989 (ref. 6).

possible binding site for the factor. Information about the ideal binding site sequence can be obtained by the gel mobility shift assay, DNase I footprinting experiments, mutational analysis, and by sequence comparison of homologous binding sites. Although DNA fragments generated from plasmid DNA have been used effectively, it is more common to employ synthetic double-stranded oligonucleotides. Oligonucleotides with lengths of 14–50 bp have been used successfully in DNA affinity columns. In some cases, different sequences that are recognized by a single factor have been used to generate two affinity matrices. The extract is chromatographed on the first column, then the eluate is applied to the second column. This approach has the advantage in that any other factors that bind to the flanking oligonucleotide sequences in the first column, will probably not bind to the second.

4.4.1 Preparation of oligonucleotides for DNA affinity chromatography

In this procedure (see *Protocol 12*), two complementary oligonucleotides are synthesized, purified, then annealed to one another. The oligonucleotides should be designed so that the resulting double-stranded oligonucleotide bears 5′ overhanging 'sticky' ends that promote efficient concatemerization. For example, the sequence 5′-GATC-3′ might be used as the 5′ terminus of each of the two complementary single-stranded oligonucleotides. The 5′ ends of the double-stranded oligonucleotide are then phosphorylated by polynucleotide kinase, enabling it to be concatemerized by DNA ligase treatment. Concatemerization minimizes steric hindrance of factor binding by the affinity matrix and may also sometimes be of benefit in cases where cooperativity in DNA binding occurs between molecules of the protein bound to adjacent sites on the DNA. Finally, the ligated concatemerized DNA is covalently attached to Sepharose, and this matrix is then used in DNA affinity chromatography.

Protocol 12. Preparation of double-stranded concatemerized oligonucleotides

Equipment and reagents

- Equipment for preparative denaturing gel electrophoresis (see Chapter 6, *Protocol 12*)
- Heating block or water baths for incubations at 15°C, 37°C, 65°C, and 90°C
- Vacuum desiccator
- T4 polynucleotide kinase (10 U/μl)
- T4 DNA ligase (1 U/μl: Boehringer Mannheim)
- 4 M LiCl
- 10 M ammonium acetate
- Two single-stranded complementary oligonucleotides bearing the DNA-binding site (see Section 4.4)
- Denaturing urea polyacrylamide gel (see Chapter 6, *Protocol 12*) in TBE buffer
- Non-denaturing 1.5% agarose gel containing ethidium bromide in TBE buffer (see Chapter 2, *Protocol 6*)
- Nucleotide mixture (20 mM ATP containing 20 mM DTT)
- 10 × polynucleotide kinase buffer (0.5 M Tris–HCl pH 7.5, 100 mM $MgCl_2$, 10 mM spermidine, 10 mM EDTA)
- 10 × ligation buffer (0.7 M Tris–HCl pH 7.5, 100 mM $MgCl_2$)
- Nucleotide mixture containing trace radiolabel (optional) (20 mM ATP, 20 mM DTT, 5 μCi [γ-^{32}P]ATP at 3000 Ci/mmol).

Method

1. Purify approximately 1 mg of each single-stranded oligonucleotide by electrophoresis on a denaturing urea polyacrylamide gel in TBE buffer. For oligonucleotides of 15–30 nucleotide (nt) residues, use a 16% polyacrylamide gel. Measure the oligonucleotide concentration in the final solution by determination of its A_{260}.

2. Mix together in a microcentrifuge tube approximately 400 μg each of the two complementary oligonucleotides, add 20 μl of 10 × polynucleotide kinase buffer and make the final reaction up to 150 μl with distilled water.
3. Incubate the tube at 90°C for 5 min in a heating block or water bath, then switch off the heating unit and allow the tube to cool slowly to room temperature. When cool, briefly spin in a microcentrifuge to collect water droplets from the inside of the tube lid.
4. Add 30 μl of nucleotide mixture containing trace radiolabel and mix well by gently pipetting up and down. Then add 20 μl (200 U) T4 polynucleotide kinase, mix and incubate at 37°C for 2 h. The small amount of radiolabel used here (optional) helps determine the coupling efficiency to the Sepharose (see below).
5. Incubate at 65°C for 10 min to inactivate the kinase, then extract once with phenol/chloroform (1:1) and once with chloroform alone.
6. Distribute the sample equally into two microcentrifuge tubes. To each add 50 μl 10 M ammonium acetate, 100 μl distilled water, and 0.75 ml of absolute ethanol, mix and incubate at −20°C for 15 min. Pellet the DNA in a microcentrifuge for 15 min, and briefly dry under vacuum.
7. Redissolve each DNA pellet in 200 μl distilled water and then add 25 μl 3 M sodium acetate pH 5.6, 2.5 μl 1 M $MgCl_2$, and 0.68 ml of absolute ethanol. Mix and incubate at −20°C for 15 min.
8. Pellet the DNA by centrifugation in a microcentrifuge for 15 min. Wash the DNA pellet with 75% ethanol and dry under vacuum.
9. Redissolve each DNA pellet in 65 μl distilled water and pool together. Add 20 μl of 10 × ligation buffer and 40 μl nucleotide mixture (lacking radiolabel). Mix well by gently pipetting up and down. Then add 10 μl T4 DNA ligase, mix, and incubate overnight at 15°C.
10. Analyse 1 μl of the ligation reaction by electrophoresis alongside DNA markers on a non-denaturing 1.5% agarose gel containing ethidium bromide in TBE buffer (Chapter 2, *Protocol 6*). The majority of the ligation-products should be at least 10mers. Unsatisfactory ligation can be caused by improper pairing of the oligonucleotide strands (examine the sequence for the possibility of hairpin-formation). Different ligation temperatures may solve this problem (the optimal incubation temperature can vary between 4°C and 30°C).
11. To the ligation reaction, add 66 μl 10 M ammonium acetate, then extract once with phenol and once with chloroform.
12. Add 266 μl propan-2-ol and incubate at −20°C for 15 min to precipitate the DNA. Pellet the DNA by centrifugation in a microcentrifuge for 15 min at 4°C and wash it twice with 1 ml of 70% ethanol.

Protocol 12. *Continued*

13. After drying under vacuum, redissolve the DNA in 100 μl distilled water. *Do not* dissolve the DNA in a buffer containing Tris, since Tris will interfere with subsequent coupling of the DNA to the matrix.

4.4.2 Preparation of the DNA binding site affinity matrix

Although alternative approaches are possible (14, 26, 33), the most commonly used method for generating a DNA affinity matrix is to covalently crosslink multimerized double-stranded oligonucleotides to Sepharose. In *Protocol 13* a method is presented for generating activated Sepharose that gives very high coupling efficiencies. However, in many cases, commercially available activated supports such as cyanogen-bromide activated Sepharose (Pharmacia) are quite satisfactory when used according to the manufacturer's instructions. *Note*: when generating the activated Sepharose as described in *Protocol 13*, the cyanogen bromide is highly toxic. *Therefore utmost care must be taken during this procedure*; all steps (including the weighing of the cyanogen bromide) must be performed in a fume hood with adequate airflow and all materials coming into contact with solution containing cyanogen bromide should be immersed in 1.0 M glycine in a large beaker in the fume hood and left overnight before removing from the fume hood.

Protocol 13. Coupling of oligonucleotides to a Sepharose matrix

Equipment and reagents

- Motorized rotating wheel
- Sintered glass funnel (60 ml, coarse)
- Sepharose CL2B (Pharmacia)
- Ligated oligonucleotides (see *Protocol 12*)
- 1 M glycine
- 1 M ethanolamine–HCl pH 8.0
- Cyanogen bromide (solid; Sigma)
- Dimethylformamide
- Column storage buffer (0.3 M NaCl, 10 mM Tris–HCl pH 7.5, 1 mM EDTA, 0.02% NaN_3)
- 5 M NaOH
- 10 mM potassium phosphate buffer pH 8.0
- 1 M potassium phosphate buffer pH 8.0
- 1 M KCl.

Method

All steps must be performed in a fume hood with adequate airflow, and all materials coming into contact with cyanogen bromide should be immersed in 1.0 M glycine in a large beaker in the fume hood and left overnight before discarding.

1. Decant 10 ml (settled volume) of the Sepharose CL2B into a sintered-glass funnel, then wash with 500 ml distilled water. Transfer the Sepharose in a 20 ml slurry into a 150 ml glass beaker with a magnetic stirrer. Keep the temperature of the slurry at 15°C by placing the beaker in a water bath.

2. While stirring gently, add 2 ml dimethylformamide containing 1.1 g freshly dissolved cyanogen bromide slowly over a period of 1 min. *Caution, cyanogen bromide is extremely toxic!*
3. Add 1.8 ml 5M NaOH slowly over a period of 10 min (approximately 30 μl every 10 sec).
4. Add 100 ml ice-cold distilled water, then transfer the slurry into a 60 ml coarse sintered-glass funnel.
5. Wash the matrix three times with 100 ml ice-cold distilled water, then with 100 ml ice-cold 10 mM potassium phosphate (pH 8.0). These washes can be speeded up by the application of slight suction. If suction is used, care must be taken not to suck the matrix dry since sucking air through it may destroy the reactive groups in the matrix.
6. With a glass rod, transfer the resin from the funnel into a 15 ml screw-capped plastic tube. Add 4 ml 10 mM potassium phosphate (pH 8.0) and the ligated oligonucleotides. After briefly mixing, immediately take out 50 μl buffer (this is to be used to determine the coupling efficiency of the DNA to the Sepharose).
7. Incubate the rest of the slurry on a rotating wheel at room temperature overnight (or for at least 16 h).
8. Transfer the slurry into a sintered-glass funnel. Collect the first few ml of buffer flowing through.
9. Wash the Sepharose twice with 100 ml of distilled water, then with 100 ml 1M ethanolamine–HCl pH 8.0.
10. Transfer the Sepharose to a 15 ml screw-capped tube, add 4 ml 1M ethanolamine–HCl pH 8.0, and incubate on a rotating wheel at room temperature for 4 h.
11. Transfer the slurry to a sintered-glass funnel and wash the resin successively with:

    100 ml 10 mM potassium phosphate buffer pH 8.0
    100 ml 1 M potassium phosphate buffer pH 8.0
    100 ml 1 M KCl
    100 ml distilled water
    100 ml column storage buffer.
12. Store the matrix at 4°C. In this form, it is stable for at least two years.
13. To assess the coupling efficiency of the DNA to the matrix, determine by scintillation counting the relative amount of radioactivity in the buffer before and after coupling (from steps 6 and 8). The incorporation is normally greater than 90%.

4.4.3 DNA affinity chromatography of transcription factors

Protocol 14 describes a procedure that works well for the purification of many sequence-specific transcription factors. For certain DNA-binding proteins, however, other conditions may be optimal. The parameter that differs most often from one factor to another is the choice of competitor DNA; changes in either the quantity or type of competitor can have dramatic effects on factor purification. As a rule, it is wise to use the competitor DNA that works well in DNA binding assays such as the gel mobility shift assay. Competitor DNAs that have been used successfully are synthetic double-stranded poly(dI-dC)·poly(dI-dC), poly(dA-dT)·poly(dA-dT) (both from Pharmacia) and natural DNAs from calf thymus, salmon sperm, or *E. coli*. To prepare the synthetic polymers for use as competitors, dissolve them in TEN buffer (10 mM Tris–HCl pH 7.5, 1 mM EDTA, 100 mM NaCl) to a concentration of 10 A_{260} units/ml, heat to 90°C for 10 min, then allow to cool down slowly to room temperature to renature. If the average length of these competitors exceeds 1 kb, reduce this by sonication.

When deciding on the quantity of competitor to use, a balance must be struck between two opposing considerations. First, sufficient competitor is required to sequester non-specific DNA binding proteins so that they do not attach to the affinity column. Second, if too much competitor DNA is used, then a large proportion of the sequence-specific DNA binding protein of interest will also flow through the column, since it will bind to the competitor DNA to some degree. If in doubt, it is advisable first to try lower amounts of competitor rather than higher. If analysis of the resulting fractions indicates that insufficient competitor was used, then it is easy to re-chromatograph the material generated from the first purification on another DNA affinity column. Although the conditions given in *Protocol 14* should work satisfactorily for most factors, the best purification for individual factors will be achieved by determining the optimal competitor quantities empirically. Finally, it is important to note that the optimal quantity of competitor depends on the amount and purity of the sample. If the factor has already been substantially purified, considerably less competitor should be used.

Another parameter that may vary somewhat from factor to factor is the optimal salt concentrations for running and eluting the DNA affinity columns. In most instances, satisfactory results are achieved by applying the sample in 0.1 M KCl, washing with this salt concentration, then eluting with 1.0 M KCl. In some cases, however, an intermediate salt concentration wash can be used to remove weakly interacting proteins before the factor of interest is recovered at a higher salt concentration.

Because of the existence of non-specific DNA-binding proteins in crude cell extracts, it is usually necessary to perform several consecutive cycles of sequence-specific DNA affinity chromatography in order to purify a protein to homogeneity (see *Figure 7*). To do this, the eluate from the first DNA

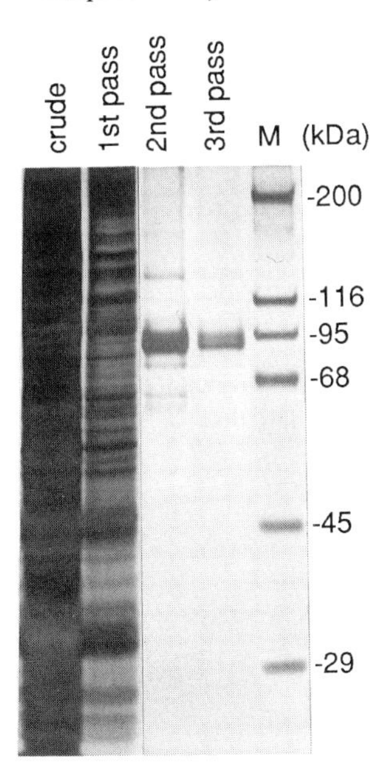

Figure 7. Purification of Sp1 by DNA affinity chromatography. A silver-stained SDS-polyacrylamide gel containing various fractions generated during a standard Sp1 purification. Lanes from left to right contain: crude factor partially purified on Sephacryl S-300 (50 μg), eluate from the 'first pass' DNA affinity column (10 μg), eluate from the 'second pass' DNA affinity column (400 ng), and eluate from the 'third pass' DNA affinity column (100 ng). Lane M contains protein molecular weight standards (sizes in kDa).

affinity column is diluted down to decrease the salt concentration, additional competitor DNA is added, then the sample is subjected to a second round of affinity chromatography. Since a proportion of the factor is lost upon every pass over an affinity column, it is usually best to analyse the eluted material from the second column by SDS-PAGE (see *Protocols 3* and *4*) and staining (see *Protocols 16* and *17*), then go on to the third DNA affinity column only if it is deemed necessary. Sometimes, it is advantageous if the second or third column bears an oligonucleotide different to that used in the first column. In this way, contaminating proteins that may have bound to the sequences flanking the DNA recognition element used in the first column will not be co-purified further.

The volume of the DNA affinity matrix used depends on the amount and purity of the extract applied (the purer and smaller amount of protein applied, the smaller the column needed). Generally, small plastic columns are used containing 1 ml of DNA affinity matrix. If larger column volumes are required, it is often better to run multiple columns in parallel rather than a single larger one. For example, for the first round of affinity chromatography (first pass) four columns can be used, for the second pass two columns, and for the third pass one column (see *Protocol 14*). When running these small columns, it does not matter if the buffer sinks down to the surface of the matrix; they will not run dry. Finally, it is worth noting that several different factors can conveniently be purified simultaneously by stacking a set of different DNA affinity columns on top of each other, so that the flow through for the first becomes the input into the second. The columns are separated after the first wash and then washed and eluted individually.

Protocol 14. DNA recognition site affinity chromatography

Equipment and reagents

- Small glass or plastic columns; 10 ml capacity (e.g. BioRad 'Econocolumns')
- DNA affinity matrix prepared as in *Protocol 13*
- Partially purified nuclear extract (see *Protocols 9–11*)
- Poly(dI-dC)·poly(dI-dC) (Pharmacia)
- Calf thymus DNA (Sigma) [Dissolve this DNA in distilled water, deproteinize it by repeated phenol extraction, sonicate to reduce the average DNA length to <1 kb, ethanol precipitate the DNA, then dissolve it at 5 mg/ml in distilled water]
- Z′ buffer (25 mM Hepes-KOH pH 7.6, 12.5 mM $MgCl_2$, 10 μM $ZnSO_4$, 20% glycerol, 0.1% Nonidet-P40, 1 mM DTT) [Add the DTT just before use. Adjust the pH to pH 7.6 with KOH. Generally, Z′ buffer is prepared in two forms; one containing 1 M KCl, the other containing no KCl. By combining these two solutions, all the necessary intermediate salt concentrations can be achieved]
- Column regeneration buffer (2.5 M NaCl, 10 mM Tris–HCl pH 7.5, 1 mM EDTA)
- Column storage buffer (0.3 M NaCl, 10 mM Tris–HCl pH 7.5, 1 mM EDTA, 0.02% NaN_3).

Method

The procedure given below is for a partially purified nuclear extract originally derived from 36 litres of HeLa cell suspension culture (approximately 60 mg total protein), although it can easily be scaled up or down as appropriate. Perform all steps at 0–4°C.

1. Prepare a set of four 10 ml capacity columns, each containing 1 ml DNA affinity matrix, and wash them four times with 5 ml of Z′ buffer containing 0.1 M KCl.
2. Mix the nuclear extract with 2 ml of 10 A_{260} units/ml poly(dI-dC)·poly(dI-dC) and 100 μg calf thymus DNA. Incubate for 10 min on ice, then centrifuge for 10 min at 10 000 *g* (e.g. in a fixed angle Sorvall SS34 rotor at 10 000 r.p.m.). Recover the supernatant. This spin removes the aggregated material (pellet) which would otherwise lead to reduced column flow rates and suboptimal purification.
3. Apply equal volumes of the supernatant to each of the four columns and run them at a flow rate of approximately 15 ml/h per column. If Sephacryl S-300 chromatography was used as an initial fractionation (see *Protocol 9*), the total sample volume at this stage is usually about 120 ml.

4. Wash the columns four times with 2 ml Z′ buffer containing 0.1 M KCl; discard the washings.

5. Elute the bound protein from the column by applying 2 × 1.25 ml of Z′ buffer containing 1.0 M KCl. Pool the eluates from the four columns.

6. For the second round of purification, dilute the first pass eluate with 6 vol. Z′ buffer lacking KCl to reduce the overall salt concentration, then add 50 μl of 10 A_{260} units/ml of poly(dI-dC)·poly(dI-dC). Load on to two new columns of DNA affinity matrix, then wash and elute as for the first pass (steps 4 and 5), except collect the eluted material in five 0.5 ml fractions so that the eluted factor can be obtained as concentrated as possible. Check for factor activity and purity as described in Section 5.

7. If a third round of purification is necessary, dilute the pooled second pass eluate with 6 vol. Z′ buffer lacking KCl, and add 15 μl of 10 A_{260} units/ml poly(dI-dC)·poly(dI-dC). Load on to a single fresh column of DNA affinity matrix and repeat steps 4 and 5; collect the eluted material as five 0.5 ml fractions.

8. (Optional.) Dialyse the fractions against Z′ buffer containing 0.1 M KCl using either conventional small dialysis bags or a commercially available microdialysis system (e.g. BioRad).

9. Take a small aliquot of each fraction and assay for transcription factor activity as described in Section 5. Freeze the fractions in small aliquots in liquid nitrogen and store at −70°C. Most purified transcription factors are stable in this form for over 1 year. Since multiple freeze–thaw cycles can destroy factor activity, it is usual practice to divide the fractions into many small aliquots before freezing.

10. Regenerate the DNA affinity columns by washing each successively twice with 10 ml column regeneration buffer then twice with 10 ml column storage buffer. Apply 5 ml column storage buffer to each column, stopper the columns and store indefinitely at 4°C

---

5. Determining the purity and activity of a purified transcription factor

Having subjected the transcription factor preparation to DNA affinity chromatography, the next step is to determine the purity and activity of the resulting fractions. The sample purity must be assessed for two reasons:

(a) It is important to know whether significant quantities of contaminating proteins are present that may complicate further experiments. For example, if one wishes to clone the cDNA encoding the factor of interest through the protein microsequencing approach (Section 7), one must be certain that the sequence obtained is not that of a contaminant.

(b) If functional analysis of the factor is intended, one must know whether the activities measured are conferred by a single component or by a collection of related or unrelated factors. This consideration is particularly important where multiple proteins can each bind individually to the same DNA sequence. A good example of this is provided by the AP-1 family of transcription factors. Initial attempts to purify this 'factor' to homogeneity were deemed unsuccessful since multiple polypeptides were present in the final preparations. It soon became clear, however, that these multiple polypeptides corresponded to members of the Fos and Jun families of transcription factors that all function through the AP-1 DNA consensus sequence (8). One approach to determine which proteins in the preparation actually contact the DNA recognition element is to perform UV crosslinking (Section 3.1) or South-Western blotting (Section 3.2). Determining factor purity may also provide insight into whether the factor of interest is composed of single or multiple polypeptides.

It is also important to determine the transcriptional and/or DNA binding activities of a purified transcription factor preparation. In particular, it is essential to show that the purified factor displays all the activities expected, such as the correct DNA binding specificity and an ability to activate transcription of the appropriate genes *in vitro*. Furthermore, by comparing the activities of the purified fraction with the original and partially purified extracts, one can assess how efficient the purification scheme has been.

5.1 Assaying the purity of a transcription factor preparation

To determine the purity of the transcription factor preparation, a sample should be electrophoresed on an SDS polyacrylamide gel. Full details of SDS-PAGE are given in Section 3.1.3. As is sometimes the case when the purified protein is dilute, limitations on the volume of sample that can be applied to the SDS-polyacrylamide gel may mean that an aliquot of the sample must first be concentrated. This is accomplished by precipitating the protein with trichloroacetic acid (TCA) (*Protocol 15*). TCA precipitation also serves to remove much of the KCl which otherwise can cause gels to run anomolously. In order to estimate the size and amount of the purified protein, co-electrophorese a known amount of a mixture of protein size standards on an adjacent gel lane.

After electrophoresis, the individual protein bands are visualized by either Coomassie Blue staining (*Protocol 16*) or silver staining (*Protocol 17*). Silver staining has the advantage that it is more sensitive than Coomassie Blue staining (the lower limits of detection are approximately 1 ng and 50 ng per protein band, respectively). Because it is very easy to perform, Coomasssie Blue staining is the method of choice when there is a sufficiently large quantity of protein. Conversely, silver staining is the method of choice when limited material is available. Due to the extreme sensitivity of silver staining, it is important to keep the gel apparatus and all tubes scrupulously clean. This is usually achieved by wearing disposable plastic or latex gloves at all times when handling these items and by cleaning the gel apparatus thoroughly before use.

Protocol 15. Concentration of proteins by TCA precipitation

Reagents

- TCA solution [100% (w/v) TCA in distilled water, containing 0.4% sodium deoxycholate]
- 2 × SDS sample buffer (10% glycerol, 5% 2-mercaptoethanol, 3% SDS, 62.5 mM Tris–HCl, pH 6.8, 0.2% bromophenol blue). Store in aliquots at −20°C.

Method

1. Add 0.25 vol. TCA solution to a sample of the protein preparation, mix by vortexing, then incubate on ice for 15 min.
2. Spin in a microcentrifuge for 15 min and, without disturbing the pellet, remove the supernatant with a micropipette.
3. Add 1 ml acetone, mix well by vortexing, and spin in microcentrifuge for 5 min. Carefully remove the supernatant. This acetone wash removes residual TCA in the sample.
4. Dry the pellet under vacuum for 5 min.
5. Add an appropriate quantity of 2 × SDS sample buffer, vortex, then spin briefly.
6. Pierce a hole in the tube top, place in a boiling water bath for 5 min, then load on to an SDS-polyacrylamide gel. If the bromophenol blue in the sample turns green due to residual TCA, before loading add a small volume of 2 M Tris–HCl pH 6.8 until the blue colour of the loading dye returns, then load on to the gel as usual.

Protocol 16. Coomassie staining proteins in SDS-polyacrylamide gels

Reagents

- Coomassie staining solution (0.03% Coomassie Brilliant Blue R250, 25% 2-propanol, 10% acetic acid). Mix well to dissolve the Coomassie Blue dye
- Mixture of protein size standards (Sigma)
- Destaining solution (10% acetic acid, 10% methanol in water) [An alternative destaining solution is 10% acetic acid, 25% methanol].

Protocol 16. *Continued*

Method

Carry out all incubations at room temperature in a clean plastic box with constant gentle shaking on a shaking platform.

1. Remove the electrophoresed gel from the glass plates and incubate it in Coomassie staining solution (sufficient to cover the gel) for 30 to 60 min.
2. Rinse the gel twice with a small volume of destaining solution.
3. Add a liberal volume of destaining solution and incubate with gentle shaking until the background staining has been reduced sufficiently to visualize the protein bands (normally approximately 2 h).
4. Soak the gel for 15 min in 5% glycerol before drying on a heated vacuum dryer. The glycerol helps prevent gel cracking during the drying step.

Protocol 17. Silver staining of SDS-polyacrylamide gels

Reagents

- 1 M DTT
- Citric acid (solid)
- Silver solution (0.1% $AgNO_3$ in distilled water)
- Developing solution [Dissolve 7.5 g anhydrous Na_2CO_3 in 250 ml distilled water, then add 0.125 ml of 37% formaldehyde and stir for 2 min].

Method

Given the sensitivity of silver staining, wear disposable plastic or latex gloves at all stages of gel manipulation. Carry out all incubations at room temperature in a clean plastic box with gentle shaking on a shaking platform. Prepare all solutions with glass distilled water, just before use.

1. Remove the electrophoresed gel from the glass plates and incubate twice for 15 min each time with 250 ml of 50% methanol.
2. Transfer the gel into 250 ml of 5% methanol and incubate for 15 min.
3. Incubate for 15 min in 250 ml of distilled water containing 8 μl of 1M DTT.
4. Wash the gel briefly with 100 ml distilled water, twice.
5. Rinse the gel with a little silver solution, pour this away, then add the remaining silver solution and incubate for 15 min.
6. Wash briefly with distilled water, twice.
7. Rinse the gel twice with a little developing solution, pour away, then add the remaining developing solution. Incubate until the brown-staining protein bands which appear are of the desired intensity.

8. Pour off most of the developing solution, sprinkle solid citric acid into the solution containing the gel and swirl. Continue adding the citric acid slowly until the fizzing ceases. Add a little distilled water and incubate for 15 min.
9. Incubate the gel three times for 15 min each time with 250 ml of distilled water.
10. Soak the gel in 5% glycerol for 15 min before drying on a heated vacuum drier.

5.2 Assaying the activity of a purified transcription factor preparation

The DNA binding properties of a purified transcription factor can be measured using the gel mobility shift assay and DNase I footprinting procedures covered in this chapter (Section 2) and in Chapter 6. This information will allow a qualitative and quantitative assessment of the purified preparation. To measure the transcriptional properties of a purified factor, one requires an assay system that is depleted of this factor, but has the full complement of general transcription factors (RNA polymerase II, TFIIA, B, D, etc.). One obvious approach is to generate a transcriptional extract from a cell or tissue type that does not contain the factor of interest. In many cases, however, this is not feasible, and one must find a way to deplete the factor from a transcription system. The most common way to achieve this is to use a DNA affinity chromatography matrix to deplete the factor from a crude transcription extract. This is normally performed in batch (*Protocol 18*), although an alternative is to deplete the extract by passing it through a small column containing the DNA affinity matrix.

Another approach that can sometimes be used if the transcription factor of interest is glycosylated is to deplete this factor using WGA affinity chromatography (see Section 4.3.3). Although this can be carried out batchwise as with DNA affinity depletion, often the most straightforward approach is to use the flow-through fraction from the WGA column used in factor purification (*Protocol 11*) as a factor-depleted transcription extract.

Other alternatives to generating systems that lack the factor of interest but contain all the general factors include:

- conventional chromatographic fractionation (28)
- the addition of oligonucleotides that are recognized by the factor (34)
- depletion of the factor using antibodies directed against it.

Hopefully, one of these approaches will yield an extract that is depleted in the factor of interest and is therefore unable to efficiently utilize promoters that are responsive to this factor (but remain active for promoters that only

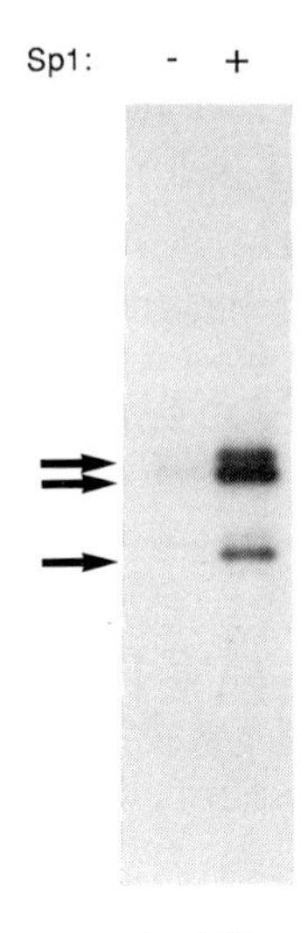

Figure 8. *In vitro* transcriptional activation of the SV40 early promoter by Sp1. Standard chromatographic fractionation of a HeLa nuclear extract generated fractions 'CL-225' and 'Sp2' that, together, contain RNA polymerase II and all general transcription factors (5, 28). These components were tested for the ability to transcribe the SV40 virus early promoter, either in the absence (−), or in the presence (+) of purified Sp1. After incubation at 30°C for 1 h, any transcripts produced were analysed by primer extension analysis (see Chapter 1, Section 7). The radiolabelled primer extension products corresponding to correctly initiated transcripts are indicated by arrows. The addition of Sp1 results in an approximately 20-fold increase in transcription indicating that the SV40 promoter is responsive to this factor.

require factors that are not depleted). Addition of the purified factor preparation should then restore transcriptional activity. *Figure 8* shows one such analysis.

Protocol 18. Depletion of sequence-specific transcription factors from extracts

Equipment and reagents

- Motorized rotating wheel
- Crude nuclear extract
- Z′ buffer containing 0.1 M KCl (25 mM Hepes-KOH pH 7.6, 100 mM KCl, 12.5 mM $MgCl_2$, 20% glycerol, 0.1% Nonidet-P40, 10 μM $ZnSO_4$, 1 mM DTT). Add the DTT just before use.

Method

1. Adjust the KCl concentration of the nuclear extract to 0.05–0.1 M KCl.

2. To a volume of extract (e.g. 300 μl), add 1/3 vol. of the DNA affinity matrix that has been pre-equilibrated with Z′ buffer containing 0.1 M KCl. Incubate at 4°C on a rotating wheel for 30 min.
3. Spin the slurry in a microcentrifuge for 1 min, then transfer the supernatant into another tube, taking care to avoid removing any matrix. Repeat the centrifugation to remove any remaining traces of matrix.
4. Test the transcriptional activity of the extract both in the presence and absence of purified transcription factor.
5. If sufficient depletion of the factor has not been achieved, steps 2 and 3 can be repeated with fresh DNA affinity matrix. The efficiency of depletion is most readily assessed by the gel mobility shift assay (Section 2.1).

5.3 Renaturation of transcription factors from an SDS-polyacrylamide gel

In this method, the purified sample is electrophoresed on an SDS polyacrylamide gel (Section 3.1.3), then the various protein bands are located, cut from the gel, and the individual polypeptides eluted. The proteins are then denatured and renatured by the successive addition then removal of guanidinium chloride. Finally, the renatured proteins are tested for activity (28, 35). These procedures are all described in *Protocol 19*. Although this protocol is often successful, it will not work in the substantial proportion of cases where the protein does not recover its activity upon renaturation, and will not work for proteins that need to heteromultimerize in order to bind to DNA. Because of this, a negative result does not necessarily indicate that a particular polypeptide is not the factor of interest.

The usual method to measure the activity of a transcription factor recovered from SDS gel slices is the gel mobility shift assay (Section 2.1). Since the final yield of activity is typically 5–20% of the input, fairly large quantities of the factor should be run on the SDS gel; usually at least 50 times the amount required to give a reasonably strong signal in the gel mobility shift assay. Furthermore, because staining procedures irreversibly fix proteins in the gel, the factor must be localized indirectly. The best way to do this is to electrophorese, on one half of the gel, a small quantity of the transcription factor preparation alongside a small quantity of pre-stained marker proteins. This part of the gel is then silver-stained (*Protocol 17*) and the location of the various polypeptides determined with respect to the markers. On the other half of the gel, which is not stained, a large aliquot of the transcription factor preparation is run alongside an easily visible amount of pre-stained markers. Regions containing the desired bands are then localized by comparison with the other gel half.

For the renaturation protocol, SDS polyacrylamide gel electrophoresis is performed as described in *Protocols 3* and *4*, except for the following alterations:

(a) All buffers that contain SDS should contain the highest quality SDS since certain impurities can inhibit protein renaturation.

(b) The gel running buffer should contain thioglycolate to scavenge the gel for any free radicals that may damage the protein.

(c) Samples should not be precipitated with TCA before electrophoresis if this can possibly be avoided.

(d) The sample should be incubated at 65°C rather than 100°C for 5 min before electrophoresis.

At this point, it is worth noting that abundant non-specific DNA binding proteins are frequently present in a 'purified' transcription factor preparation. In HeLa extracts, two common contaminants are poly (ADP-ribose) polymerase (116 kDa; refs 36, 37) and the Ku antigen (subunits of 70 kDa and 80 kDa; ref. 38). To help avoid mis-identification of a non-specific DNA binding protein as the factor of interest, it is often worthwhile to demonstrate that the protein tentatively identified as the desired factor is not purified when a DNA affinity column is used that lacks the DNA binding site of interest.

Protocol 19. Renaturation of transcription factors from SDS-polyacrylamide gels

Equipment and reagents

- Equipment and reagents for SDS-PAGE (section 3.1.3) [SDS containing solutions should be prepared using SDS of the highest possible grade (e.g. 'Surfact-Amps' SDS purified detergent solution, Pierce Chemicals). The gel running buffer should be supplemented with thioglycolate to 0.1 mM]
- *Either* a microdialysis system (BioRad) *or* Sephadex G50 spun columns (Pharmacia)
- Prestained protein molecular weight markers (e.g. Rainbow markers; Amersham)
- Elution buffer (0.1% SDS, 50 mM Tris–HCl pH 7.9, 0.15 M NaCl, 5 mM DTT, and 0.1 mg/ml BSA)
- Z′ buffer containing 0.1 M KCl (25 mM Hepes-KOH pH 7.6, 100 mM KCl, 12.5 mM $MgCl_2$, 20% glycerol, 0.1% Nonidet-P40, 10 μM $ZnSO_4$, 1 mM DTT). (Add the DTT just before use.)
- Z′ buffer containing 0.1 M KCl and 6 M guanidinium chloride.

Method

1. Electrophorese the transcription factor preparation on an SDS-polyacrylamide gel (Section 3.1.3), and cut out of the gel a series of gel slices that contain the various polypeptide species (a region that lacks any protein should also be chosen as a control).
2. Crush the gel slices with a small pestle (an ideal pestle is the type designed to fit neatly inside a microcentrifuge tube).

3. Add 1 ml elution buffer and incubate for 2–4 h at room temperature with occasional agitation.
4. Spin for 2 min in a microcentrifuge. Whilst taking care to avoid the polyacrylamide, transfer the supernatant into a siliconized Corex centrifuge tube.
5. Add 5 vol. cold (−20°C) acetone, mix, and incubate at −70°C for 30 min.
6. Spin at 10 000 *g* for 30 min, then decant the supernatant.
7. Wash the pellet by adding 2 ml cold 80% acetone, then re-centrifuge and carefully remove the supernatant with a Pasteur pipette.
8. After allowing the protein pellet to dry at room temperature for 10 min, resuspend it in 100 μl of Z′ buffer containing 0.1 M KCl and 6 M guanidinium chloride. Incubate at room temperature for 15 min.
9. Remove the guanidinium chloride by *either* of two methods:
   (a) Dialyse against Z′ buffer containing 0.1 M KCl for 2–4 h at 4°C in a microdialysis system,

   or

   (b) centrifuge the sample through a 1 ml Sephadex G50 spun column.
10. Assay 1–10 μl aliquots of the renatured protein by the gel mobility shift assay (section 2.1).

6. Problems and trouble-shooting

The purification procedures listed above will work fairly well for many transcription factors. However, since transcription factors can differ significantly from one another in their chromatographic properties, the standard purification procedures often give sub-optimal results. Problems with factor purification fall into two major categories; low yield and low purity. Possible causes and remedies for these problems are discussed below. It should be noted that yield and purity are not independent of one another. Indeed, in many cases they are reciprocally related (see Section 4.4.3); changes in parameters to increase factor purity, for instance, will often result in lower yields of the factor.

6.1 Low yield of transcription factor

Although all transcription factors appear to be of relatively low abundance, the actual amount present in a particular cell or tissue varies considerably from factor to factor. As a general rule, however, one can expect to isolate between a few micrograms to a few tens of micrograms of factor from around 20 g of cells. If antibodies against the factor of interest are available, one way of estimating yields is to compare the total amount of factor present in the purified preparation with that in the original extract using an immunological assay. Alternatively, an estimate of factor recovery can be obtained by assaying the DNA binding activity of the various fractions by the gel mobility shift assay.

Low yields are most often the result of the factor not binding efficiently to the DNA affinity matrix. There are a number of possible reasons for this:

(a) Too much competitor DNA has been added. This is the most common reason. Try using less competitor or another type of competitor DNA.
(b) The oligonucleotide sequence is not optimal for binding. Try another oligonucleotide.
(c) The factor may be unstable in purified form. In this case, proceed through the purification procedure as rapidly as possible. Addition of non-specific carrier protein, such as BSA, to highly purified fractions may also help stabilize the transcription factor. Minimize freeze-thawing of the factor.
(d) The factor is being destroyed by proteolytic enzymes in the extract. Although this is not generally a problem with extracts from HeLa cells, many other cells and tissues are notoriously rich in degradative enzymes (see Chapter 3). If proteolytic activity is considered a possible problem, include other protease inhibitors in addition to PMSF and sodium metabisulphite (Section 4.2) such as 1 mM benzamidine-HCl, aprotinin, pepstatin, and leupeptin (see Chapter 3, Section 3.2.1).

6.2 Transcription factor preparation is heterogeneous

Ideally, the purification procedures described in this chapter will yield a homogeneous preparation of a single polypeptide. In practice, however, this is almost never accomplished, and it is generally unreasonable to expect even the most highly purified fraction to be more than 95% pure.

If the factor preparation contains multiple major polypeptides, there are a number of possible reasons for this:

(a) Too little competitor DNA was used in the DNA affinity purification, allowing non-specific DNA binding proteins to attach to the DNA affinity matrix. Try using more competitor DNA.
(b) The chosen oligonucleotide contains two or more distinct binding sequences. Define the consensus binding motif for the factor of interest in more detail and/or use an oligonucleotide with different flanking sequences for factor purification.
(c) More than one protein binds to the chosen DNA recognition element. Check this by UV crosslinking (Section 3.1).
(d) The factor of interest may be multimeric (many factors bind to DNA as heterodimers) or may exist in a variety of differentially spliced forms due to alternative splicing of its RNA transcript (39).

(e) The preparation may be purer than you think; the contaminating polypeptides may not be derived from the nuclear extract at all, but may be proteins such as keratins derived from the experimenter's fingers! Wear gloves throughout the purification and analysis and clean the electrophoresis apparatus used for SDS-PAGE thoroughly.

(f) The factor may have been degraded partially during purification. Try including additional protease inhibitors such as aprotinin, pepstatin, and leupeptin (see Chapter 3, Section 3.2.1) at all stages of the purification.

(g) The factor might exist in more than one electrophoretically distinct form due to alternative routes of post-translational modification (see Section 8). Inclusion of phosphatase inhibitors (e.g. 1 mM NaF and 10 mM sodium phosphate) during factor purification may prevent removal of phosphate residues from the protein by endogenous phosphatases.

Whatever the reason for there being multiple polypeptides in the final transcription factor preparation, it is imperative to identify which polypeptide(s) actually binds to the DNA recognition element of interest. This can be achieved by UV crosslinking or South-Western blotting (Sections 3.1 and 3.2). Alternatively, one can electrophorese the purified sample on an SDS-polyacrylamide gel, cut out and elute the various polypeptide bands, renature them and assay their activity (see Section 5.3).

7. Isolation of cDNA clones encoding transcription factors

One of the most significant advances in recent years in the study of transcription has been the isolation of cDNA clones encoding transcriptional regulatory molecules. By sequencing these clones and comparing them with one another, it has become apparent that transcription factors exist in a series of families that are related in primary protein sequence. Manipulation and mutation of the cDNA coding sequences, together with sequence analysis, have revealed a variety of motifs such as zinc fingers and homeodomains which are DNA-binding domains, leucine zippers and helix–loop–helix motifs that define dimerization domains, and 'acid blob', glutamine-rich regions, and proline-rich regions that constitute transcriptional activation domains (1–4). In turn, the identification of conserved sequences in these functional units have allowed the development of directed approaches to isolate other members of the same transcription factor family. Although it is beyond the scope of this chapter to discuss the approaches used to isolate transcription factor cDNA clones in detail, a brief overview of the two main approaches is given below.

In the first approach to cloning a transcription factor cDNA, the factor is first purified to homogeneity. It is then cleaved into a series of peptides by proteolytic enzymes, and these peptides are resolved from one another chromatographically [usually by high-performance liquid chromatography (HPLC)]. N-terminal microsequencing is then performed on a group of these peptides. Next, 'guess-mer' oligonucleotides are synthesized that correspond to the determined protein sequence (39–42). Finally, these oligonucleotides are used to isolate clones from a cDNA library (usually in phage lambda as vector). If the open reading frame of the cloned cDNA contains the original peptide sequences, then this usually indicates that the desired cDNA has indeed been isolated. The advantage of this approach is that one can be certain that the clone isolated encodes the protein that has been purified. The main limitation, however, is that a considerable quantity of pure factor is required to obtain reliable sequence information. Although reliable microsequence data can be obtained with less than 1 μg pure protein, frequently at least 5–10 μg is required in order to optimize proteolytic cleavage and because the recovery of many peptides on HPLC is low. Another important point is that only factor preparations of the highest purity can be used in this approach; otherwise, one could easily end up isolating the cDNA for a contaminating protein.

The other strategy commonly used to isolate the cDNA for a transcription factor takes advantage of the factor's sequence-specific interaction with DNA (see Chapter 6; 43–46). In this approach, a phage lambda cDNA expression library is plated out and the expressed proteins transferred on to a nitrocellulose membrane. Next, the proteins are denatured and subsequently renatured to regain biological activity. The membrane is then probed with a radiolabelled oligonucleotide containing the sequence of interest. If autoradiography reveals a lambda plaque which binds the labelled oligonucleotide, this is picked, grown up, and its cDNA analysed. The advantages of this method are that it is a relatively easy to perform and does not require the factor to be purified. Disadvantages, however, include the fact that the procedure is unable to clone factors that do not recover activity upon denaturation and renaturation or that are composed of more than one polypeptide species. Furthermore, it is unlikely to be successful if the factor of interest has a high dissociation rate from the DNA.

8. Post-translational modification of transcription factors

Many transcription factors have their activity regulated by post-translational modification. The most prevalent form of modification seems to be phosphorylation, although a considerable number of transcription factors are glycosylated (32). In some cases, phosphorylation is known to regulate the DNA binding potential of the transcription factor, whereas for others it regulates transcriptional potential (47, 48). The function(s) associated with transcription factor glycosylation have not yet been determined. Given below is a brief description of some of the methods used to analyse transcription factor modification. For detailed coverage of the various techniques employed, the reader is directed elsewhere (49, 50).

8.1 Transcription factor phosphorylation

The first step in analysing the post-translational phosphorylation of a transcription factor is usually to determine whether the factor is normally phosphorylated *in vivo*. The most straightforward approach to do this is to grow tissue culture cells expressing the factor in the presence of [^{32}P]orthophosphate, then to isolate the factor from these cells and determine whether it is radiolabelled. This is best achieved by immunoprecipitation from crude cell extract, although analytical-scale factor purification can also be used (13).

Once a factor is known to be normally phosphorylated *in vivo*, the experimenter may wish to determine whether dephosphorylation of the factor affects its activity. Dephosphorylation is most commonly achieved by incubating a purified preparation of the factor with either alkaline phosphatase or acid phosphatase. Before analysing the transcription factor in DNA-binding or transcriptional assays, it is advisable to remove the phosphatase. This can be achieved either through chromatography (13) or dephosphorylating the transcription factor using a phosphatase attached to a readily removable insoluble support (51). When performing these experiments, one should be careful to include essential controls such as performing the dephosphorylation reaction in the presence of phosphatase inhibitor. In the longer term, the experimenter may wish to locate precisely the sites of phosphorylation *in vivo*, test the effect of mutating these residues, and determine the kinase and signal transduction pathway that mediate its regulation (47, 48, 52–55).

8.2 Transcription factor glycosylation

Many transcription factors are glycosylated. In all characterized examples, this involves GlcNAc residues attached to serine and/or threonine residues of the protein (32). Since proteins bearing GlcNAc residues are often bound by the lectin WGA (6, 31), the easiest way to test for such glycosylation is to determine whether the factor is retained on a WGA column and can be eluted subsequently by the competing sugar, GlcNAc (see Section 4.3.3). Although a positive result does indicate glycosylation, a negative result does not necessarily indicate lack of glycosylation since only a fraction of GlcNAc-bearing proteins (apparently those that are multiply modified) bind sufficiently tightly to WGA. A much more definitive test for glycosylation is an *in vitro* labelling procedure that utilizes the enzyme galactosyl transferase to attach

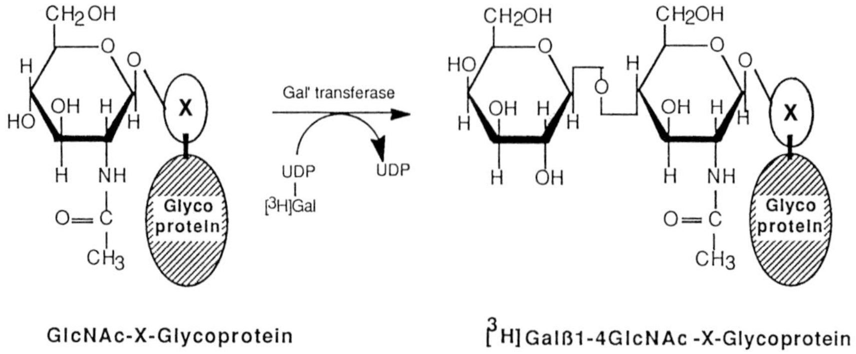

GlcNAc-X-Glycoprotein

[^{3}H]Galβ1-4GlcNAc-X-Glycoprotein

Figure 9. *In vitro* labelling of glycoproteins bearing terminal GlcNAc residues using bovine galactosyltransferase. The GlcNAc residue of a glycoprotein transcription factor is attached to a serine or threonine residue (X). The enzyme galactosyl transferase transfers the tritiated galactose moiety of a UDP-galactose substrate on to the glycoprotein such that it now has a disaccharide attached to it. When analysed by SDS-PAGE and autoradiography, the transcription factor will be found to be radiolabelled.

radiolabelled galactose moieties covalently on to any protein that bears terminal GlcNAc residues (*Figure 9*). This procedure is described below.

8.2.1 Testing for transcription factor glycosylation using galactosyl transferase

In this procedure, a purified preparation of the transcription factor is incubated with bovine milk galactosyl transferase (Gal transferase) in the presence of UDP-[^{3}H]galactose (32, 56). During this reaction, the enzyme transfers the tritiated galactose moiety on to any terminal GlcNAc residues present on the transcription factor. The reaction products are then separated by SDS-PAGE and the extent of radiolabelling determined by autoradiography. Knowing the specific radioactivity of the UDP-[^{3}H]galactose used and the quantity of factor analysed, one can estimate the number of moles of GlcNAc present per mole of protein. However, since the Gal transferase preparation itself contains glycosylated proteins, it must first be subjected to autogalactosylation in the presence of unlabelled UDP-galactose before it can be used (*Protocol 20*).

The Gal transferase assay for transcription factor glycosylation is described in *Protocol 21*. Note that, despite the precaution of using Gal transferase which has been autogalactosylated (*Protocol 20*), an 'enzyme alone' control should be included in *Protocol 21* to monitor for any low level residual autogalactosylation which may occur. As an alternative to *Protocols 20* and *21*, it is now possible to obtain a commercial *O*-GlcNAc detection kit (Oxford Glycosystems).

Protocol 20. Autogalactosylation of the gal transferase enzyme

Reagents

- Bovine colostrum galactosyl transferase (Sigma)
- Autogalactosylation buffer [50 mM Tris–HCl pH 7.3, 0.4 mM UDP-galactose, 5 mM $MnCl_2$, 1 mM 2-mercaptoethanol, 1% aprotinin (Sigma)]
- Saturated ammonium sulphate (in distilled water)
- Enzyme storage buffer (25 mM Hepes-NaOH pH 7.3, 5 mM $MnCl_2$, 50% glycerol).

Method

1. Dissolve 1–5 units of the galactosyl transferase in 100 μl autogalactosylation buffer.
2. Incubate at 37°C for 30 min.
3. Slowly add 0.57 ml saturated ammonium sulphate, whilst mixing by gentle pipetting (making the final solution 85% saturated). Incubate on ice for 30 min.
4. Spin in a microcentrifuge for 30 min. Remove the supernatant, then re-centrifuge the tube briefly and carefully remove the remaining supernatant.
5. Gently wash the pellet with 0.5 ml of 85% saturated ammonium sulphate to remove residual UDP-galactose. Carefully remove all traces of supernatant.
6. Redissolve the enzyme in enzyme storage buffer at 0.05 units/μl (50 mU/μl). Store at −20°C. Enzyme preparations stored in this way are stable for at least six months.

Protocol 21. Galactosyl transferase assay for transcription factor glycosylation

Reagents

- Gal transferase buffer (10 mM Hepes-NaOH pH 7.3, 0.15 M NaCl, 0.3% Nonidet P-40, 1 mM 2-mercaptoethanol)
- Autogalactosylated galactosyl transferase (from *Protocol 20*)
- Enzyme storage buffer (see *Protocol 20* reagents for composition)
- Galactose buffer (0.1 M galactose, 0.1 M Hepes-NaOH pH 7.3, 0.15 M NaCl, 50 mM $MgCl_2$, 5% Nonidet P-40)
- UDP-[^{3}H]galactose (1 mCi/ml, 17.3 Ci/mmol; Amersham)
- 25 mM 5′-AMP
- 2 × SDS sample buffer (10% glycerol, 5% 2-mercaptoethanol, 3% SDS, 62.5 mM Tris–HCl pH 6.8, 0.2% Bromphenol Blue)
- ^{14}C-methylated protein markers and 'Amplify' fluorographic reagent (Amersham)
- Fixing solution (see *Protocol 4*).

Protocol 21. *Continued*

Method

1. Dry down in a vacuum desiccator sufficient UDP-[^{3}H]galactose (5 μl per reaction). Redissolve this in 25 mM 5′-AMP to the original volume.
2. Mix up to 10 μl of protein sample with an amount of Gal transferase buffer such that the final volume is 37.5 μl.
3. Add, in the following order: 10 μl (50 mU) autogalactosylated Gal transferase freshly diluted to 5 mU/μl in enzyme storage buffer, 2.5 μl galactose buffer, and 5 μl UDP-[^{3}H]galactose. Mix and incubate at 37 °C for 30 min.
4. Add an equal volume of 2 × SDS sample buffer, incubate at 100 °C for 5 min and electrophorese on an SDS-polyacrylamide gel (Section 3.1.3). If the sample volume is too large, precipitate the proteins with TCA first (*Protocol 15*). In order to determine the relative molecular mass of radiolabelled species, co-electrophorese ^{14}C-methylated protein markers on the same gel.
5. Incubate the gel in fixing solution (*Protocol 4*) for 15 min with gentle agitation at room temperature, then with 'Amplify' for 15 min.
6. Dry the gel and expose it to X-ray film. Develop the exposed film according to the manufacturer's instructions.

Acknowledgements

I thank Dirk Bohmann, Jim Kadonaga, Karen Perkins, and Frank Pugh for providing protocols, Mike Taylor for providing the gel mobility shift assay figures, Robert Tjian and Mike Briggs for the S-300 fractionation figure, and Jim Kadonaga, Mike Taylor, and Bob White for their comments on this manuscript. Finally, thanks to Ruth Dendy for help with preparing the manuscript. S. P. Jackson is supported by a project grant from the Cancer Research Campaign (UK) and is a member of the Department of Zoology, Cambridge University.

References

1. Mitchell, P. J. and Tjian, R. (1989). *Science,* **245,** 371–8.
2. Johnson, P. F. and McKnight, S. L. (1989). *Annu. Rev. Biochem.,* **58,** 799–839.
3. Saltzman, A. G. and Weinmann, R. (1989). *FASEB J.,* **3,** 1723–33.
4. (1991) *Trends Biochem. Sci.,* **16,** 393–447.
5. Dynan, W. S. and Tjian, R. (1983). *Cell,* **35,** 79–87.
6. Jackson, S. P. and Tjian, R. (1989). *Proc. Natl Acad. Sci. USA,* **86,** 1781–5.
7. Hai, T., Liu, F., Allegretto, E. A., Karin, M., and Green, M. R. (1988). *Genes Dev.,* **2,** 1216–26.

8. Kouzarides, T. and Ziff, E. B. (1989). *Cancer Cells,* **1,** 71–6.
9. Mohun, T. J., Chambers, A. E., Towers, N., and Taylor, M. V. (1991). *EMBO J.,* **10,** 933–40.
10. White, R. and Jackson, S. P. (1992). *Cell,* **71** (in press).
11. Taylor, M., Treisman, R., Garrett, N., and Mohun, T. (1989). *Development,* **106,** 67–78.
12. Taylor, M. V., Gurdon, J. B., Hopwood, N. D., Towers, N., and Mohun, T. J. (1991). *Genes Dev.,* **5,** 1149–60.
13. Jackson, S. P., MacDonald, J. J., Lees-Miller, S., and Tjian, R. (1990). *Cell,* **63,** 155–65.
14. Chodosh, L. A. (1991). In *Current protocols in molecular biology*, Vol. 2 (ed. F. M. Asubel *et al.*), 12.5.1–12.5.8. J. Wiley & Sons, Chichester.
15. Chodosh, L. A., Carthew, R. W., and Sharp, P. A. (1986). *Mol. Cell Biol.,* **6,** 4723–33.
16. Wu, C., Wilson, S., Walker, B., Dawid, I., Paisley, T., Zimarino, V., and Ueda, H. (1987). *Science,* **238,** 1247–53.
17. Larson, J. S., Schuetz, T. J., and Kingston, R. E. (1988). *Nature,* **335,** 372–5.
18. Laemmli, U. K. (1970). *Nature,* **227,** 680–5.
19. Towbin, H., Staehelin, T., and Gordon, J. (1979). *Proc. Natl Acad. Sci. USA,* **76,** 4350–4.
20. Miskimins, W. K., Roberts, M. P., McClelland, A., and Ruddle, F. H. (1985). *Proc. Natl Acad. Sci. USA,* **82,** 6741–4.
21. Rigby, P. W. J., Dieckmann, M., Rhodes, C., and Berg, P. (1977). *J. Mol. Biol.,* **113,** 237–51.
22. Feinberg, A. P. and Vogelstein, B. (1983). *Anal. Biochem.,* **132,** 6–13.
23. Silva, C. M., Tully, D. B., Petch, L. A., Jewell, C. M., and Cidlowski, J. A. (1987). *Proc. Natl Acad. Sci. USA,* **84,** 1744–8.
24. Kadonaga, J. T. and Tjian, R. (1986). *Proc. Natl Acad. Sci. USA,* **83,** 5889–93.
25. Rosenfeld, P. J. and Kelly, T. J. (1986). *J. Biol. Chem.,* **261,** 1398–408.
26. Kasher, M. S., Pintel, D., and Ward, D. C. (1986). *Mol. Cell. Biol.,* **6,** 3117–27.
27. Dignam, J. D., Lebovitz, R. M., and Roeder, R. G. (1983). *Nucleic Acids Res.,* **11,** 1475–89.
28. Briggs, M. R., Kadonaga, J. T., Bell, S. P., and Tjian, R. (1986). *Science,* **234,** 47–52.
29. March, F. C., Parikh, I., and Cuatrecasas, P. (1974). *Anal. Biochem.,* **60,** 149–52.
30. Davidson, B. L., Leighton, T., and Rabinowitz, J. C. (1979). *J. Biol. Chem.,* **254,** 9220–26.
31. Lichtsteiner, S. and Schibler, U. (1989). *Cell,* **57,** 1179–87.
32. Jackson, S. P. and Tjian, R. (1988). *Cell,* **55,** 125–33.
33. Chodosh, L. A. (1991). In *Current protocols in molecular biology*, Vol. 2 (ed. F. M. Asubel *et al.*), 12.6.1–12.6.9. Wiley, Chichester.
34. White, R., Jackson, S. P., and Rigby, P. W. J. (1992). *Proc. Natl Acad. Sci. USA,* **89,** 1949–53.
35. Hager, D. A. and Burgess, R. R. (1980). *Anal. Biochem.,* **109,** 76–86.
36. Slattery, E., Dignam, J. D., Matsui, T. and Roeder, R. G. (1983). *J. Biol. Chem.,* **258,** 5955–59.
37. Ueda, K. and Hayaishi, O. (1985). *Annu. Rev. Biochem.,* **54,** 73–100.

38. Zhang, W.-W. and Yaneva, M. (1992). *Biochem. Biophys. Res. Commun.*, **186,** 574–9.
39. Santoro, C., Mermod, N., Andrews, P. C., and Tjian, R. (1988). *Nature,* **334,** 218–24.
40. Kadonaga, J. T., Carner, K. R., Masiarz, F. R., and Tjian, R. (1987). *Cell,* **51,** 1079–90.
41. Lathe, R. (1985). *J. Mol. Biol.,* **183,** 1–12.
42. Martin, F. H., Castro, M. M., Aboul-ela, F., and Tinoco, I., Jr. (1985). *Nucleic Acids Res.,* **13,** 8927–38.
43. Singh, H., LeBowitz, J. H., Baldwin, A. S., and Sharp, P. A. (1988). *Cell,* **52,** 415–23.
44. Huynh, T. V., Young, R. A., and Davis, R. W. (1985). In *DNA cloning,* Vol. 1, *a practical approach* (ed. D. M. Glover), pp. 49–78. IRL Press, Oxford.
45. Vinson, C. R., LaMarco, K. L., Johnson, P. F., Landschulz, W. H., and McKnight, S. L. (1988). *Genes Dev.,* **2,** 801–6.
46. Singh, H. (1991). In *Current protocols in molecular biology*, (ed. F. M. Asubel *et al.*), 12.7.1–12.7.10. Wiley, Chichester.
47. Jackson, S. P. (1992). *Trends Cell Biol.*, **2,** 104–8.
48. Hunter, T. and Karin, M. (1992). *Cell*, **70,** 375–87.
49. Hunter, T. and Sefton, B. M. (ed.) (1991). *Methods in enzymology,* Vol. 200. Academic Press, New York.
50. Hunter, T. and Sefton, B. M. (ed.) (1991). *Methods in enzymology,* Vol. 201. Academic Press, New York.
51. Raychaudhuri, P., Bagchi, S., and Nevins, J. R. (1989). *Genes Dev.,* **3,** 620–7.
52. Gonzalez, G. A. and Montminy, M. R. (1989). *Cell,* **59,** 675–80.
53. Pulverer, B. J., Kyriakis, J. M., Avruch, J., Nikolakaki, E., and Woodgett, J. R. (1991). *Nature,* **353,** 670–4.
54. Boyle, W. J., Smeal, T., Defize, L. H. K., Angel, P., Woodgett, J. R., Karin, M., and Hunter, T. (1991). *Cell,* **64,** 573–84.
55. Smeal, T., Binetruy, B., Mercola, D. A., Birrer, M., and Karin, M. (1991). *Nature,* **354,** 494–6.
56. Holt, G. D. and Hart, G. W. (1986). *J. Biol. Chem.,* **261,** 8049–57.

6

Analysis of protein–DNA interactions

MICHAEL J. GARABEDIAN, JOSHUA LaBAER, WEI-HONG LIU, and JAY R. THOMAS

1. Introduction

An important finding of contemporary studies in molecular biology is that much of the process of cellular differentiation is regulated at the level of gene transcription. Analyses of cloned genes have revealed that in a given tissue or cell type, gene expression is determined largely by proteins that bind to specific DNA sequences linked to those genes. This chapter considers in detail numerous procedures for characterizing these important protein–DNA interactions.

2. The mobility shift DNA-binding assay for protein–DNA interactions

The gel mobility shift assay is a simple and sensitive method for determining interactions between protein and DNA. This assay was developed by Fried and Crothers (1) and Garner and Revsin (2) for analysing protein–DNA interactions. This assay separates protein–DNA complexes from free DNA by non-denaturing polyacrylamide gel electrophoresis. Unlike footprinting techniques that rely on the loss of a signal to determine protein–DNA interactions (negative assay), the gel mobility shift assay yields a positive signal; the appearance of a DNA fragment with an altered mobility. An unexpected property of the gel assay is its ability to detect and physically separate alternative conformational states of complexes that rapidly equilibrate in solution. However, the gel shift assay does not give a direct readout of the DNA nucleotides that the protein is recognizing. For this type of information, a higher resolution technique such as DNase I footprinting or methylation interference is necessary. Fortunately, a combination of these assays can often be coupled with considerable advantage.

The mobility shift assay can be divided into several parts:

(a) *Preparing the DNA probe*: the DNA containing the recognition site of interest prepared either from a plasmid or in the form of a synthetic oligonucleotide, is radiolabelled to high specific activity and then purified.

(b) *Preparing the gel*: low crosslinking, low ionic strength, vertical polyacrylamide gels are cast and allowed to polymerize.

(c) *Carrying out the DNA binding reactions*: the DNA binding proteins, either purified proteins or contained in nuclear or whole cell extracts, are incubated with the labelled DNA fragment.

(d) *Gel electrophoresis and autoradiography*: protein–DNA complexes are separated from free DNA by polyacrylamide gel electrophoresis and visualized by autoradiography.

2.1 Preparation of the DNA probe

Either a restriction fragment from a recombinant plasmid containing the DNA binding site of interest (50–300 bp) or a synthetic oligonucleotide (20–50 bp) may be used as the probe. Simple procedures for radiolabelling and purifying DNA fragments and for labelling synthetic oilgonucleotides are described below. Unincorporated nucleotides can be removed by ethanol precipitation or 'spun column' chromatography (see Chapter 1, *Protocol 3*; ref. 3).

2.1.1 End labelling DNA binding sites in recombinant plasmid DNA

The 3′-end of the DNA may be labelled using the 5′–3′ polymerase activity of the Klenow fragment of DNA polymerase I to fill in from a recessed 3′ end produced by restriction endonuclease cleavage, using the corresponding 5′ overhang as the template. This labelling procedure (see *Protocol 1*) can generally be used after restriction digestion without intermediate purification of the restricted DNA and yields a relatively high specific activity probe.

Protocol 1. 3′-end-labelling using DNA polymerase I (Klenow fragment)

Reagents

- DNA containing the protein recognition site
- 10 × restriction enzyme buffer[a] [50 mM–0.5 M NaCl, 0.5 M Tris–HCl pH 7.5, 0.1 mM $MgCl_2$, 10 mM dithiothreitol (DTT)]
- 10 × nucleotide mixture (containing each deoxyribonucleoside triphosphate at 2.5 mM except those to be used as label) in TE buffer

- TE buffer (10 mM Tris–HCl, 1 mM EDTA pH 8.0)
- Appropriate restriction enzyme(s)
- [α-^{32}P]dNTP(s) at 3000 Ci/mmol (110 TBq/mmol), 10 mCi/ml (370 MBq/ml)[b]
- DNA polymerase I (Klenow fragment)
- 0.5 M EDTA pH 8.0
- Chase solution (1 mM each of dATP, dGTP, dTTP, dCTP in TE buffer).

A. *Restriction digestion*

1. Excise a DNA fragment containing the binding site of interest by digesting the plasmid DNA with restriction enzyme(s) that create a 5′ overhang.

2. Mix together in a microcentrifuge tube the following:

DNA to be digested	1 μg
10 × restriction enzyme buffer	1 μl
H_2O	to a final volume of 10 μl
restriction enzyme(s)	5 units.

3. Incubate at 37°C for 1–2 h or until the DNA has been fully digested.

B. *Labelling reaction*

1. Mix together in a microcentrifuge tube the following:

restriction digested DNA	1 μg in 10 μl
10 × restriction enzyme buffer	1.5 μl
10 × nucleotide mixture	2.5 μl
H_2O	to a final volume of 25 μl
[α-^{32}P]dNTP[b]	10 μl
DNA polymerase I (Klenow fragment)	20 units.

2. Incubate at room temperature for 10 min.

3. To ensure that all end-labelled molecules are of the same length, add 2 μl chase solution and incubate for an additional 5 min at room temperature (optional).

4. Terminate the reaction by adding 1 μl 0.5 M EDTA and remove the unincorporated radioactivity by ethanol precipitation.

5. Resuspend the labelled DNA in 20 μl TE buffer and resolve the DNA fragments by electrophoresis in a low melting temperature agarose or polyacrylamide gel (see *Protocols 5* and *6*).

[a] Most restriction enzyme reaction buffers are compatible with DNA polymerase I, which requires $MgCl_2$ and DTT for activity.
[b] Commonly one radiolabelled dNTP is used (e.g. [α-^{32}P]dCTP) but several may be used if desired. The choice of radiolabelled dNTP(s) dictates the composition of the 10 × nucleotide mixture which contains 2.5 mM of each dNTP except those used as radiolabelled dNTPs.

DNA may be 5′-end-labelled using T4 polynucleotide kinase. The enzyme catalyses the transfer of the γ-phosphate of ATP to the 5′-hydroxyl group of a

DNA fragment. Most fragments, however, have 5′-phosphate groups which must be removed using alkaline phosphatase before they can be used as substrates (see *Protocol 2*). The dephosphorylated DNA fragment can be labelled at its 5′ end as described in *Protocol 3*.

Protocol 2. Removal of 5′-terminal phosphate groups using alkaline phosphatase

Reagents

- Alkaline phosphatase, calf intestine [Pharmacia (No. 27-0620)]
- 10 × phosphatase buffer (0.5 M Tris–HCl pH 9.0, 0.1 M $MgCl_2$, 10 mM $ZnCl_2$)
- TE buffer (10 mM Tris–HCl, 1 mM EDTA pH 8.0)
- Appropriate restriction enzyme(s)
- Restriction enzyme buffer (see *Protocol 1*)
- DNA to be digested
- 0.5 M EDTA pH 8.0
- Phenol:chloroform:isoamyl alcohol (24:24:1 by vol.)
- Chloroform:isoamyl alcohol (24:1 by vol.).

Method

1. Restriction digest the DNA in a microcentrifuge tube using restriction enzyme(s) that create a 5′ overhang (if possible). Mix together:

DNA to be digested	10 μg
10 × restriction enzyme buffer (see *Protocol 1*)	10 μl
H_2O	to a final volume of 100 μl
restriction enzyme(s)	10–20 units.

Incubate at 37°C for 1–2 h or until the DNA has been fully digested.

2. Set up the phosphatase reaction by mixing:

restriction digested DNA	100 μl
10 × phosphatase buffer	10 μl
calf intestinal phosphatase	0.1 units of per μg DNA.

Let the reaction proceed for 30 min at 37°C.

3. Stop the reaction by adding 10 μl 0.5 M EDTA pH 8.0 and heating to 68°C for 10 min.

4. Cool the reaction and extract the DNA once with 100 μl phenol:chloroform:isoamyl alcohol (24:24:1), then once with 100 μl chloroform:isoamyl alcohol (24:1).

5. Ethanol precipitate the DNA.

6. Dissolve the DNA precipitate in 50 μl TE buffer and store in the −20°C freezer.

Protocol 3. 5′-end-labelling using T4 polynucleotide kinase

Reagents

- Dephosphorylated DNA (0.2 mg/ml)
- 10 × kinase buffer (0.5 M Tris–HCl pH 7.5, 0.1 M $MgCl_2$ 50 mM DTT, 1 mM spermidine, 1 mM EDTA)
- [γ-^{32}P]ATP; 7000 Ci/mmol (257 TBq/mmol) 160 mCi/ml (5.92 GBq/ml)
- TE buffer (10 mM Tris–HCl, 1 mM EDTA pH 8.0)
- T4 polynucleotide kinase
- Phenol:chloroform:isoamyl alcohol (24:24:1 by vol.).

Method

1. Mix in a microcentrifuge tube:

dephosphorylated DNA (0.2 mg/ml)	10 μl
10 × kinase buffer	2 μl
H_2O	to 20 μl
[γ-^{32}P]ATP	1–2 μl
T4 polynucleotide kinase	10 units.

2. Incubate for 30–60 min at 37 °C.
3. Stop the reaction by adding 80 μl TE buffer and phenol extract the DNA once with 100 μl phenol:chloroform:isoamyl alcohol (24:24:1).
4. Ethanol precipitate the DNA to remove unincorporated radiolabelled nucleotides.

2.1.2 End-labelling of synthetic oligonucleotides

Synthetic oligonucleotides are synthesized without a 5′-phosphate and therefore can be directly labelled using T4 polynucleotide kinase and [γ-^{32}P]ATP (see *Protocol 4*).

Protocol 4. 5′-end-labelling of synthetic oligonucleotides using T4 DNA polynucleotide kinase

Reagents

- Oligonucleotide (10 pmol/μl)
- 10 × kinase buffer (see *Protocol 3*)
- [γ-^{32}P]ATP; 7000 Ci/mmol (257 TBq/mmol) 160 mCi/ml (5.92 GBq/ml)
- TE buffer (10 mM Tris–HCl, 1 mM EDTA pH 8.0)
- T4 polynucleotide kinase
- Yeast tRNA carrier or Sephadex G50 spun columns (see Chapter 1, *Protocol 3*) for removal of unincorporated [γ-^{32}P]ATP.

Protocol 4. *Continued*

Method

1. Mix in a microcentrifuge tube:

oligonucleotide	1 μl
10 × kinase buffer	2 μl
H_2O	to 20 μl final volume
[γ-^{32}P]ATP	1–2 μl
T4 polynucleotide kinase	10 units.

2. Incubate for 30–60 min at 37 °C.
3. Stop the reaction by adding 80 μl of TE buffer and heat to 68 °C for 10 min.
4. Remove the unincorporated [γ-^{32}P]ATP. For short oligonucleotide probes (<50 bp), removal of unincorporated label by ethanol precipitation gives poor recovery of the labelled probe, unless a carrier nucleic acid, such as yeast tRNA, is included. Spun column chromatography (3, 4) is a simple alternative for the removal of unincorporated radionucleotides; see Chapter 1, *Protocol 3*.

2.1.3 Purification of DNA probes

Separation and purification of the DNA restriction fragments labelled for use as probes in gel mobility shift or footprinting assays are easily accomplished by electrophoresis through acrylamide or agarose gels. Agarose gels have a lower resolving power than polyacrylamide gels and should be used to isolate fragments >200 bp in size (see *Protocol 5*). A higher resolution non-denaturing polyacrylamide gel is used to isolate small DNA fragments (<200 bp) and oligonucleotides (see *Protocol 6*).

Protocol 5. Purification of DNA using low melting temperature agarose gels

Equipment and reagents

- Horizontal gel electrophoresis apparatus
- UV transilluminator (long wavelength, 366 nm)
- Low melting temperature agarose (e.g. Sea Plaque No. 50102; FMC Bio Products)
- Tris-acetate electrophoresis (TAE) buffer (40 mM Tris-acetate, 1 mM EDTA) is prepared as a concentrated stock solution (50 × TAE) as follows. For 50 × TAE stock, mix 242 g Tris base, 57.1 ml glacial acetic acid, 100 ml 0.5 M EDTA pH 8.0 and add H_2O to one litre final volume.
- 0.2 M NaCl, 20 mM EDTA
- Ethidium bromide (10 mg/ml in water) [This stock solution should be stored in a foil-wrapped bottle at room temperature. *Note*: Ethidium bromide is a potent mutagen and gloves should be worn when working with this dye]
- 10 × DNA loading buffer (50% glycerol, 0.5% Bromophenol Blue, 0.5% xylene cyanol)
- TE buffer (10 mM Tris–HCl, 1 mM EDTA pH 8.0)
- Phenol:chloroform:isoamyl alcohol (24:24:1 by vol.).

Method

1. Combine in a 250 ml Erlenmeyer flask or glass bottle:

low melting agarose	1–2 g
1 × TAE buffer	100 ml.

2. Heat in a microwave oven until the agarose is dissolved.

3. Let the agarose cool to 65°C and then add 2 μl 10 mg/ml ethidium bromide; mix thoroughly.

6. Pour the agarose into the gel mould and let the agarose set completely (~30 min). To set gels more rapidly, cast them in the cold room.

7. Add enough 1 × TAE buffer to just cover the gel.

8. Add 0.1 vol. of 10 × DNA-loading buffer to the labelled DNA fragments and electrophorese at 6 V/cm for about 1 h.

9. To visualize the labelled DNA fragments, place the gel on a long wavelength UV transilluminator. Using a razor blade or scalpel, excise the portion of the gel containing the relevant fragment.

10. To melt the agarose, place the gel slice at 68°C and add 3 vol. 0.2 M NaCl, 20 mM EDTA.

11. Extract the agarose containing the labelled DNA twice with phenol buffered at pH 8.0 and once with phenol:chloroform:isoamyl alcohol (24:24:1).

12. Ethanol precipitate the DNA by adding 2 vol. 95% ethanol. Centrifuge for 5 min at 12 000 *g* to pellet the DNA.

13. Rinse the DNA pellet with 1 ml of 70% ethanol and dry.

14. Dissolve the purified labelled DNA in TE buffer. It can be stored at 4°C for up to two weeks.

Protocol 6. Purification of DNA probes using polyacrylamide gels

Equipment and reagents

- Vertical gel electrophoresis tank, 15 cm × 20 cm glass plates, 1.5 mm spacers and combs
- 40% acrylamide stock (38:2 acrylamide: bisacrylamide) [Mix 38.0 g acrylamide, 2 g *N,N'*-methylene bisacrylamide, and water to 100 ml. To dissolve the chemicals, heat the solution to 37°C with constant stirring. Filter through a 2 μm membrane filter and store in a foil wrapped bottle. Acrylamide stock is stable for several weeks at 4°C]
- 10% (w/v) ammonium persulphate (e.g. Bio Rad Laboratories, No. 161-0700)
- Tris-borate electrophoresis (TBE) buffer (90 mM Tris-borate, 2 mM EDTA) is prepared as a concentrated stock solution (5 × TBE) [To prepare 5 × TBE stock, mix 54.0 g Tris base, 27.5 g boric acid, 20 ml 0.5 M EDTA pH 8.0, and add water to one litre final volume]
- (TEMED) (e.g. BioRad Laboratories, No. 161-0800)
- 10 × DNA loading buffer (see *Protocol 5*)
- TE buffer (10 mM Tris–HCl, 1 mM EDTA pH 8.0)

Protocol 6. *Continued*

- Reagents for visualizations of DNA band of interest after gel electrophoresis [ethidium bromide for staining (see *Protocol 5*) or X-ray film for autoradiography]
- Gel elution buffer (20 mM Tris–HCl pH 8.0, 2 mM EDTA pH 8.0, 0.4 M NaCl, 0.05% SDS)
- DEAE-Sephacel equilibrated in TE buffer in a syringe column (4, 5) or prepacked Elutip-d microcolumn (Schleicher & Schuell, No. 27360)
- 50 mM NaCl in TE buffer
- 1.0 M NaCl in TE buffer
- Yeast tRNA (previously phenol-extracted).

A. *Gel electrophoresis*

1. Depending on the sizes of the DNA fragments to be separated (see *Appendix 2*) choose an appropriate gel concentration from the following table and mix the indicated volumes of reagents (for 100 ml gels).

	Polyacrylamide gel concentration (%)			
Stock solutions (for 100 ml gels)	**3.5**	**5**	**8**	**12**
40% acrylamide (ml)	8.75	12.5	20.0	30.0
Water (ml)	70.55	66.8	59.3	49.3
5 × TBE buffer (ml)	20.0	20.0	20.0	20.0
10% ammonium persulphate (w/v) (ml)	0.7	0.7	0.7	0.7
TEMED (μl)	40	40	40	40

2. Clean the gel plates thoroughly and assemble the gel plates with the spacers. Pour the mixture into the gel mould and let it polymerize for 30–60 min with a sample well comb in place.
3. Remove the comb and flush out the wells with 1 × TBE buffer or water.
4. Attach the gel to the electrophoresis tank.
5. Fill the buffer reservoir tanks with 1 × TBE buffer. Flush out the wells and remove any air bubbles trapped beneath the gel with a bent Pasteur pipette or syringe needle.
6. Mix the labelled DNA with an appropriate amount of DNA loading buffer and load the mixture into a sample well using a long pipette tip.
7. Run the gel at 6 V/cm for long enough to separate the labelled DNA fragments (determined empirically).
8. Separate the gel plates using a spatula to prise them apart, leaving the gel on one of the plates.

B. *Isolation of the DNA fragment(s)*

1. Locate the DNA band(s) of interest by staining with ethidium bromide or by autoradiography. For autoradiography, cover the gel and plate with

plastic wrap (cling film). In the darkroom, tape a piece of X-ray film on to the plastic wrap and puncture the film and the gel with a needle (in one movement) to mark the orientation. Do this in several peripheral locations. Carefully remove the tape and film from the plastic wrap after 1 min exposure and develop the film.

2. Cut out from the X-ray film the autoradiographic image of the required fragment and align the film over the gel using the needle marks. Remove the segment of the gel underneath the gap in the film (where the autoradiographic image of the fragment band used to be) with a scalpel or razor blade.
3. Place the gel slice containing the labelled DNA in a microcentrifuge tube and crush the acrylamide gel into small pieces against the wall of the tube using a disposable pipette tip. Add 0.5 ml gel elution buffer.
4. Incubate for a time period from 4 h to overnight at 37°C to elute the labelled DNA from the gel.
5. Centrifuge the sample for 1 min at 12 000 *g* in a microcentrifuge and transfer the supernatant to a fresh tube; avoid as much of the polyacrylamide fragments as possible.
6. Dilute the supernatant into 5 ml 50 mM NaCl in TE buffer and bind to 0.2 ml DEAE-Sephacel equilibrated in TE in a syringe column or use a prepacked Elutip-d microcolumn. For maximal recovery, reload the flow-through on to the column.
7. Wash the column once with 2 ml 50 mM NaCl in TE buffer.
8. Elute the labelled DNA with 1 ml 1.0 M NaCl in TE buffer. (Typically >90% of the radioactivity is eluted.)
9. For fragments less than 50 bp, add 10 μg yeast tRNA. Divide the labelled DNA (~500 μl) between two microcentrifuge tubes and fill to the top with 95% ethanol. Let the DNA mixture sit on ice for 5–10 min and then spin for 5–10 min in a microcentrifuge at 12 000 *g* to precipitate the DNA.
10. Remove the supernatant and wash the DNA pellet once with 70% ethanol. Dry the pellet.
11. Resuspend the DNA pellet in 100 μl TE buffer. Cerenkov count 1 μl of the DNA solution. Typically a labelled DNA fragment isolated in this manner will yield between 10 000–100 000 c.p.m./μl.

2.2 The binding reaction and electrophoresis of protein–DNA complexes

A wide variety of parameters can influence protein–DNA interactions, and so affect the results of the mobility shift assay. For example, monovalent (Na^+

and K^+) and divalent (Mg^{2+}) cations, pH, non-ionic detergents (e.g. Nonidet P-40), binding temperature, binding time, protein concentration, and the type and concentration of competitor DNA can influence protein–DNA complex formation. In addition, the gel composition (both the percentage and degree of crosslinking) as well as the electrophoretic conditions (ionic strength and temperature) can significantly alter the mobililty of a given protein–DNA complex, often to one's advantage. Thus, the systematic alteration of both DNA binding and gel electrophoresis conditions can ensure optimal protein–DNA interaction. What follows are general conditions for binding and electrophoresis of protein–DNA complexes. Each parameter should be titrated for every protein–DNA interaction to be studied.

2.2.1 The protein–DNA binding buffer

Low salt (<150 mM NaCl or KCl) tends to favour protein–DNA interactions by decreasing the dissociation rate of the complex. Conversely, higher salt facilitates dissociation. In addition, low amounts of non-ionic detergents, such as Nonidet P-40 (0.5–0.05%), carrier proteins like bovine serum albumin (BSA) or insulin (100–500 μg/ml), and polyamines, such as spermidine (5 μM), can stabilize certain protein–DNA interactions. Two examples of protein–DNA binding buffers commonly used are given in *Table 1*.

Table 1. Protein–DNA binding buffers

2 × binding buffer A
40 mM Tris–HCl pH 7.9
100 mM NaCl
20% glycerol
0.2 mM DTT

2 × binding buffer B
40 mM Tris–HCl pH 7.9
4 mM $MgCl_2$
100 mM NaCl
2 mM EDTA
20% glycerol
0.2% Nonidet P 40
2 mM DTT
100 μg/ml BSA

The addition of non-specific competitor DNA to the binding reaction is important when one is trying to select a single DNA binding protein from a vast number of proteins, such as in crude extracts. The most commonly used competitor DNAs are synthetic copolymers, such as poly(dI-dC) or poly(dA-dT). These copolymers provide a vast excess of low affinity binding sites which absorb DNA binding proteins non-specifically, thus allowing detection of specific protein–DNA complexes. Heterologous sequences, such as *E. coli* DNA, calf thymus, or salmon sperm DNA can also be used, but often fortuitously contain binding sites which will reduce the amount of specific protein–DNA complexes observed. A procedure for preparing poly(dI-dC) is given in *Protocol 7*.

Protocol 7. Preparing poly(dI-dC): shearing and renaturing the copolymer

Equipment and reagents

- Sonicator (e.g. Branson Sonifier model 250 with a microtip probe)
- 90°C water bath/or heat block
- 50 mM NaCl in TE buffer (see *Protocol 6*)
- Poly(dI-dC) (Sigma, No. P 9514)
- Renaturation buffer (50 mM NaCl, 10 mM Tris–HCl pH 8.0, 1 mM EDTA pH 8.0).

Method

1. Resuspend the copolymer at a concentration of 10 mg/ml in 50 mM NaCl in TE buffer and place in a 1.5 ml microcentrifuge tube.
2. Briefly sonicate the DNA (10–20 sec on output control setting 4) to ensure that the copolymer is of uniform length.
3. Heat the DNA to 90°C for 10 min and allow to cool slowly to room temperature over several hours.
4. Dilute a portion of the poly(dI-dC) to 1 mg/ml in 50 mM NaCl in TE buffer and store at 4°C. Store the 10 mg/ml stock at −20°C in 100 μl aliquots.

2.2.2 The polyacrylamide gel

The mobility of a protein-DNA complex through a non-denaturing polyacrylamide gel is determined by the size, charge, and conformation of the protein bound to the DNA fragment. The low ionic strength (0.5–0.25 × TBE; see *Protocol 6*) as well as the 'caging effect' of the gel matrix helps to stabilize the protein–DNA complexes. Changing the acrylamide gel percentage, crosslinking, pH, ionic strength, or temperature may alter the mobility of a protein–DNA complex or reveal novel interactions. The preparation of a suitable gel is described in *Protocol 8*.

Protocol 8. Preparation of low crosslinking/low ionic strength polyacrylamide gels for mobility shift assay

Equipment and reagents

- Vertical electrophoresis tank, 15 cm × 20 cm glass plates, 1.5 mm spacers and combs
- 30% Acrylamide stock[a] (60:1 acrylamide: bisacrylamide) [Mix 29.0 g acrylamide, 0.5 g *N,N'*-methylene bisacrylamide and water to 100 ml. To dissolve the chemicals, heat the solution to 37°C with constant stirring. Filter through a 2 μm membrane and store at 4°C in a foil-wrapped bottle. Stable for several weeks]
- 0.5 × TBE electrophoresis buffer (45 mM Tris-borate, 1 mM EDTA) is prepared as a 5 × TBE stock as follows: Mix 54.0 g Tris base, 27.5 g boric acid, 20 ml 0.5 M EDTA pH 8.0, and add water to 1 litre final volume
- 10% (w/v) ammonium persulphate (e.g. BioRad, No. 161-0700)
- TEMED (e.g. BioRad, No. 161-0800).

Protocol 8. *Continued*

Method

1. Choose an appropriate gel concentration from the following table and mix the indicated volumes of reagents (for 40 ml gels).

	Polyacrylamide gel concentration (%)		
Stock solutions (for 40 ml gels)	**4.0**	**6.0**	**8.0**
30% acrylamide stock (ml)	5.3	8.0	10.75
H_2O (ml)	29.7	27.0	25.25
5 × TBE buffer (ml)	4.0	4.0	4.0
10% ammonium persulphate (ml)	0.4	0.4	0.4
TEMED (μl)	40	40	40

2. Pour the acrylamide mixture into the gel mould. Let the gel polymerize for 1 h before using it to fractionate the protein–DNA complexes (see *Protocol 9*).

[a] The acrylamide:bisacrylamide ratio (and hence the degree of crosslinking of the gel) may be varied anywhere between 80:1 to 30:1.

2.2.3 Protein–DNA binding and gel electrophoresis

The usual order of events is to prepare the polyacrylamide gel and, while it is undergoing pre-electrophoresis, set up the protein–DNA binding reactions. After the alloted incubation period, the binding reaction mixtures are loaded on to the pre-electrophoresed gel and electrophoresis is continued. Finally, the DNA bands are visualized by autoradiography. The detailed procedure for each of these steps is described in *Protocol 9*.

Protocol 9. Binding reactions and gel electrophoresis of protein–DNA complexes

Reagents

- Protein-DNA binding buffer (see *Table 1*)
- poly(dI-dC) (see *Protocol 7*)
- Low crosslinking/low ionic strength polyacrylamide gel (see *Protocol 8*)
- 5 × TBE buffer stock (see *Protocol 6*)
- Sample loading buffer [0.5 × TBE (see *Protocol 6*), 10% glycerol,[a] 0.25% Bromophenol Blue, 0.25% xylene cyanol]
- DNA binding protein (1–10 μl containing 0.5–100 ng purified binding protein or 1–10 μg total protein when using nuclear or whole cell extracts)
- ^{32}P-labelled DNA fragment containing the binding site of interest (1–10 pmol: 5000–50 000 c.p.m.).

Method

1. Pour the gel and let it polymerize for 1 h (see *Protocol 8*). After the gel has polymerized, remove the comb and rinse the wells with water.
2. Attach the gel to the electrophoresis tank.
3. Fill the buffer tanks with the electrophoresis buffer (e.g. 0.5 × TBE buffer). Flush out the wells and remove any air bubbles trapped beneath the gel with a bent Pasteur pipette or syringe needle.
4. Pre-electrophorese the gel for 60–90 min at 200 V at room temperature or at 350 V for 60–90 min at 4°C. (The current should drop from about 25 mA to less than 10 mA during the pre-electrophoresis when using the 0.5 × TBE buffer system.)

   [During electrophoresis, a pH gradient can form across the system, leading to uneven migration and banding patterns. Recirculating the buffer between the buffer tanks using a pump ensures uniform pH and ionic strength throughout the system and is especially important during extended electrophoretic runs.]
5. While the gel is pre-electrophoresing, set up the binding reactions as follows. Mix together in a microcentrifuge tube:

   10 μl of 2 × protein–DNA binding buffer
   1–2 μl of 1 mg/ml poly(dI-dC) (competitor DNA)
   1–10 μl of DNA binding protein (~0.5–100 ng purified protein or 1–10 μg total protein when using a nuclear or whole cell extract)
   H_2O to 20 μl final volume.

   Incubate for 5–10 min at the desired temperature (e.g. 22°C).
6. Add 1 μl ^{32}P-DNA fragment(s) (~1–10 pmol; typically 5000–50 000 c.p.m.) and incubate for 5–30 min at the desired temperature (e.g. 22°C).
7. Add 2 μl sample loading buffer to each binding reaction.
8. Load one-half of each sample (10 μl) on to the gel and electrophorese at 200 or 350 V at room temperature or 4°C, respectively for 1–2 h or until the Bromophenol Blue reaches the bottom of the gel.
9. Remove the gel plates from the electrophoresis tank. Insert a spatula and slowly prise apart the glass plates.
10. With the gel still attached to one of the plates, place a similar sized piece of Whatman 3MM paper on top of the gel.
11. Carefully peel the paper with the gel attached to it from the plate and cover the exposed surface of the gel with plastic wrap.
12. Dry the gel under vacuum at 80°C for 30 min.

Protocol 9. *Continued*

13. Autoradiograph the dried gel for several hours to visualize the DNA bands. A hypothetical autoradiogram is shown in *Figure 1a* to illustrate the type of data expected.

[a] Ficol 6000 may be substituted for glycerol.

2.3 The specificity of protein–DNA interactions

The specificity of a protein–DNA interaction can be easily measured by adding increasing amounts of unlabelled competitor DNA to a fixed amount of the labelled fragment. The DNA-binding proteins are then added and the

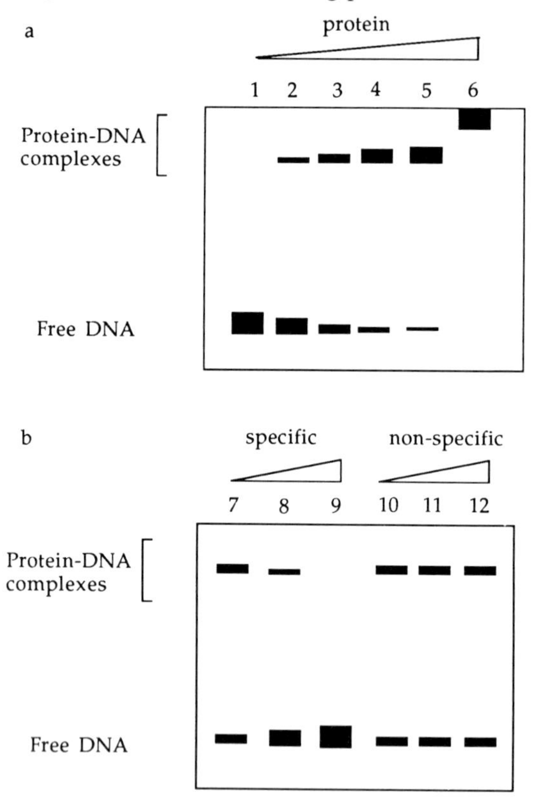

Figure 1. Hypothetical autoradiogram of a gel mobility shift experiment. (a) Protein–DNA binding. A DNA fragment containing a protein recognition site was ^{32}P-labelled and incubated with increasing amounts of a DNA binding protein (lanes 1–6). Lane 1 contains only the DNA probe (no added protein). Note that at extremely high protein concentrations, the protein–DNA complexes do not enter the gel appreciably. (b) Specificity of protein–DNA interaction. Lanes 7–9 include increasing concentrations of an unlabelled competitor DNA, identical to the probe in the binding reaction, whereas lanes 10–12 contain increasing amounts of a non-specific competitor DNA.

complexes resolved by electrophoresis on a non-denaturing gel. If a vast excess of unlabelled competitor DNA is added which contains a binding site identical to the probe, it should bind the protein and abolish the complex(es) observed from the trace amount of labelled DNA in the reaction. However, if the unlabelled competitor DNA does not contain a high-affinity binding site for the protein, then it will not affect the protein–DNA interaction (*Figure 1b*).

The gel mobility shift assay is also an ideal method for determining equilibrium and kinetic constants between proteins and DNA, since protein–DNA complexes can be easily separated from free DNA. An apparent equilibrium dissociation constant (K_{eq}) can be determined experimentally by producing a standard binding curve; a known quantity of labelled DNA is mixed with increasing concentrations of protein, with the DNA concentration held constant at a level well below the concentration of the protein. The resulting protein–DNA complexes are quantified by scintillation counting or densitometry. The point at which 50% of the labelled DNA is bound with protein corresponds to the apparent K_{eq} as shown by the following.

Protein–DNA interactions can be represented as an equilibrium reaction:

$$[\text{protein}_{\text{free}}] + [\text{DNA}_{\text{free}}] \underset{k_2}{\overset{k_1}{\rightleftharpoons}} [\text{protein-DNA complex}]$$

$$K_{eq} = \frac{[\text{protein}_{\text{free}}]\,[\text{DNA}_{\text{free}}]}{[\text{protein-DNA complex}]} = \frac{k_1}{k_2}.$$

where K_{eq} is the apparent equilibrium dissociation constant (K_D app), k_1 is the dissociation rate constant, and k_2 is the association rate constant.

When 50% of the protein is bound to DNA,

$$[\text{DNA}_{\text{free}}] = [\text{protein–DNA complex}],$$

and the above equation simplifies to:

$$K_{eq} = [\text{protein}_{\text{free}}].$$

For proteins with high affinity, the protein concentration used, and therefore the DNA concentration used, must be very low (10^{-8}–10^{-12} M). Thus, the DNA fragments employed must be radiolabelled to high specific activity.

2.4 Altered DNA conformations as detected by gel mobility shift assay: analysis of protein-induced DNA bending

The conformation of DNA can be altered by interaction with regulatory proteins (5). Proteins from both prokaryotes and eukaryotes, such as catabolite-activating protein (6) and the oncoproteins Fos and Jun (7), respectively, can induce bends in the DNA helix, potentially affecting the interactions of other regulatory proteins bound at separate sites. Bent DNA fragments can be detected because they migrate more slowly than linear DNA molecules of the same size during electrophoresis. In addition, when a

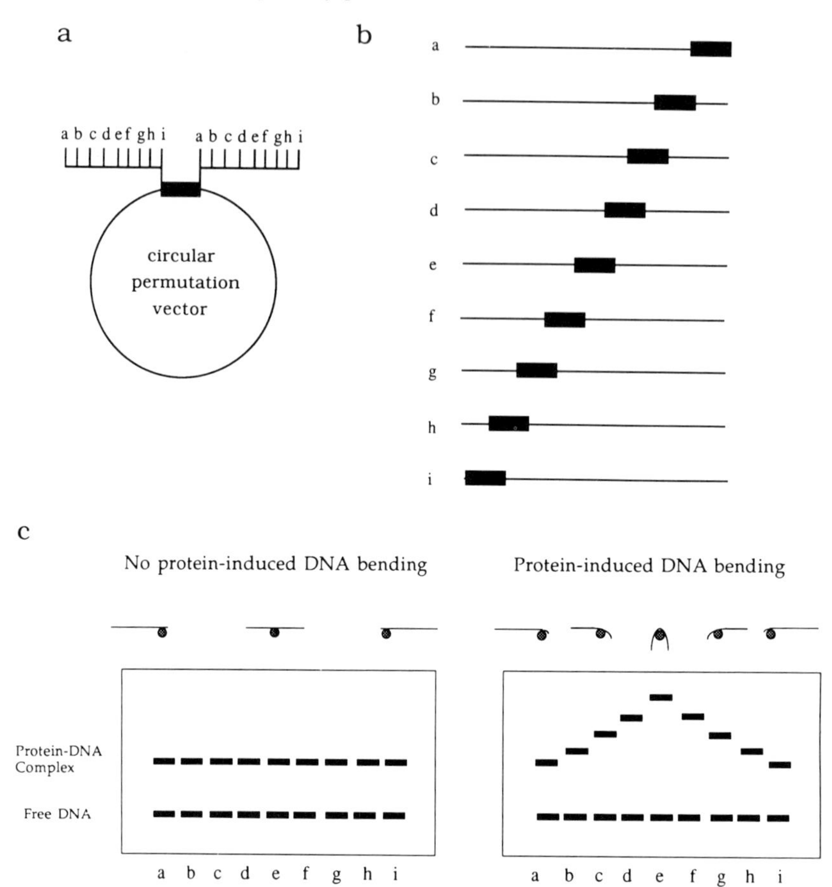

Figure 2. Protein-induced DNA bending. (a) Circular permutation vector (10). The prototypic vector for analysing DNA bending contains two identical DNA segments with restriction sites (denoted by letters a–i) in a direct repeat spanning a central region containing a protein binding site (black rectangle). (b) Constructing sets of DNA fragments with binding site permutations. Fragments of identical length which place the protein-binding site in permutated order are generated by digesting the circular permutation vector with restriction enzymes a–i. (c) Hypothetical autoradiograms of protein–DNA complexes as a function of binding site position. No change in the electrophoretic mobility of the protein–DNA complex is observed if the protein does not bend DNA. If a protein bends DNA, however, a reduction in the electrophoretic mobility of the protein–DNA complex is observed as a function of the position of the binding site.

bend occurs at the centre of a DNA molecule, it will affect the shape of that molecule to a greater degree than if the bend occurs near an end (see *Figure 2*). Fortunately, protein-induced DNA bending can be easily monitored in the gel mobility shift assay by using circularly permuted DNA fragments (see *Figure 2*) that are identical in length, but place the protein-binding site at different distances from the centre of the molecule. If a protein bends the

DNA, then binding sites in the centre of a DNA fragment will lead to the protein–DNA complex being retarded more severely than if the binding site lies close to an end. However, the mobility of the protein–DNA complex should not be affected by binding site location within the fragment if bending does not occur (see *Figure 2*). Thus, the relative mobility of the protein–DNA complex as a function of the distance from the centre of the binding site to the end of the probe can be used to monitor protein induced bending (8). Methods for the analysis of protein-induced DNA bending using circularly permutated DNA are described in *Protocol 10*. The degree and direction of bending can also be estimated by utilizing intrinsically bent DNA as a standard (7, 9).

Protocol 10. Analysis of protein-induced DNA bending using circularly permuted DNA and the gel mobility shift assay

Reagents

- Circular permutation vector (10)
- DNA binding protein (see *Protocol 9*)
- End-labelling reagents and DNA fragment isolation reagents (see *Protocols 1–6*)
- Materials for gel mobility shift assay (see *Protocols 7–9*)
- Suitable restriction enzymes.

Method

1. Clone the protein binding site of interest into the circular permutation vector (see *Figure 2a* and ref. 10).
2. Digest the DNA with restriction enzymes chosen to generate DNA fragments that contain the protein binding site at different distance from the centre of the molecule (see *Figure 2a* and *b*).
3. End label the DNA probes (see *Protocols 1–3*) and purify them by gel electrophoresis (see *Protocols 5* and *6*).
4. Mix the binding protein of interest with the DNA and resolve the complexes using the gel mobility shift assay (see *Protocol 9*).

2.5 Dimerization of DNA-binding proteins

Many eukaryotic and prokaryotic transcriptional regulatory proteins recognize DNA-binding sites with a twofold axis of symmetry (6). This observation prompted speculation that dimerization may be a general feature of DNA-binding proteins, with each protein subunit contacting one-half of the DNA recognition site. Indeed, X-ray diffraction analysis of crystallized protein–DNA complexes from bacterial repressor proteins and steroid hormone receptors demonstrate directly that these proteins bind DNA as dimers (6, 11).

The gel mobility shift assay is a useful tool for determining the number of protein molecules bound per DNA fragment. In principle, dimerization can be assessed by using two different-sized derivatives of the same protein, incubating these proteins with specific DNA sequences and resolving the protein–DNA complexes by gel electrophoresis. Individually these proteins will exhibit a particular electrophoretic mobility when bound to DNA. Together, however, if the protein binds DNA as a dimer, then an additional complex of intermediary electrophoretic mobility is observed (see *Figure 3a*). This novel intermediary species represents one molecule of each protein bound simultaneously to the DNA. The appearance of additional complexes may suggest higher order structures, such as trimers or tetramers.

Figure 3b shows an actual analysis of this type for the glucocorticoid receptor. As shown in *Figure 3b*, incubating the target DNA with increasing concentrations of a 1:1 mixture of two different-sized derivatives of the same protein (over-expressed and purified from *E. coli*) results in a series of protein–DNA complexes visualized by non-denaturing gel electrophoresis. At low protein concentrations, two retarded species are observed, representing one of each protein molecule bound per DNA fragment. With increasing concentrations of protein, additional species appear, which correspond to two protein molecules bound. Novel intermediate complexes represent one molecule of each of the two different proteins binding to the DNA fragment. As expected for dimers, the intermediate complexes were present in roughly twice the abundance of the other two complexes. Footprinting in the gel can be performed (see Section 3) to locate the exact nucleotides bound by the protein molecule(s) in each of the specific complexes.

Several methods are available for creating DNA-binding proteins of different sizes from a single cloned gene. Hope and Struhl pioneered the use of *in vitro* synthesized proteins for DNA binding studies (12). Two different-sized derivatives of the same protein are co-synthesized *in vitro* by translation of two mRNAs of different coding length produced from *in vitro* transcription using bacteriophage RNA polymerase. The two proteins of different size are mixed with the target DNA and resolved by non-denaturing gel electrophoresis.

Expression of cloned transcription factors *in vivo* can also be used to assess dimer formation. Extracts of mammalian cells transfected with a cDNA expression vector for the intact transcription factor protein can be mixed with extracts from cells transfected with a series of deletion derivatives (13). As above, the two different-sized protein derivatives are mixed together in the presence of DNA and assayed for their ability to form complexes of intermediate mobility, indicating dimer formation.

3. Assays for sites of protein–DNA contact: footprinting techniques

3.1 Nuclease protection assays

Nuclease protection experiments furnish a direct readout of the DNA nucleotides recognized by DNA-binding proteins. The technique is very simple in

(a)

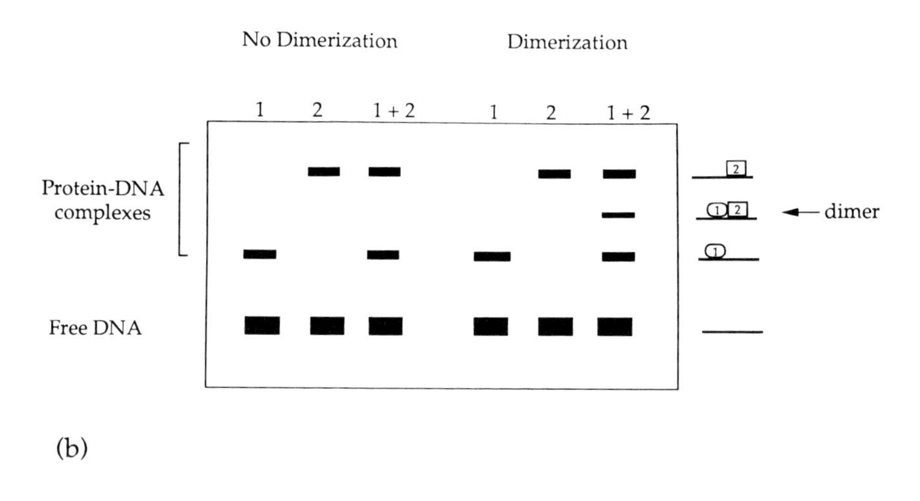

(b)

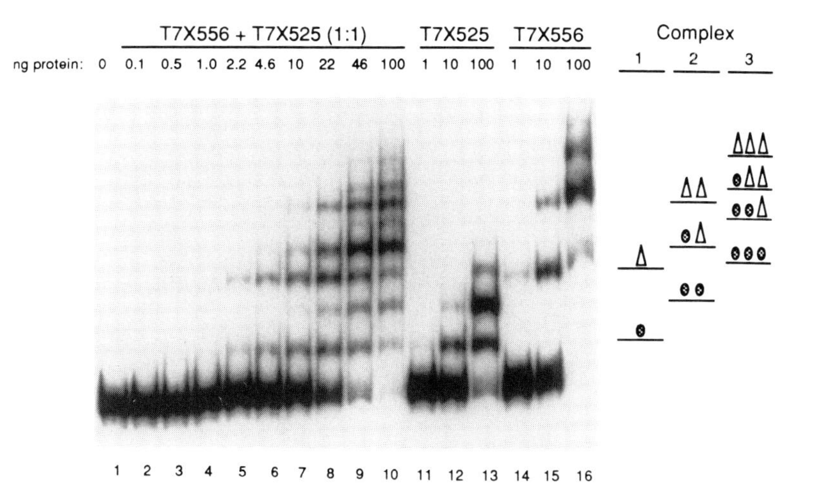

Figure 3. Assay for dimerization of DNA-binding protein. (a) A hypothetical autoradiogram of a gel mobility shift assay of dimerization. Two different-sized derivatives of the same DNA-binding protein (designated 1 and 2) were individually or simultaneously mixed with DNA and protein–DNA complexes resolved on non-denaturing polyacrylamide gels. The figure illustrates the band profile expected in each case. (b) A DNA fragment containing a binding site for the glucocorticoid receptor was ^{32}P-labelled and incubated with the indicated amounts of two protein derivatives of different size from the glucocorticoid receptor DNA binding domain T7 ×556 (mol. wt. ~17 kDa) and T7 ×525 (mol. wt. ~14 kDa). These proteins were produced in *E. coli* and purified to homogeneity (11). The identities of the different protein–DNA complexes are schematized on the right. T7 × 556 is represented by open triangles and T7 ×525 is represented by filled ovals.

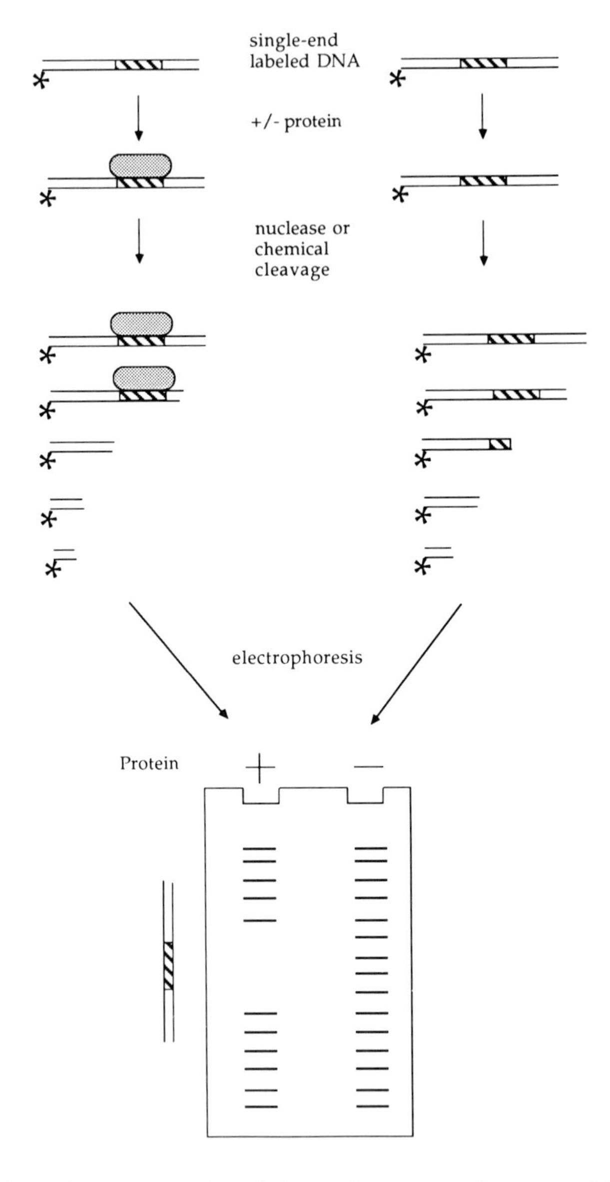

Figure 4. Schematic representation of the nuclease protection assay. DNA fragments containing the protein-binding site (hatched region) are labelled at one end with ^{32}P (denoted by an asterisk). One aliquot of the DNA is incubated with protein (filled ellipse). Both the protein-bound DNA and free DNA mixtures are reacted with chemical or enzymatic reagents under conditions that lead to random cleavage of the DNA once on average. This produces a succession of DNA fragments separated by single nucleotide gaps, termed a ladder. The protein-bound DNA and free DNA samples are electrophoresed side by side on a denaturing polyacrylamide gel. The protein binding site(s) or 'footprint(s)' appear(s) as an interruption in the ladder of DNA fragments and is visualized by autoradiography.

principle; a sequence-specific DNA binding protein 'protects' the nucleotides involved in DNA binding from enzymatic or chemical attack. A probe containing a binding site is labelled at one end and allowed to equilibrate with DNA binding protein(s). The complex is then digested with enzymatic nucleases such as DNase I or with chemical reagents which cleave DNA, such as the hydroxy radical or copper-phenanthroline ions (15). Finally, the cleaved products are resolved on denaturing polyacrylamide gels and autoradiographed.

The level of cleavage by either chemicals or nucleases is chosen to result in one cleavage per DNA molecule, on average, at random sites and produces a population of radiolabelled fragments that differ by a single nucleotide (termed a 'ladder'). Regions in which DNA-bound proteins protect the DNA backbone from chemical or enzymatic hydrolysis result in a nuclease resistant region called a 'footprint'. The location and extent of the DNA being bound by protein can be determined by electrophoresing the products of this digestion on a denaturing polyacrylamide gel. The hydrolysis products of another aliquot of the DNA which was not incubated with the protein(s) are electrophoresed in a neighbouring well of the same gel to act as size markers (see *Figure 4*).

3.1.1 Preparation of the single-end labelled DNA probe

The DNA probe is prepared by restriction digestion of an appropriate recombinant plasmid DNA, then radiolabelling the exposed 3′ end, and finally restricting with a second enzyme which releases the end-labelled fragment from the plasmid. The experimental procedures are given in *Protocol 11*.

Protocol 11. Preparation of the single-end-labelled DNA probe

Reagents

- Recombinant plasmid DNA (1 mg/ml in TE buffer)
- Appropriate restriction enzyme(s) (10 units/μl)
- 10 × restriction enzyme buffer (see *Protocol 1*)
- 3′-end-labelling reagents (see *Protocol 1*)
- An agarose or acrylamide gel for DNA fragment isolation (see *Protocols 5* and *6*)
- TE buffer (10 mM Tris–HCl, 1 mM EDTA pH 8.0).

A. *First restriction digest*

1. Mix together in a 1.5 ml microcentrifuge tube:

plasmid DNA (1 mg/ml)	2 μl
10 × restriction enzyme buffer	2 μl
H_2O	to 20 μl final volume
first restriction enzyme (10 units/μl)	1 μl.

2. Incubate at 37°C for 2 h.

Protocol 11. *Continued*

3. Label the DNA at the 3′ end as described in *Protocol 1*.
4. Ethanol precipitate the labelled DNA.
5. Resuspend the double-end-labelled DNA in 20–50 μl TE buffer (see *Protocol 1*).

B. *Second restriction digest*

1. Mix together in a 0.5 ml microcentrifuge tube:

Double-end-labelled DNA	18 μl
10 × restriction enzyme buffer	2 μl
Second restriction enzyme	2 μl.

2. Incubate at 37°C for 1 h until the digest is complete.

C. *Isolation of the single-end-labelled DNA probe*

Gel purify the single-end-labelled DNA probe by electrophoresis on a 1.5% low melting temperature agarose gel or on a 5% polyacrylamide as described in *Protocols* 5 and 6, respectively.

Note: the quality of the single-end labelled DNA is essential to successful footprinting; contaminants from agarose or acrylamide gels can adversely affect the quality and reproducibility of the footprint. Therefore, DNA probes isolated by gel electrophoresis should be further purified by DEAE chromatography (Elutip-d, Schleicher & Schuell) or by the glass powder method (Gene-clean Bio101) to ensure consistent results.

3.1.2 Preparation of denaturing polyacrylamide gels

The denaturing (urea) polyacrylamide gel used to separate the radiolabelled DNA products of the protection experiments are prepared as described in *Protocol 12*. These gels can be prepared the day before the protection experiment and left overnight.

Protocol 12. Preparation of denaturing (urea) polyacrylamide gels

Equipment and reagents

- Vertical electrophoresis tank, 15 cm × 60 cm glass plates, 0.4 mm spacers and comb
- Electrophoresis buffer; 5 × TBE buffer stock (see *Protocol 6*)
- Urea, ultra pure; e.g. ICN/Schwarz/Mann Biotech, No. 821527)
- 10% (w/v) ammonium persulphate
- TEMED
- 40% acrylamide stock (38:2, acrylamide: bisacrylamide) [Mix 38.0 g acrylamide, 2 g *N,N*′-methylene bisacrylamide and water to 100 ml final volume. Dissolve the chemicals by heating the solution to 37°C with constant stirring. Filter the solution through a 2 μm membrane filter and store it at 4°C in a foil-wrapped bottle. Acrylamide stock is stable for several weeks].

Method

1. Choose an appropriate gel concentration from the following table, for 50 ml gels.

Stock solutions (for 50 ml gels)	Polyacrylamide gel concentration (%)			
	5	**6**	**7**	**8**
40% acrylamide (ml)	6.25	7.5	8.75	26.6
H_2O (ml)	16.25	15.0	13.75	12.5
5 × TBE buffer (ml)	10.0	10.0	10.0	10.0
Urea (g)	24	24	24	24
10% ammonium persulphate (ml)	0.25	0.25	0.25	0.25
TEMED (μl)	40	40	40	40

2. Clean the gel plates thoroughly. Siliconize one of the plates with 5% dichlorodimethylsilane in chloroform. Wash the siliconized plate again and assemble the gel plates with the spacers.
3. Weigh out the urea and place it into a 250 ml beaker or flask. Add the required volume of acrylamide stock, water, and 5 × TBE buffer. Gently heat and stir the gel mixture until all of the urea is in solution (~10 min on low heat).
4. Filter the mixture by gravity through a funnel lined with Whatman No. 1 filter paper. (This removes any undissolved urea particles and results in a better gel.)
5. After filtering the mixture, add the ammonium persulphate and TEMED and immediately pour the gel. Insert the comb and let the gel mixture polymerize for 1 h (gels can be left overnight).
6. Once the gel has polymerized, remove the comb and assemble the gel apparatus. Pre-electrophorese the gel for 30–60 min in 1 × TBE at 45 W constant power (the gel plates should be quite warm just before the gel is loaded with samples in *Protocol 13*, step 11).

3.1.3 The nuclease protection assay using DNase I

When the single-end-labelled DNA probe and a suitable denaturing (urea) polyacrylamide gel have been prepared (see *Protocol 12*), the nuclease protection assay can be carried out as described in *Protocol 13*. The results of a typical assay are shown in *Figure 5*.

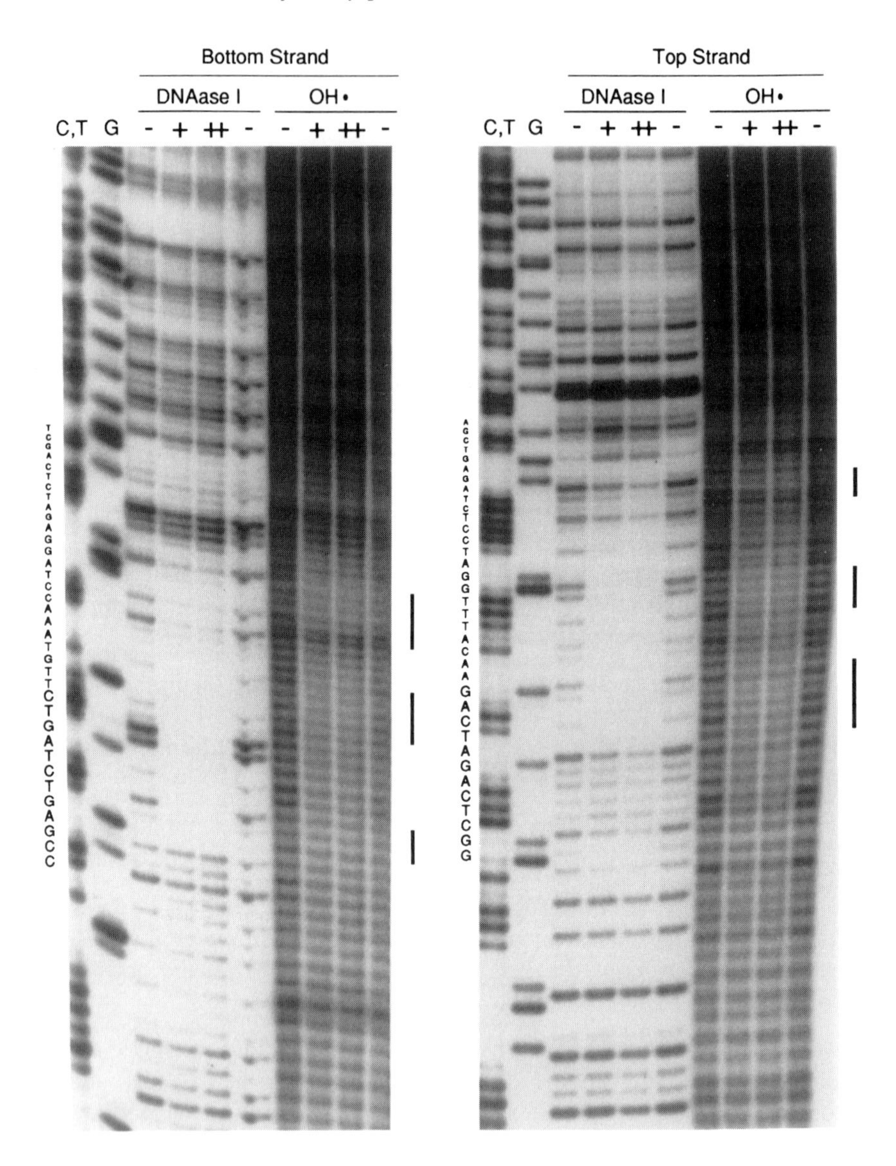

Figure 5. DNase I and hydroxyl radical footprinting. A DNA fragment containing a binding site for the glucocorticoid receptor was ^{32}P-labelled on either the top or bottom strand, mixed with increasing concentrations (+) (++) of the DNA binding domain of the glucocorticoid receptor over-expressed and purified from *E. coli* (11) or left without protein (−). The protein-bound DNA and free DNA were cleaved with either DNase I or hydroxyl radicals (OH·). Bases protected from hydroxyl radical cleavage are indicated by solid bars. Lanes G and C,T are probe DNA cleaved with guanosine-specific or cytidine and thymidine-specific reagents respectively as described elsewhere for chemical sequencing (4). The nucleotide sequence in the footprint region is shown on the left.

Protocol 13. Nuclease protection assay using DNase I[a]

Materials

- Single-end-labelled DNA probe (see *Protocol 11*)
- DNA binding protein (see *Protocol 9*)
- 2 × binding buffer (40 mM Tris–HCl pH 7.9, 2 mM EDTA, 20% glycerol, 0.2 ml Nonidet P-40, 4 mM $MgCl_2$, 2 mM DTT)
- Salts mixture (10 mM $MgCl_2$, 5 mM $CaCl_2$)
- DNase I stock [DNase I (Worthington or United States Biochemical; 2182 units/mg) in 50% glycerol, 1 × binding buffer] This stock is stable indefinitely at −20°C
- 1 mg/ml poly(dI-dC) (see *Protocol 7*)
- 20% polyvinyl alcohol (Sigma, No. P 8136)
- TE buffer (10 mM Tris–HCl, 1 mM EDTA pH 8.0)
- 1 × TBE buffer (see *Protocol 6*)
- Phenol:chloroform:isoamyl alcohol (24:24:1 by vol.)
- DNase I dilutions
  Dilute the DNase I stock into a 1:1 mixture of 2 × binding buffer and salts mixture just prior to use: dilutions can range from 1:1000–1:10 000 and should be optimized for each protein–DNA interaction studied. As a simple guide, use an amount of DNase I that will cleave only about 50% of the labelled DNA probe used in the reaction
- DNase I stop mixture (0.2 M NaCl, 40 mM EDTA, 1% SDS, 125 μg/ml tRNA, 100 μg/ml Proteinase K). [*Note*: the Proteinase K must be added just prior to use]
- Pre-electrophoresed denaturing (urea) polyacrylamide gel (see *Protocol 12*)
- Formamide loading dye for denaturing gels (95% formamide, 4% EDTA pH 8.0, 0.1% Bromophenol Blue, 0.1% xylene cyanol).

Method

Note: cleavage by DNase I is not completely random. Therefore, to control for sequence-specific cleavage by the enzyme, 'naked' DNA is also digested and run alongside the protein bound products.

Set up separate 'protein' and 'DNA' mixtures as follows:

1. Protein mixture (50 μl per reaction):
In a single microcentrifuge tube mix together the following and store on ice:

2 × binding buffer	25 μl
10–1000 ng DNA-binding protein	1–25 μl
H_2O	to 50 μl final volume.

2. DNA mixture (50 μl per reaction):
Mix the following together and leave at room temperature:

Single-end-labelled DNA fragment	1–10 μl (~20 000 c.p.m.)
20% polyvinyl alcohol	10 μl
1 mg/ml poly(dI-dC)	1 μl
TE buffer	to 50 μl final volume.

3. Combine 50 μl of the protein mixture and 50 μl of the DNA mixture; incubate at room temperature for 15 min. This is the protein-binding reaction. Also set up the 'naked DNA' tube by mixing another 50 μl DNA mixture with 25 μl of 2 × binding buffer and 25 μl water.

Protocol 13. *Continued*

4. Add 100 μl of the salts mixture to each tube and incubate for another 1 min.
5. Add 10 μl of the diluted DNase I solution to each tube and incubate for exactly 1 min. (The concentration of DNase I that gives ~50% digestion should be titrated for each DNA fragment; see *Reagents*.)
6. Add 200 μl DNase I stop mixture to each tube and incubate at 37°C for at least 15 min.
7. Add 200 μl phenol:chloroform:isoamyl alcohol (24:24:1) to each tube and vortex for 10 sec.
8. Spin each tube in a microcentrifuge at 12 000 *g* for 5 min. Transfer the aqueous (upper) phase to a fresh tube and precipitate the DNA by adding 1 ml ice-cold 95% ethanol. Pellet the DNA by centrifugation at 12 000 *g* for 5 min. Rinse the pellet with cold 70% ethanol and dry under vacuum.
9. Resuspend each pellet in 10 μl formamide-loading dye and heat to 95°C for 2–3 min.
10. Determine the total radioactivity recovered by Cerenkov counting each sample for 1 min in a scintillation counter.
11. Load equal amounts of radioactivity (about 5000–8000 c.p.m.) per lane on to a pre-electrophoresed denaturing urea polyacrylamide gel. Electrophorese in 1 × TBE buffer (see *Protocol 6*) at 45 W constant power for 2–3 h or until the Bromophenol Blue is near the bottom of the gel.
12. Remove the gel plates from the box. Insert a spatula and slowly prise apart the glass plates.
13. With the gel still attached to one of the plates, place a similar sized piece of Whatman 3MM paper on top of the gel.
14. Carefully peel the Whatman 3MM paper with the gel attached to it from the plate and cover the exposed surface of the gel with plastic wrap.
15. Dry the gel under vacuum at 80°C for 30 min.
16. Autoradiograph the dried gel at −70°C for several hours with an intensifying screen to visualize the DNA bands.

[a] From ref. 14.

3.2 Hydroxyl radical footprinting

This footprinting technique utilizes the highly reactive hydroxyl radical to cleave the DNA (16, 17). Unlike most cleavage reagents, the hydroxyl radical is the size of a water molecule and is small enough to attack the exposed sites on the 'backside' of the DNA helix to which a protein is bound. This method (see *Protocol 14*) can identify the precise nucleotides that are contacting the bound protein. In addition to its small size, the free radical, which is generated by the reduction of hydrogen peroxide with iron (II), shows virtually no sequence dependence in its ability to cleave DNA (see *Figure 5*).

Protocol 14. Hydroxyl radical footprinting

Reagents

- Single-end-labelled DNA fragment (see *Protocol 11*)
- DNA binding protein (see *Protocol 9*)
- 2 × binding buffer (without glycerol; see *Protocol 13* for composition)
- Stop mixture (same composition as the DNase I stop mixture; see *Protocol 13*)
- Ferrous ammonium sulphate $[Fe(NH_2)_2(SO_4)_2{\cdot}6H_2O]$ (Aldrich)
- 0.5 M EDTA pH 8.0
- Sodium ascorbate
- 30% hydrogen peroxide stock
- Poly(dI-dC) (see *Protocol 7*)
- Polyvinyl alcohol
- TE buffer (10 mM Tris–HCl, 1 mM EDTA pH 8.0)
- 1 × TBE buffer (see *Protocol 6*)
- Phenol:chloroform:isoamyl alcohol (24:24:1 by vol.)
- Formamide-loading dye (see *Protocol 13*)
- Pre-electrophoresed denaturing (urea) polyacrylamide gel(s) (see *Protocol 12*).

Prepare the following reagents *fresh* before each experiment:

- 80 μM $Fe(NH_2)_2(SO_4)_2{\cdot}6H_2O$
- 160 μM EDTA
- 20 mM sodium ascorbate
- 0.06% hydrogen peroxide (diluted from the 30% stock).

Method

1–3. Set up separate 'protein' and 'DNA' mixtures as in the DNase I footprinting assay (see *Protocol 13*, steps 1–2) and combine them as in *Protocol 13*, step 3.

4. Generate an iron (II)/EDTA complex by mixing the following reagents in a 1.5 ml microcentrifuge tube:

80 μM $Fe(NH_2)_2(SO_4)_2{\cdot}6H_2O$	200 μl
160 μM EDTA	200 μl.

5. To generate the hydroxyl radical, carefully add 10 μl of the iron/EDTA solution, 10 μl 0.6% hydrogen peroxide and 10 μl 20 mM sodium ascorbate to the inside wall of the microcentrifuge tube containing the protein–DNA complex.

6. Mix the droplets suspended on the walls of the tube with the protein DNA complex by vortexing to begin the DNA cleavage.

Although the sodium ascorbate helps to stabilize the free radicals, they are extremely short-lived. It is therefore important to initiate and mix the reaction as quickly as possible.

Protocol 14. *Continued*

The final concentration in a typical reaction is 10 μM iron (II), 20 μM EDTA, 0.03% hydrogen peroxide, 1 mM sodium ascorbate.

7. Allow the reaction to proceed for 1 min at room temperature.
8. Terminate the reaction by adding 170 μl DNase I stop mixture and incubate at 37°C for 15 min.
9. Purify the DNA by phenol extraction and analyse the products by electrophoresis as described in *Protocol 13*, steps 7–16.

This footprinting procedure can be technically challenging. For example, hydrogen peroxide may adversely affect the DNA-binding ability of certain proteins, possibly by disrupting the protein's structure. This problem can be partially overcome by reducing the concentration of hydrogen peroxide in the reaction by one-half. In addition, since free radicals are very short-lived, it is often helpful to initiate the hydroxyl radical reaction by mixing each individual reagent with the protein–DNA complex simultaneously, thereby avoiding the decay of the free radical that occurs when the chemicals are premixed. This can be performed in a slightly different way than described in *Protocol 14*. Aliquots of 5 μl 80 μM Fe $(NH_2)_2$ $(SO_4)_2 \cdot 6H_2O$, 5 μl 160 mM EDTA, 10 μl 20 mM sodium ascorbate, and 10 μl 0.06% hydrogen peroxide are placed in a Petri dish, taking care to keep the reagents separate from one another. Each reagent is then taken up, one by one, into the tip of an adjustable pipetting device, allowing a small gap of air after each to act as a partition, except for the final two chemicals, ferrous ammonium sulphate and EDTA, which are allowed to mix. Then the cleavage reaction is initiated by injecting all of the reagents into the binding reaction. Initiating the reaction in this way, (i.e. without pre-mixing the reagents) maximizes the exposure of the protein–DNA complex to the short-lived free radical and can help to achieve reproducible footprints.

3.3 Footprinting protein–DNA complexes following mobility shift assay

3.3.1 Copper-phenanthroline footprinting

This method (see *Protocol 15*) combines the technical simplicity of the gel mobility assay with the detailed sequence information obtained from protection experiments. Like the hydroxyl radical, the 1,10 phenanthroline-copper ion has the ability to cleave DNA (15). This DNA cleavage reagent has been successfully used to footprint protein–DNA complexes embedded within the gel matrix following mobility shift assay (18). After cleavage is completed, the DNA is eluted from the gel and resolved by electrophoresis on denaturing gels.

Protocol 15. *O*-Phenanthroline-copper footprinting of DNA-protein complexes *in situ* following the mobility shift assay

Reagents

- Single-end-labelled DNA (see *Protocol 11*)
- DNA-binding protein (see *Protocol 9*)
- 9 mM copper (II) sulphate (cupric sulphate) in water (Aldrich, No. 20,917-1)
- 40 mM 1,10-phenanthroline in 100% ethanol (Aldrich, No. 13,137-7)
- 3-mercaptopropionic acid (Aldrich, No. M 580-1)
- 1.2% 2,9 dimethyl 1,10 phenanthroline in 100% ethanol (Aldrich, No. 12,189-4)
- Gel elution buffer [0.2 M NaCl, 20 mM EDTA pH 8.0, 1% SDS, 1 mg/ml tRNA]
- Phenol:chloroform:isoamyl alcohol (24:24:1 by vol.)
- Formamide loading dye (see *Protocol 13*)
- Pre-electrophoresed denaturing (urea) polyacrylamide gel(s) (see *Protocol 12*).

Method

1. Incubate the DNA binding protein of interest with a single-end-labelled DNA probe (prepared as in *Protocol 11*) and resolve the resulting complexes on a non-denaturing polyacrylamide gel as described in *Protocol 9*, except do not use tracking dye in lanes containing protein–DNA complexes. The standard reaction should be scaled up 5–10 fold and run over several lanes of the gel.
2. After electrophoresis is complete, remove the top plate and transfer the gel (still attached to the other plate) to a glass dish containing 200 ml 10 mM Tris–HCl pH 8.0. Allow the gel to equilibrate in this buffer for 10 min at room temperature.
3. While the gel is equilibrating, mix the following reagents together in a small beaker or flask:

9 mM $CuSO_4$	1 ml
40 mM 1,10-phenanthroline	1 ml.

   Allow this mixture to stand for 1 min at room temperature (turns blue). Then add 18 ml water.
4. Add the entire 20 ml of the $CuSO_4$/phenanthroline solution to the equilibrated gel.
5. Initiate the reaction by adding 100 μl 3-mercaptopropionic acid diluted in 20 ml water (the solution should turn brown).
6. Allow the cleavage reaction to proceed for 1 min at room temperature without shaking. (The time of incubation with the cleavage reagent should be optimized for each protein–DNA complex studied.)
7. Stop the reaction by adding 10 ml 1.2% 2,9-dimethyl-1,10-phenanthroline (the solution should turn yellow).

Protocol 15. *Continued*

8. After 2 min, pour off the solution and rinse the gel once with water.
9. Using the bottom gel plate for support, remove the gel from the glass dish and cover with plastic wrap.
10. Expose the wet gel to X-ray film for several hours or overnight at 4°C.
11. Excise that portion of the gel corresponding to the protein–DNA complex and likewise excise the area of the gel containing the free DNA probe.
12. In each case, elute the DNA from the gel by soaking in 500 µl gel elution buffer overnight at 37°C.
13. Discard the gel fragments and extract each supernatant with an equal volume of phenol:chloroform:isoamyl alcohol (24:24:1).
14. Remove each supernatant to a fresh tube and precipitate the DNA by adding 2.5 vol. 95% ethanol. Rinse each pellet once with 70% ethanol and dry.
15. Dissolve each of the DNA pellets in 10–20 µl formamide-loading dye, denature them by heating to 90°C for 3 min, and then load equal amounts of radioactivity on to a pre-electrophoresed denaturing (urea) polyacrylamide gel (see *Protocol 13*, steps 9–11).
16. Visualize the DNA band profile by autoradiography (see *Protocol 13*, steps 12–16).

An example of the use of copper-phenanthroline footprinting is shown in *Figure 6* for glucocorticoid receptor-DNA binding.

4. Methylation interference assay

Footprinting or protection experiments use nucleases or chemical cleavage reagents to analyse DNA sequences that already have bound proteins and thus are protected by them. For interference assays, the actual protein binding is sterically blocked by chemical modifications on the DNA (see *Figure 7*).

The interference technique (see *Protocol 16*) again exploits the gel mobility

---

Figure 6. 1,10-Phenanthroline-copper ion (OP–Cu) footprinting. A DNA fragment containing a binding site for the glucocorticoid receptor was ^{32}P-labelled on either the top or bottom strand, mixed with protein, and resolved on a non-denaturing polyacrylamide gel. Protein–DNA complexes (monomer and dimer, complexes 1 and 2, respectively) were footprinted *in situ* with 1,10-phenanthroline-copper ion, eluted from the gel, and resolved on a 6% denaturing polyacrylamide gel. Lanes G and C,T are probe DNA cleaved with guanosine-specific or cytidine and thymidine-specific reagents respectively as described elsewhere for chemical sequencing (4). The nucleotide sequence in the footprint region is shown at the left.

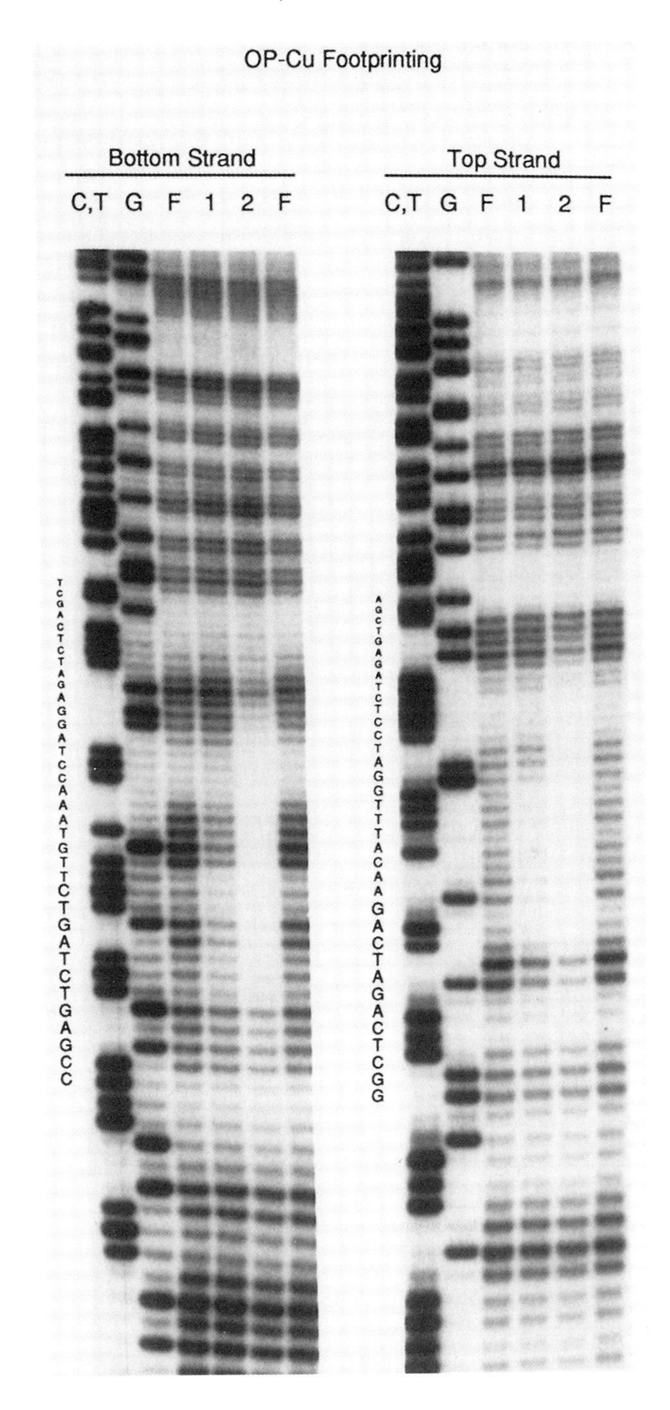
OP-Cu Footprinting
Bottom Strand
C,T G F 1 2 F
Top Strand
C,T G F 1 2 F
TCGACTCTAGAGGATCCAAATGTTCTGATCTGAGCC
AGCTGAGATCTCCTAGGTTTACAAGACTAGACTCGG

shift assay. A single-end-labelled DNA fragment is treated with dimethylsulphate, methylating the N-7 position of deoxyguanosine residues (19, 20). On average, each DNA molecule is methylated once at random sites. The modified DNA is incubated with the protein of interest and protein-DNA complexes are separated from free DNA by gel electrophoresis. Both protein-bound DNA and free DNA are eluted, cleaved at the sites of modification and electrophoresed on denaturing (urea) polyacrylamide gels along with size standards. The protein-complexed DNA will exhibit cleavages at all positions *except* those involved in DNA binding. The free DNA will display an increased cleavage at sites where the modification of the DNA interferes with protein binding, compared to the DNA-protein complex situation. Steric inhibition at modified DNA sites is interpreted as contacts or interaction between protein and those DNA sites. This approach can identify specific guanosine nucleotides involved in binding the protein.

Protocol 16. Methylation interference assay using the gel mobility shift assay: identifying guanine-specific contacts

Reagents

- Single-end-labelled DNA (see *Protocol 11*)
- DNA-binding proteins (see *Protocol 9*)
- 99% dimethyl sulphate (DMS) (Aldrich, No. D18,630-9)
- 20% piperidine in water (Sigma, No. P-5881)
- DMS stop solution [1.5 M sodium acetate pH 7.0, 200 μg/ml tRNA (see *Protocol 4*), 7% (v/v) 2-mercaptoethanol]
- TE buffer (10 mM Tris–HCl, 1 mM EDTA pH 8.0)
- Gel elution buffer (0.2 M NaCl, 20 mM EDTA, pH 8.0, 1% SDS, 1 mg/ml tRNA)
- Phenol:chloroform:isoamyl alcohol (24:24:1 by vol.)
- Formamide-loading dye (see *Protocol 13*)
- Pre-electrophoresed denaturing (urea) polyacrylamide gel(s) (see *Protocol 12*).

Method

1. Partial methylation of DNA: mix together in a microcentrifuge tube the following reagents:

Single-end-labelled DNA	$1–2 \times 10^6$ c.p.m.
TE buffer	200 μl
99% DMS	1 μl.

   Incubate at room temperature for 4 min.[a]
2. Terminate the methylation reaction by adding 50 μl DMS stop mixture.
3. Ethanol precipitate the DNA by adding 750 μl 95% ethanol (−20°C).
4. Centrifuge for 5–10 min in a microcentrifuge (12 000 *g*) to pellet the partially methylated DNA.
5. Remove the supernatant. Rinse the tube once with 70% ethanol and dry the pellet.

6. Resuspend the DNA pellet in 10–20 μl TE buffer to give about 20 000 c.p.m./μl.
7. Incubate the DNA-binding protein with DNA and resolve the protein-DNA complexes from free DNA by non-denaturing gel electrophoresis (see *Protocol 9*).
8. After electrophoresis, remove the gel from the apparatus and take off the top plate, making sure that the gel remains attached to the remaining plate. Cover the exposed gel surface with plastic wrap.
9. Expose the wet gel to X-ray film for several hours or overnight at 4°C.
10. Excise that portion of the gel corresponding to the protein-DNA complex and likewise excise the gel containing the free DNA probe.
11. In each case, elute the DNA from the gel by soaking in 500 μl gel elution buffer overnight at 37°C.
12. Discard the gel fragments and extract each supernatant with an equal volume of phenol:chloroform:isoamyl alcohol (24:24:1).
13. In each case, transfer the aqueous (upper) supernatant to a fresh tube and precipitate the DNA by adding 2.5 vol. 95% ethanol. Rinse the pellet once with 1 ml 70% ethanol and dry.
14. Redissolve each DNA pellet in 50 μl water.
15. Cleave the sugar-phosphate backbone at sites of methylated guanine residues by adding 50 μl 20% piperidine to 50 μl of each of the eluted DNAs (both protein-bound and free). Place the tubes at 90°C for 30 min. (Use Sarstedt screw-capped microcentrifuge tubes lined with gaskets to prevent the tops from popping open.)
16. Stop the reaction by precipitating the DNA. To do this, add to each tube 100 μl 0.6 M sodium acetate and 0.5 ml 95% ethanol (−20°C). Pellet the DNA by centrifuging for 5–10 min at 12 000 *g*.
17. Remove the supernatant completely using a drawn out Pasteur pipette. Rinse each tube once with 70% ethanol and dry the pellets in a Speed Vac for 5 min.
18. Dissolve each of the DNA pellets in 10 μl formamide-loading dye, and denature them by heating to 90°C for 3 min.
19. Load equal amounts of radioactivity on to a pre-electrophoresed denaturing (urea) polyacrylamide gel, resolve the fragments by electrophoresis, and visualize by autoradiography (see *Protocol 13*, steps 10–16).

[a] The time of incubation with DMS should be optimized for each DNA fragment.

An experimental example of the use of the methylation interference assay is shown in *Figure 8* for the glucocorticoid receptor.

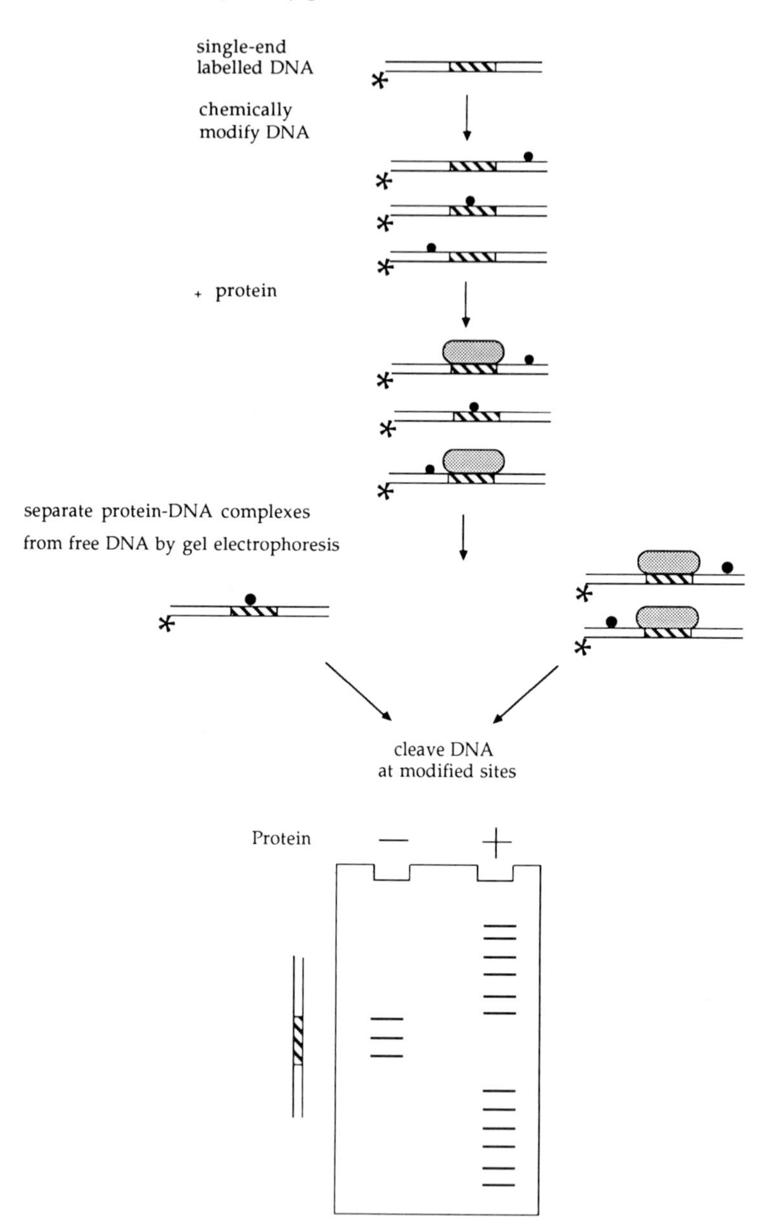

Figure 7. Schematic representation of the methylation interference assay. DNA fragments containing the protein-binding site (diagonal stripes) are labelled at one end with ^{32}P (shown by an asterisk) and randomly methylated with dimethylsulphate (filled circles) such that each molecule is randomly methylated once on average. The modified DNA is incubated with protein (filled ellipse) and allowed to bind DNA. Alterations at nucleotides essential for protein–DNA interaction will interfere with binding, while fragments modified at non-binding DNA sites will complex with protein. The two populations of DNA

5. Random mutagenesis of DNA sequences

Transcription factors and the genetic elements to which they bind have been studied primarily by mutating cloned DNA segments and assessing the phenotypic consequences *in vitro* and/or *in vivo*. A variety of techniques are available for introducing single-base changes into cloned DNA sequences (4). Generally, these mutagenesis procedures fall into two categories: specific and random.

Specific mutagenesis alters the DNA sequence in a defined way by using a single oligonucleotide with a specific base change. This method is extremely useful for determining the effects of an individual alteration on a prescribed function. However, in situations where it is necessary to produce many mutations, the technique's high degree of specificity becomes a major limitation, requiring a different oligonucleotide for each alteration.

Random mutagenesis, on the other hand, can efficiently and economically generate a large number of mutations within a defined region of DNA. Random or saturation mutagenesis approaches include the synthesis of degenerate or 'doped' oligonucleotides (21, 22) as well as chemical mutagenesis (23).

- *'Doped' oligonucleotides* are mixtures of oligomers synthesized by including small amounts of the other three nucleotides at concentrations that maximize single base substitutions. The mutation frequency is set by the amounts of 'non-wild-type' nucleotides included during the synthesis.
- *'Chemical mutagens'*, such as nitrous acid, formic acid, and hydrazine, generate single-base substitutions at every position within a given DNA sequence as well.

With either 'doped' oligonucleotides or chemical mutagens, the altered sequences are cloned and amplified as a population of 'mutant' molecules (termed a 'library'). Rapid DNA sequencing procedures make it practical to identify every possible substitution within the population by simply sequencing random isolates from the library. Alternatively, a functional assay may be coupled with random mutagenesis to isolate both gain or loss of function mutants (24). Methodologies are described below for saturation mutagenesis using both 'doped' oligonucleotides and chemical mutagens.

---

fragments are separated by non-denaturing gel electrophoresis. Both protein-bound and free DNA are eluted from the gel and the recovered DNA fragments are chemically cleaved at sites of DNA modification. Samples are electrophoresed on denaturing gels and autoradiographed. Bands will appear at positions corresponding to protein contact points for free DNA. Samples prepared from protein–DNA complexes will be cleaved and display bands at all positions *except* those contacted by protein.

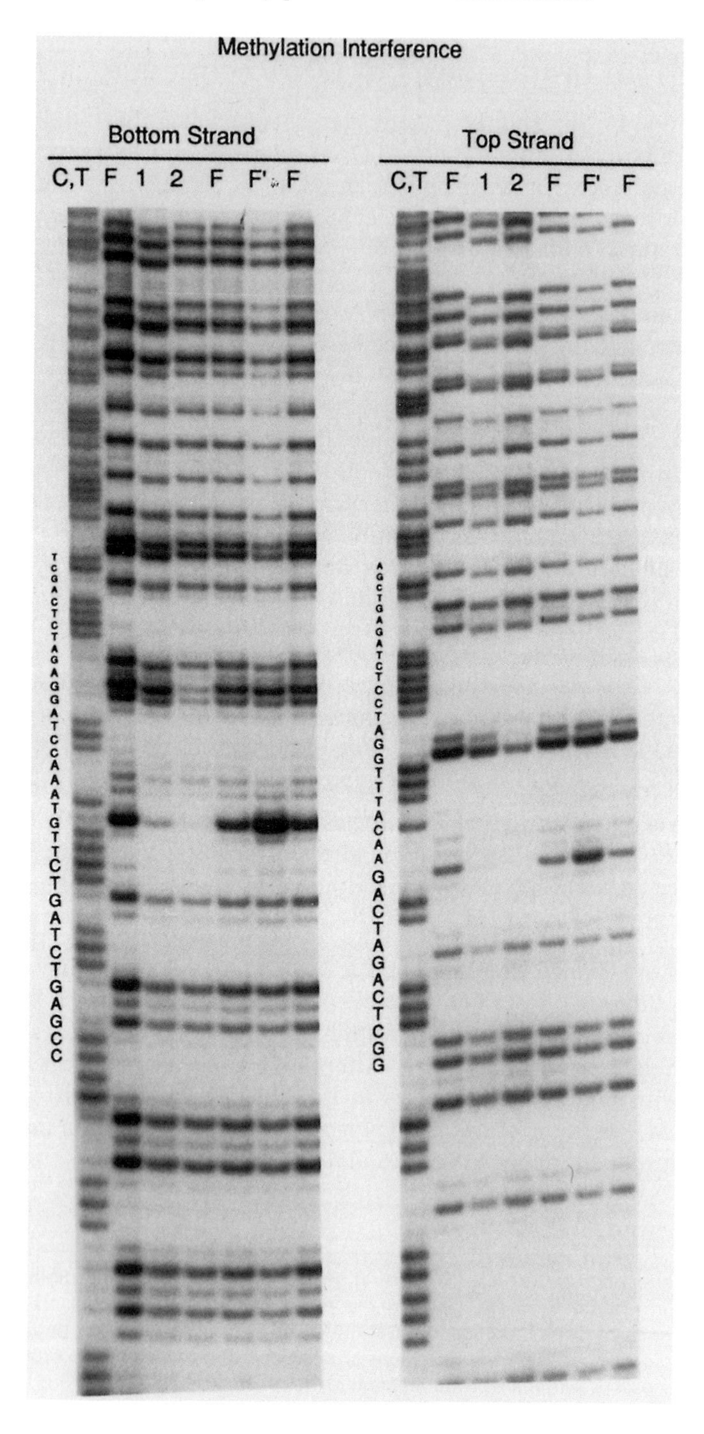
Methylation Interference
Bottom Strand
C,T F 1 2 F F' F
TCGACTCTAGAGGATCCAAATGTTCTGATCTGAGCC
Top Strand
C,T F 1 2 F F' F
AGCTGAGATCTCCTAGGTTTACAAGACTAGACTCGG

Figure 8. Methylation interference. A DNA fragment containing the binding site for the glucocorticoid receptor was ^{32}P-labelled on either the top or bottom strand, partially methylated, incubated with protein, and resolved on a non-denaturing polyacrylamide gel. Both protein-bound (complexes 1 and 2) and free DNA (F) were eluted from the gel, cleaved at the modified bases with piperidine, and resolved on a denaturing polyacrylamide gel. Lanes indicated by F prime (F′) represent the probe remaining in the free DNA band when, under non-methylated conditions, most of the probe was completely complexed with protein. Methylated molecules should be enriched at guanine positions that inhibit DNA binding. Lane C,T is DNA probe cleaved with cytidine- and thymidine-specific reagents respectively as described elsewhere for chemical sequencing (4). The nucleotide sequence in the footprint region is shown at the left.

5.1 Saturation mutagenesis of a defined region using degenerate oligonucleotides

This approach (see *Protocol 17* and *Figure 9*) efficiently creates numerous point mutations within a given region by using products of a single oligonucleotide synthesis. A blend of degenerate oligonucleotides with single base-pair substitutions is generated by including low concentrations of the other three nucleotide precursors during synthesis. Converting this mixture of single-stranded oligonucleotides into double-stranded DNA is achieved by mutually primed synthesis. By designing the oligonucleotides with palindromic 3′ ends, two molecules of the oligomer can hybridize and act as primers for synthesis of the complementary strand. The double-stranded DNA is cleaved with restriction enzymes, ligated into an appropriate vector and introduced into *E. coli*. Each transformant represents a single oligonucleotide from the library of random mutants. The DNA is then sequenced to determine the nature of the mutations.

Protocol 17. Saturation mutagenesis using degenerate oligonucleotides

Equipment and reagents

- Automated DNA synthesizer
- 10 × annealing/synthesis buffer (1.5 M NaCl, 0.3 M Tris–HCl pH 7.5, 0.3 M $MgCl_2$, 0.15 M DTT, 1 mg/ml BSA)
- dNTP mixture (2.5 mM of each of the four dNTPs in TE buffer)
- DNA polymerase I (Klenow fragment)
- [α-^{32}P]dGTP at 3000 Ci/mmol (220 TBq/mmol), 10 mCi/ml (370 MBq/ml)
- TE buffer (10 mM Tris–HCl, 1 mM EDTA pH 8.0)
- Synthetic oligonucleotide (see steps 1–3 below)
- 0.5 M EDTA pH 8.0
- Phenol:chloroform:isoamyl alcohol (24:24:1)
- 3.0 M sodium acetate pH 5.2
- Appropriate restriction enzyme
- 10 × restriction enzyme buffer (see *Protocol 1*)
- Non-denaturing gel for DNA isolation (see *Protocol 6*)
- Vector for subcloning
- DNA sequencing gel.

Protocol 17. *Continued*

A. *Synthesis of degenerate oligonucleotide*

1. Design the oligonucleotide with an 8 bp palindrome at its 3′ end and restriction endonuclease sites at both ends (if possible), with the nucleotides to be altered sandwiched between. A sample oligonucleotide is shown below.

*Xba*I base substitutions *Xba*I

5′ GCCGTCTAGACTCGAGGCTCAGATCAGAACATTTGGATCC**CTCTAGAG** 3′

palindrome

2. Use homogeneous nucleotides during the synthesis of positions where the restriction enzymes are located (such as the 5′ and 3′ end) but a mixture of nucleotides at positions where the base substitutions are desired. If the DNA synthesizer cannot be programmed for mixtures, combine the appropriate amounts of nucleotides before solubilization in anhydrous acetonitrile. For the above oligonucleotide, a single base substitution frequency of about 35% was obtained when the nucleoside precursors were 'doped' to a final concentration of 1.7% of each three non-designated bases.

   Purify the degenerate oligonucleotides by either HPLC or by electrophoresis on a 12% denaturing gel. The isolation of oligonucleotides from gels is described elsewhere (4).

B. *Synthesis of the complementary strand*

Second strand synthesis uses the 8 bp palindrome at the 3′ end of the synthetic oligonucleotides. Under annealing conditions, two oligomers will hybridize and act as mutual primers for complementary strand synthesis; effectively converting single-stranded DNA into a double-stranded duplex, suitable for cloning.

1. Mix in a 0.5 ml microcentrifuge tube:

purified 'doped' oligonucleotide (0.5 mg/ml)	5 μl
10 × annealing/synthesis buffer	1 μl
H_2O	to 10 μl final volume.

   Incubate at 37 °C for 15 min then cool slowly to room temperature over several hours.

2. Add to the annealing reaction:

H_2O	16 μl
dNTP mixture (2.5 mM of each)	3 μl
[α-^{32}P]dGTP	1 μl
DNA polymerase I (Klenow fragment)	5 units.

   Incubate for 1 h at 37°C

3. To terminate the reaction, add 80 μl TE buffer and extract the reaction mixture with 100 μl phenol:chloroform:isoamyl alcohol (24:24:1).
4. To the aqueous phase, add 5 μl 3.0 M sodium acetate and 250 μl 95% ethanol. Centrifuge for 10 min at 12 000 *g* in a microcentrifuge to pellet the DNA.
5. Rinse the pellet with 1 ml 70% ethanol and allow it to dry under vacuum. Resuspend the DNA in 200 μl TE buffer.

C. *Digestion of mutant DNA*

1. Mix the following:

double-stranded-mutant DNA in TE buffer	90 μl
10 × restriction buffer	10 μl
restriction enzyme	80 units.

Incubate for 2 h at 37°C

D. *Cloning of substitution mutants*

1. Gel purify the DNA fragment containing the degenerate oligonucleotides using a non-denaturing gel (see *Protocol 6*).
2. Clone this DNA fragment into an appropriate vector.
3. Isolate mutations and characterize the site of mutation in each case by sequencing random clones.

5.2 Chemical mutagenesis

Another procedure for generating random single-base substitutions is to treat single-stranded DNA with various chemical mutagens (23). Mutations occur during second strand synthesis when the DNA polymerase encounters a damaged base and misincorporates a nucleotide. The resulting DNA duplex containing the random single-base mutations is cloned, amplified as a population and sequenced, selected or screened for an altered phenotype (see *Figure 10*).

This type of chemical mutagenesis requires single-stranded DNA. Methods for cloning target DNA into M13 plasmids [or comparable single-stranded vectors, such as the phagemid BlueScript (Stratagene)] and preparation of single-stranded DNA templates are described elsewhere (4).

Using chemical mutagens with different nucleotide target specificities maximizes single-base substitutions at all positions. The concentration of the DNA, mutagen, and temperature are held constant, with the extent of mutagenesis controlled by the time of the chemical treatment. Therefore, by varying the reaction time, the extent of mutagenesis for a given fragment can be increased or decreased. Suggested incubation times are given in *Protocol 18* for each chemical mutagen. These times are determined empirically to

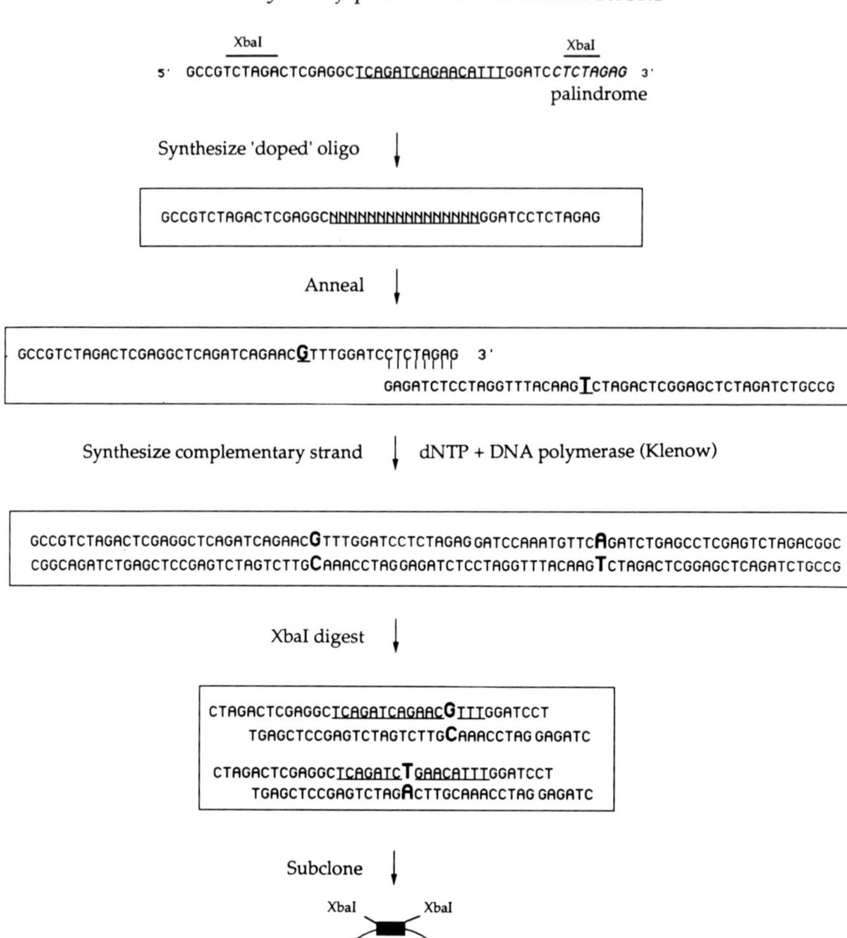

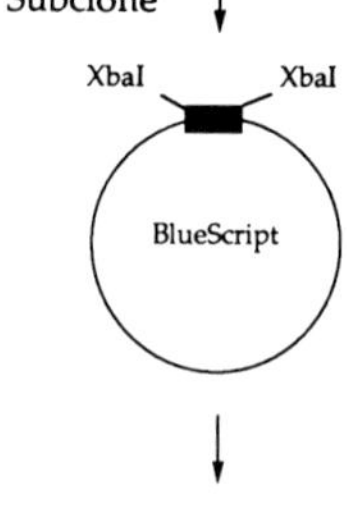

Figure 9. Saturation mutagenesis using degenerate oligonucleotides. The top line shows a 'wild-type' oligonucleotide designed with a 3′ end palindrome and flanked by 5′ and 3′ restriction sites. The 17 central bases (underlined) are synthesized with 95% of the indicated base and 1.7% of the other three nucleotides to produce a mixture of mutant oligonucleotides. The oligonucleotides are annealed at their complementary 3′ end and double-stranded DNA is synthesized by extension of the 3′ ends with DNA polymerase I (Klenow fragment) and the four deoxyribonucleoside triphosphates. The double-stranded DNA is suitable for cloning into a *Xba*I cut vector. Mutations are identified by sequencing random clones.

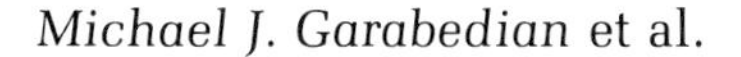

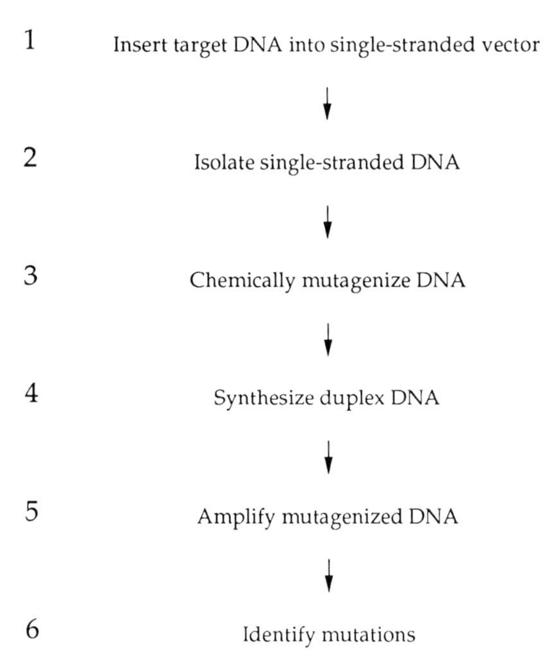

Figure 10. Steps involved in chemical mutagenesis. Target DNA is inserted into a single-stranded vector (such as M13) and single-stranded DNA is prepared (steps 1, 2). The single-stranded DNA is exposed to the chemical mutagens (step 3); duplex DNA is synthesized (step 4), subcloned into a non-mutagenized vector and amplified as a population (step 5). Individual clones are directly sequenced or screened for an altered phenotype (ref. 23) and then sequenced to identify the mutated nucleotides (step 6).

favour single base substitutions of a 400 bp insert in M13. A time-course of mutagenesis is a practical way to establish optimal mutagenesis conditions for a given target sequence. Complementary strand synthesis of the mutagenized DNA and the isolation of mutant clones is described in *Protocol 19*.

Protocol 18. Random chemical mutagenesis using nitrous acid, formic acid, and hydrazine

Reagents

- Single-stranded DNA containing the binding site of interest (1 mg/ml)
- 2.5 M sodium acetate pH 4.3 (Mallinckrodt, No. 7356)
- 3.0 M sodium acetate pH 5.2 (Mallinckrodt, No. 7356)
- TE buffer (10 mM Tris–HCl, 1.0 mM EDTA pH 8.0)
- 10 mg/ml yeast tRNA, phenol-extracted (see *Protocol 4*).

The following reagents are potent mutagens which must be handled with extreme care.

- 18.0 M formic acid [772 μl formic acid (88%; ~23 M; Fisher No. A118p-500) mixed with 228 μl H_2O]
- 18.0 M hydrazine [575 μl hydrazine (anhydrous 98%; ~31 M; Aldrich, No. 21,515-5) mixed with 425 μl H_2O]
- 2.0 M sodium nitrite (Sigma, No. S-2252).

Protocol 18. *Continued*

A. *Mutagenesis with nitrous acid*

This mutagen alters deoxycytidine, deoxyadenosine, and deoxyguanidine by deamination, resulting in deoxyuridine, deoxyinosine, and deoxyxanthosine, respectively. Nitrous acid is formed by the addition of sodium acetate to sodium nitrite.

1. Set up the following in a 0.5 ml microcentrifuge tube (final volume 50 μl).

single-stranded DNA (1 mg/ml)	10 μl
H_2O	10 μl
2.5 M sodium acetate pH 4.3	5 μl
2.0 M sodium nitrite	25 μl.
(added last: time = zero)	

2. Incubate at 25 °C. To establish optimal mutagenesis conditions for a given target sequence, vary the treatment time with nitrous acid. Thus, at $t = 2$, 6, and 18 min, remove 16 μl and terminate the mutagenesis immediately by adding:

3.0 M sodium acetate pH 5.2	5 μl
H_2O	27 μl
yeast tRNA 10 mg/ml	2 μl.

This is best done by mixing these reagents together to form a stop mixture which is kept ice-cold until needed.

3. Immediately add 100 μl 95% ethanol (ice cold) and centrifuge the DNA for 10 min at 12 000 *g* to pellet the mutagenized DNA.

4. Resuspend the DNA in 100 μl TE buffer.

5. Add 10 μl 3.0 M sodium acetate pH 5.2 and 250 μl 95% ethanol (ice cold). Centrifuge the DNA for 10 min at 12 000 *g* to pellet the mutagenized DNA.

6. Repeat the ethanol precipitation (steps 4–5) and finally dry the DNA pellet.

7. Resuspend the DNA in 20 μl TE buffer.

B. *Mutagenesis with formic acid and hydrazine*

Formic acid alters deoxyguanosines and deoxyadenosines in DNA by breaking the *N*-glycosyl bonds of purine nucleosides. Hydrazine attacks deoxycytidines and thymidines in DNA by introducing breaks into pyrimidine rings.

1. Set up the following in a 0.5 ml microcentrifuge tube:

single-stranded DNA (1 mg/ml)	10 μl
H_2O	30 μl
18.0 M formic acid or hydrazine	60 μl.

2. Incubate at room temperature for 1 min and 5 min for formic acid and hydrazine, respectively.
3. To terminate mutagenesis, immediately add:

3.0 M sodium acetate pH 5.2	10 μl
yeast tRNA, 10 mg/ml (phenol extracted)	2 μl
95% ethanol (ice cold)	250 μl.

   This is best done by mixing these reagents together to form a stop mixture which is kept ice-cold until needed.
4. Ethanol precipitate the DNA twice, dry the pellet, and resuspend in 20 μl TE buffer.

Protocol 19. Complementary strand synthesis of chemically mutagenized DNA and isolation of mutants

Reagents

- Mutagenized single-stranded DNA (see *Protocol 18*)
- 10 × primer annealing buffer (0.5 M NaCl, 0.2 M Tris–HCl pH 7.5, 0.2 M $MgCl_2$)
- M13 sequencing primer (10 ng/μl; New England Bio Labs, No. 1211)
- 2.5 mM dNTP mixture in TE buffer (i.e. all four dNTPs)
- AMV reverse transcriptase (Boehringer-Mannheim)
- Phenol:chloroform:isoamyl alcohol (24:24:1 by vol.)
- TE buffer (10 mM Tris–HCl, 1 mM EDTA pH 8.0)
- Non-denaturing polyacrylamide gel and reagents (see *Protocol 6*)
- Vector for subcloning.

Method

1. Mix in a 1.5 ml microcentrifuge tube the following:

mutagenized single-stranded DNA in TE buffer	20 μl
10 × primer annealing buffer	3 μl
M13 primer (10 ng/μl in water)	5 μl
H_2O	to 30 μl final volume.

2. Incubate the sample at 75°C for 5 min then at 40°C for 15 min to let the primer anneal.
3. Initiate DNA synthesis by adding:

2.5 mM dNTP mixture	5 μl
AMV reverse transcriptase	20 units.

   Let the reaction proceed for 1 h at 40°C.
4. To stop the reaction, extract once with 50 μl phenol:chloroform:isoamyl alcohol and then ethanol precipitate the DNA.

Protocol 19. *Continued*

5. Resuspend the duplex DNA in 20 μl TE buffer.
6. Excise the insert containing the mutagenized duplex DNA by restriction digestion.
7. Purify the DNA fragment of interest by gel electrophoresis using a non-denaturing polyacrylamide gel (see *Protocol 6*).
8. Clone the mutagenized DNA into an appropriate vector and identify mutations by sequencing or phenotypic screening.

6. Purification of DNA-binding proteins

Many DNA binding sites for transcriptional regulatory proteins have been identified as a direct result of simplified procedures for detecting protein–DNA interactions, such as the gel mobility shift assay. In many instances, however, the purification (and subsequent cloning) of these regulatory proteins has lagged far behind. This is often due to a low level of these factors within a cell, making purification by traditional chromatographic techniques a tedious and time consuming project. To get past these difficulties, specialized techniques have been developed for purifying and cloning DNA binding proteins, using their DNA recognition sites as affinity probes.

Kadanoga and Tjian successfully used DNA affinity columns containing large amounts of a specific binding site to purify several regulatory proteins (25). Typically, the protein is first fractionated by standard chromatographic techniques, such as ion exchange or exclusion chromatography, and then applied to the site-specific DNA affinity column. The bound proteins are eluted and fractions assayed for the presence of the protein by the gel mobility shift or footprinting assays. The active fractions are passed over the DNA affinity column several more times to achieve a high degree of purification.

A direct method for cloning DNA-binding proteins, devised by Singh, *et al.* (26), again utilizes DNA recognition sequences as affinity probes. Proteins from a cDNA expression library are immobilized on to a nitrocellulose filter and screened with ^{32}P-labelled duplex DNA containing the binding site of interest. The binding site probe is prepared by synthesizing an oligonucleotide corresponding to the sequence recognized by the protein. To ensure tight binding of the DNA to the protein, multiple binding sites are produced by ligating the oligonucleotides together into large concatenated molecules (27). These reiterated binding sites are labelled to high specific activity and used to detect clones that express the desired DNA-binding protein. This technique allows the direct cloning of the DNA-binding proteins and eliminates the need for prior protein purification. If the protein binds as a heteromer or requires a particular post-translational modification for binding, however, it may not be detected by this method.

Protocol 20 describes the techniques involved in preparing the radiolabelled concatenated DNA from a double-stranded oligonucleotide containing the binding site.

Protocol 20. Preparation of the radiolabelled concatenated DNA

Reagents

- Double-stranded DNA oligonucleotide corresponding to the recognition sequence (typically 12–20 bp in length) (100 ng/μl in water)
- 10 × kinase buffer (0.5 M Tris–HCl pH 7.8, 0.1 M $MgCl_2$, 0.1 M DTT)
- [γ-^{32}P]ATP; 7000 Ci/mmol (259 TBq/mmol) 160 mCi/ml (5.92 GBq/ml)
- T4 polynucleotide kinase (10 units/μl)
- 50 mM ATP in water
- T4 DNA ligase (400 units/μl)
- 0.5 M EDTA pH 8.0.

Method

1. Renature the double-stranded oligonucleotide recognition site as described in *Protocol 7*, but omitting the sonication step.
2. 5′-end-labelling and ligation. Mix in a 1.5 ml microcentrifuge tube the following components:

recognition site oligonucleotide	1 μl (100 ng)
10 × kinase buffer	2 μl
[γ-^{32}P]ATP	1 μl
T4 polynucleotide kinase	1 μl
H_2O	to 20 μl final volume.

   Incubate at 37°C for 60 min.
3. Ligation of oligonucleotide into concatamers. Add to the 5′-end-labelled reaction (step 2) the following:

10 × kinase buffer	1 μl
T4 DNA ligase (400 units/μl)	1 μl
50 mM ATP	1 μl
H_2O	to 30 μl final volume.

   Incubate at 15°C for 30 min or 5 min for blunt- and sticky-ended oligonucleotides, respectively.
4. Stop the reaction by adding 1 μl 0.5 M EDTA pH 8.0. Remove unincorporated nucleotides by passing the reaction mixture twice through spun columns (see Chapter 1, *Protocol 3*).

The purpose of the ligation reaction (see *Protocol 20*, step 3) is to create a series of binding sites linked in tandem and ranging in size from monomers to 12mers. This facilitates tight binding of the DNA to the protein. Under-ligated oligonucleotides (predominantly monomers and dimers) will often be lost during the high stringency wash procedures, leading to low or unreliable signals. Conversely, over-ligated molecules (dominated by concatamers >12mers) tend to obscure specific interactions. These large concatamers bind non-specifically to many proteins, leading to an increase in background and a large number of false-positive clones. Therefore, the extent of ligation should be closely monitored by electrophoresing 1 μl of the ligation reaction on an 8% denaturing polyacrylamide gel (see *Protocol 12*). A time-course of ligation is recommended to select the best conditions for generating a range of monomers to 12mers from the oligonucleotide of interest.

When the labelled oligonucleotide concatamers have been prepared, screen a suitable cDNA expression library as described in *Protocol 21*.

Protocol 21. Screening cDNA expression libraries using recognition site probes

Reagents

- λgt11 bacteriophage cDNA expression library (e.g. Stratagene)
- Host cells for the propagation of the λgt11 bacteriophage expression library
- NZCYM plates: (150 mm × 15 mm Petri dishes) [Prepare this medium by mixing 10 g NZ amine: casein hydrolysate, 5 g bacto-yeast extract (Difco Laboratories, No. 0127-01-7), 5 g NaCl, 1 g casamino acids (Difco Laboratories, No. 0127-05-3), 2 g $MgSO_4 \cdot 7H_2O$, 20 g bacto-agar (Difco Laboratories, No. 0140-01) and adding water to one litre. Sterilize by autoclaving and pour the NZCYM agar plates. These plates should be cured before use by allowing them to dry at room temperature for several days or overnight at 37°C]
- LB medium [Mix 10 g bacto-tryptone (Difco Laboratories, No. 0123-01-1), 5 g bacto-yeast extract (Difco Laboratories, No. 0127-05-3), 10 g NaCl, and water to 1 litre. Sterilize by autoclaving]
- Top agar (0.8 g bacto-agar in 100 ml LB medium)
- SM buffer (buffer for storage and dilution of bacteriophage) [Mix 5.8 g NaCl, 2 g $MgSO_4 \cdot 7H_2O$, 50 ml 1 M Tris–HCl pH 7.5, 5 ml of 2% (w/v) gelatin and add water to one litre. Sterilize by autoclaving]
- Nitrocellulose filter circles (137 mm diameter) Schleicher & Schuell, No. 68320
- 10 mM IPTG (Sigma, No. 102101)
- Non-fat dried milk powder (e.g. Carnation)
- Binding buffer (4 mM KCl, 25 mM Hepes, pH 7.9, 2 mM $MgCl_2$, 1 mM DTT) [The stock buffer prepared is 10 × binding buffer without DTT, i.e. 40 mM KCl, 0.25 M Hepes pH 7.9, 20 mM $MgCl_2$]
- 6 M guanidinium chloride in 1 × binding buffer
- DTT (1 M stock)
- Calf thymus DNA or salmon sperm DNA (stock 5 mg/ml; sonicated and denatured by boiling)
- 10 mM $MgSO_4$
- DNA binding buffer [5% (w/v) non-fat dried milk powder, 1 mM DTT, 100 μg/ml sonicated and denatured calf thymus DNA]
- Binding site probe buffer [10^6 c.p.m./ml probe,[a] 0.25% (w/v) non-fat dried milk, 1 mM DTT, 100 μg/ml sonicated and denatured calf thymus DNA, 1 × binding buffer]
- Wash buffer [0.1% Triton X-100, 0.25% (w/v) non-fat dried milk powder, 1 × binding buffer].

A. *Preparation of bacterial host cells for infection*

1. Inoculate a single colony of host cells for the bacteriophage infection, into 20 ml of LB medium.

2. Incubate at 37°C overnight with vigorous shaking.
3. Pellet the cells by centrifuging at 4000 *g* for 5 min at 4°C.
4. Resuspend the cell pellet in 8 ml of ice-cold 10 mM $MgSO_4$.

B. *Infection of bacteria with bacteriophage cDNA expression library*

1. For each plating, add 250 μl host cells and 1 μl of the bacteriophage library [containing approximately 20 000 plaque forming units (p.f.u.)] into a sterile tube.
2. Mix the host cells with the phage and incubate for 20 min at 37°C, without shaking, to infect the bacteria.

C. *Plating the bacteriophage expression library*

1. While the bacterial cells are being infected, melt top agar in microwave oven and let it cool to about 50°C.
2. Arrange a set of sterile tubes in a rack, add 8 ml molten top agar to each tube and place them in a 50°C water bath.
3. Arrange another set of sterile tubes in a rack prewarming to 50°C. Add 250 μl infected host cells to each tube.
4. Pour the top agar from one tube into a tube containing the infected cells, mix, then immediately pour the mixture on to an NZCYM plate which has been prewarmed to 37°C. Plate one tube at a time to ensure even spreading of the top agar.
5. Let the plates sit at room temperature until the top agar hardens (~5 min). Incubate the infected plates at 42°C for 2–3 h or until the plaques are visible (plaques need only be the size of a pin-prick). Then move the plates to a 37°C incubator.

D. *Induction of protein expression: preparation of nitrocellulose overlays*

Protein expression is brought about by IPTG, an inducer of the promoter controlling the expression of the cDNA insert in the bacteriophage.

1. Number the required set of nitrocellulose filter circles.
2. Soak the nitrocellulose filters in 10 mM IPTG solution for at least 5 min, making sure that the filters are completely wet.
3. Remove the filters and place them on plastic wrap and allow to air dry for 30–60 min at room temperature.
4. Once plaques become visible on the NZCYM plates, carefully overlay the IPTG-impregnated filters on to the plates and incubate at 37°C, labelling the plates with the corresponding filter markings. In addition, unambiguously mark the position of the filters on the agar plates by making several holes through the nitrocellulose with an 18-gauge needle

Protocol 21. *Continued*

near the edges of the plates. (Do this step quickly so that the temperature of the plates does not drop below 37 °C.)

5. Incubate the filter-covered plates for 6 h or overnight at 37 °C.
6. Remove each filter from the plate using blunt-ended forceps and place on Whatman 3MM paper with the side exposed to the plaques facing upward.
7. Dry the filters for 15 min at room temperature. Store the plates at 4 °C until positive plaques have been identified.

E. *Cell lysis and protein denaturation and renaturation*

1. Place the filters in a square baking dish containing 6 M guanidinium chloride in 1 × binding buffer (containing 1 mM DTT) at 4 °C; (~25 ml for each 137 mm filter).
2. Gently shake the filters for 5 min at 4 °C.
3. Replace the guanidinium chloride with a fresh batch of the same buffer and repeat the shaking incubation.
4. Pour this second batch of denaturation solution into a graduated cylinder and dilute it with an equal volume of 1 × binding buffer containing 1 mM DTT. Pour this solution into a clean glass dish and place the filters individually into the dilute denaturant, making sure that each filter is completely covered by the solution.
5. Repeat this dilution process four more times; the concentration of guanidinium chloride in the solution will be gradually reduced from 3 M to 1.5 M, to 0.75 M, to 0.375 M to 0.188 M.
6. Finally, wash the filters in 1 × binding buffer without any guanidinium chloride.
7. Repeat this final wash once more.

F. *DNA binding*

1. Place the filters in DNA-binding buffer. Agitate the filters gently for at least 60 min at 4 °C and then rinse the filters in 1 × binding buffer containing 0.25% (w/v) non-fat dried milk and 1 mM DTT.
2. Take one filter and briefly blot it on Whatman 3MM paper to remove excess buffer (do not allow the filter to dry). Place it into a 150 × 15 mm plastic Petri dish containing binding site probe buffer. Swirl the buffer over the filter until it is completely covered.
3. Repeat step 3 above until all the filters have been transferred into the Petri dish; make sure the top filter is evenly wet by the buffer.
4. Incubate all the filters overnight at 4 °C with gentle shaking or rocking.
5. Remove the filters one at a time. Each time a filter is removed, rinse it for

a few seconds in a tray with wash buffer and transfer it to a second tray of wash buffer (usually, 100–200 ml wash buffer per filter).

6. Batch wash the filters for 10 min with gentle agitation at 4°C.
7. It is important to transfer the filters individually to a second batch of wash buffer, since the filters tend to stick to one another and can give uneven background. Having transferred all the filters, wash them for a second time, again for 10 min with gentle agitation.
8. After the second wash, remove the filters and blot excess buffer from them on Whatman 3MM paper, making sure that the filters are still damp. Do *not* let the filters dry out completely.
9. Place the filters between two sheets of plastic wrap and autoradiograph for 5–10 h or overnight at −70°C.

H. *Recovery of positive clones*

1. Isolate an agarose plug corresponding to a positive signal by aligning the marks on the corresponding stored plate with the autoradiograph.
2. Place the plug containing the bacteriophage into a sterile 1.5 ml micro-centrifuge tube containing 1 ml of SM buffer.
3. Vortex, and then leave at 4°C overnight to release the bacteriophage from the agar plug.
4. Positive plaques need to be subjected to a second screen to ensure their purity. Therefore, plate secondary bacteriophage stock (~100–1000 p.f.u./100 mm plate) and screen filters as described above. Recover the true positive clones (plaques positive through the second screening).

[a] The amount of oligonucleotide probe used during the binding reaction depends on several factors, all of which can influence the strength of the signal obtained in the assay. These include, for example, the affinity of the protein for its recognition site, the amount of protein produced during induction and the ability of the protein to renature on the filter. Too much probe can lead to a high background and too little may lead to a weak and unreliable signal. We recommend a probe concentration between 2.5–25 ng/ml or 1–10 × 10^6 c.p.m./ml. If a high background persists, the probe may be suspect. Additional spun column chromatography or ethanol precipitations may be necessary to remove unincorporated radioactive nucleotides or the ligation of recognition site oligonucleotide may need to be optimized.

An example of screening a cDNA expression library with concatenated DNA binding sites is shown in *Figure 11*.

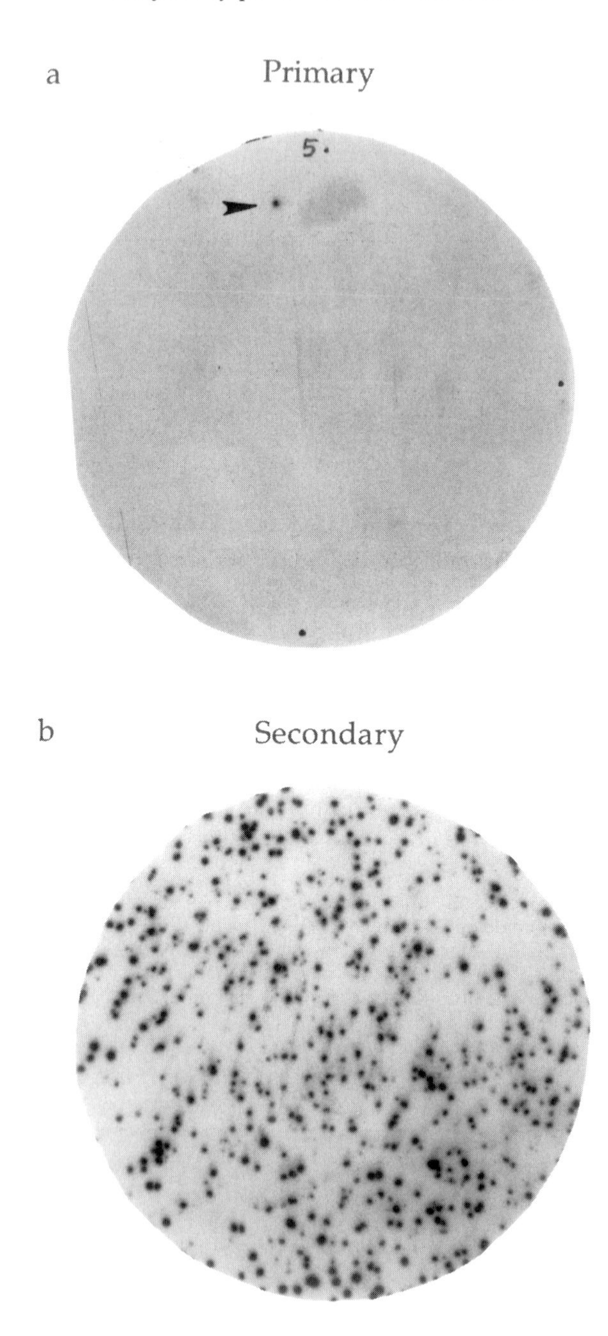
a
Primary
5.
b
Secondary

Figure 11. Screening a cDNA expression library with concatenated DNA binding sites. A *Drosophila* embryo expression library was screened with a radiolabelled concatenated oligonucleotide probe. (a) Primary screen. Each plate containing ~2000 plaques was treated with IPTG to induced protein synthesis from the phage cDNA library. The proteins were immobilized on nitrocellulose and the filters were incubated with the labelled DNA, with a positive signal visible after autoradiography. (b) Secondary screen. The positive plaque was isolated, replated and tested for its ability to bind the oligonucleotide probe once again.

References

1. Freid, M. and Crothers, D. (1981). *Nucleic Acids Res.,* **9,** 6505.
2. Garner, M. and Revsin, A. (1981). *Nucleic Acids Res.,* **9,** 3407.
3. Cunningham, M. W., Harris, D. W., and Munday, C. R. (1986). In *Radioisotopes in biology: a practical approach* (ed. R. S. Slater), p. 146. IRL Press, Oxford.
4. Maniatis, T., Fritsch, E. F., and Sambrook, J. (1989). *Molecular cloning: a laboratory manual.* Cold Spring Harbor Laboratory, New York.
5. Travers, A. (1991). *Curr. Opinion Struct. Biol.,* **1,** 114.
6. Ptashne, M. (1987). *A genetic switch, gene control and phage λ.* Blackwell Scientific Publications and Cell Press, Palo Alto, CA and Cambridge, MA.
7. Kerppola, T. K. and Curran, T. (1991). *Cell,* **66,** 317.
8. Wu, H-M. and Cruthers, D. (1984). *Nature,* **308,** 509.
9. Zinkel, S. S. and Cruthers, D. (1987). *Nature,* **328,** 178.
10. Kim, J., Zwieb, C., Wu, C., and Adhya, S. (1989). *Gene,* **85,** 15.
11. Luisi, B., Xu, X. W., Otwinowski, Z., Freedman, L. P., Yamamoto, K. R., and Sigler, P. B. (1991). *Nature,* **352,** 497.
12. Hope, I. A. and Struhl, K. (1987). *EMBO J.,* **6,** 2781.
13. Kumar, V. and Chambon, P. (1988). *Cell,* **55,** 145.
14. Galas, D. and Schmitz, A. (1978). *Nucleic Acids Res.,* **5,** 3157.
15. Sigman, D. S. and Chen, C. B. (1990). *Annu. Rev. Biochem.,* **59,** 207.
16. Tullius, T. D. and Dombroski, B. A. (1986). *Proc. Natl Acad. Sci. USA,* **83,** 5469.
17. Tullius, T. D., Dombroski, B. A., Churchill, M. E. A., and Kam, L. (1987). In *Methods in enzymology,* Vol. 155 (ed. R. Wu), p. 537. Academic Press, New York.
18. Kuwbara, M. D. and Sigman, D. S. (1987). *Biochemistry,* **26,** 7234.
19. Hendrickson, W. (1985). *BioTechniques,* **3,** 198.
20. Hendrickson, W. and Schleif, R. (1984). *J. Mol. Biol.,* **178,** 611.
21. Derbyshire, K. M., Salvo, J. J., and Grindly, N. D. F. (1986). *Gene,* **46,** 145.
22. Hill, D. E., Oliphant, A. R., and Struhl, K. (1987). In *Methods in enzymology,* Vol. 155 (ed. R. Wu), p. 588. Academic Press, New York.
23. Myers, R. M., Lerman, L. S., and Maniatis, T. (1985). *Science,* **229,** 242.
24. Schena, M. A., Freedman, L. P., and Yamamoto, K. R. (1989). *Genes Dev.,* **3,** 1590.
25. Kadanoga, J. and Tjian, R. (1986). *Proc. Natl Acad. Sci. USA,* **83,** 5889.
26. Singh, H., LeBowitz, Baldwin, A. S., Jr., and Sharp, P. A. (1988). *Cell,* **52,** 415.
27. Landschulz, W. H., Johnson, P. F., Adashi, Graves, B. J., and McKnight, S. L. (1988). *Genes Dev.,* **2,** 786.

7

Cloning and functional analysis of heterologous eukaryotic transcription factors in yeast

DANIEL M. BECKER and JOHN D. FIKES

1. Introduction

The study of transcription in yeast has provided many of our current paradigms for understanding the regulation of gene expression in higher organisms. Two things have made this possible: the ease with which, in yeast, one can combine genetic approaches with molecular genetic tools, and a striking evolutionary conservation in transcriptional mechanisms and in the transcription machinery from yeast to mammals. These same two properties make yeast an attractive surrogate organism in which to clone and characterize transcription factors from heterologous eukaryotic species, the subject of this chapter.

2. Expression cloning of heterologous transcription factors in yeast

2.1 Initial considerations

One may clone transcription factors by expressing cDNA libraries in *Saccharomyces cerevisiae* and selecting or screening for phenotypic changes. The three basic strategies are:

(a) Complementation of mutations in genes encoding yeast transcription factors. This approach, which requires the heterologous protein to substitute directly for the yeast protein, can identify cDNAs encoding proteins that are evolutionarily related, such as TFIID (1) and HAP2 (2). In addition, unlike homology-based strategies, this approach should also identify proteins of unlike sequence that serve similar functions.

(b) Suppression of defined yeast mutations. This strategy requires that the heterologous protein functions not to replace the yeast protein, but rather to bypass the need for it. Although this strategy has not yet been applied to cloning of transcription factors, it has identified components of signal transduction pathways (3).

(c) Identification of novel activities using engineered genetic screens or selections. To give an example, we have fused the E1A activation domain, which is ineffective in yeast, to the GAL4 DNA binding domain. We have introduced this fusion along with a human cDNA expression library into a *gal4*$^-$ strain. A Gal$^+$ phenotype or activation of a UAS_{gal} reporter should identify a cDNA encoding a cofactor that mediates E1A activation in mammalian cells (D. Becker and C. Fritz, unpublished data).

Table 1 lists pre-existing libraries for cDNA expression in yeast while *Table 2* lists suitable vectors for the construction of new yeast expression libraries. Practical considerations in selecting or constructing a library include:

(a) Matching the auxotrophic marker on the library vector to those available in the chosen yeast strain. This is often a major problem, especially when one or more other plasmids also need to be present, such as reporters (see Section 2.4.1) or plasmids expressing additional heterologous proteins.

(b) The level of cDNA expression required. High levels of expression may be required to permit a protein homologue to function adequately to satisfy the selection criteria. On the other hand, over-expression of some proteins, particularly strong activators, may prove toxic. In our experience, even the activation domains of types that appear not to mediate activation in yeast (e.g. the E1A activation domain and the glutamine-rich domains of Sp1) prove toxic when expressed to high levels in yeast. There are two general types of yeast promoters: constitutive promoters that differ from one another in expression level, and inducible promoters each

Table 1. Pre-existing libraries for cDNA expression in yeast

cDNA source	Vector	Primary transformants [a]	Reference or source
Human	DB20	3×10^7	(2)
Human	AB23BX	3.3×10^4	(4)
Human	SE936	3×10^7 av.	(5)
Human	ADANS	1.5×10^6	(6)
Human	FL60	2×10^5	(7)
Arabidopsis	SE936	3×10^7 av.	(5)
S. pombe	DB20	2.2×10^7	(1)
S. pombe	DB20C	6×10^6	J. D. Fikes, unpublished data
Drosophila	BL15	1×10^6	(8)
Drosophila	DB20	$>1 \times 10^6$	(9)

[a] av. = average

Table 2. Vectors for cDNA expression in *S. cerevisiae*

Vector	Selectable marker	Copy number	Promoter	Expression level	Cloning sites[a]	Source
DB10	*URA3*	High	*CYC1*	Moderate	B, N, H	(2)
DB20	*URA3*	High	*ADH*	High	B, N, H	(2)
DB20L	*LEU2*	High	*ADH*	High	N, H	P. Sugiono, personal communication
BL15	*ARG4*	High	*ADH*	High	B, N, H	(8)
DB20C	*URA3*	Low	*ADH*	Moderate	B, N, H	J. D. Fikes, unpublished data
DB20S	*URA3*	High	*ADH*[b]	Very High	B, N, H	Y. Xing, personal communication
YES2	*URA3*	High	*GAL*	Inducible	H, K, Sl, Bm, N, X, Sp, Xb, B, E, XIII	Invitrogen (San Diego)
AB23BX	*URA3*	High	*GAP*	High	B, Sl, X, Bg	(4)
SE936	*URA3*	High	*GAL*	Inducible	X, E, Xb, Sp, K	(5)
ADNS	*LEU2*	High	*ADH*	High	H, N, ScII, Sf	(3)
ADANS	*LEU2*	High	*ADH*	High	N, H	(6)
FL20	*URA3*	High	*PGK*	?	E, X	(7)

[a] Designations for restriction sites available for insertion of cDNAs are: *Bst*XI (B), *Bam*HI (Bm), *Eco*RI (E), *Not*I (N), *Hin*dIII (H), *Kpn*I (K), *Sal*I (S), *Sac*I (SI), *Sac*II (SII), *Sfi*I (Sf), *Sph*I (Sp), *Xho*I (X), *Xma*III (XIII), and *Xba*I (Xb).
[b] The *ADH* (*ADC1*) promoter fragment carried on this plasmid is smaller and more active than the fragment present on other *ADH*-containing vectors.

of which permits a range of expression levels. The inducible promoter used in current vectors is the *GAL* promoter, whose level of expression is set by varying the carbon source. This promoter may, however, be inappropriate for certain strategies [for example, the phenotype for selection of HAP2 homologues (2) requires growth on lactate, a condition under which the *GAL* promoter is not induced].

(c) The origin of the cDNAs. Clearly, the absence of an existing cDNA library from the species or tissue of interest would motivate the construction of a new library. It may be possible, however, to clone the desired transcription factor from an alternative species or tissue using the existing libraries, and then apply an homology-based approach. In several instances we have found that the cDNA encoding the *S. pombe* homologue of an *S. cerevisiae* protein is more easily cloned than the mammalian homologue (due in part to the relative complexity of each library), and that comparison of the predicted *S. pombe* and *S. cerevisiae* protein sequences provides sufficient information for design of PCR oligonucleotides to obtain counterparts from other species.

2.2 Construction of expression libraries

We designed the vector pDB20 to take advantage of the *Bst*XI cloning strategy of Aruffo and Seed (10). Six additional vectors described in *Table 2* also permit this approach. The strategy minimizes enzymatic manipulations: it obviates phosphatasing of the vector, methylation of cDNA, kinasing of linkers, and digestion of the cDNA subsequent to attachment of linkers (in this case, double-stranded adaptors). In addition, it prohibits insert multimerization. One theoretical drawback is that cDNAs insert in random orientation relative to the yeast promoter. In practice, however, this proves not to be a problem, since the *Bst*XI strategy allows one to create libraries with an enormous number of clones. In addition, pDB20 and its derivatives contain a cryptic promoter in the *ADH* terminator which drives low-level expression of cDNAs in the direction opposite to that of the *ADH* promoter (2). *Protocol 1* describes vector preparation and *Protocol 2* describes the removal of the small stuffer segment found between the *Bst*XI sites of the native vector.

Protocol 1. Purification of plasmid by centrifugation through CsCl

Equipment and reagents

- Centrifuge tubes [250 ml or 500 ml centrifuge bottles, 50 ml disposable centrifuge tube (e.g., Falcon, No. 2070), 15 ml disposable centrifuge tubes, Quick-seal ultracentrifuge tubes (13.5 ml nominal capacity) and tube sealer]
- *E. coli* host cells for transformation (see *Protocol 4*, footnote *a*)
- LB ampicillin plates [1% (w/v) tryptone, 0.5% (w/v) yeast extract, 1% (w/v) NaCl, 1.5% (w/v) agar, 100 μg/ml ampicillin]

- TBG broth[a] [1.2% (w/v) tryptone, 2.4% (w/v) yeast extract, 0.4% (v/v) glycerol, 20 mM glucose] [Autoclave. Add 0.7 vol. filter-sterilized 10 × phosphate buffer, 23.14 g KH_2PO_4 and 125.41 g K_2HPO_4 per litre; do not adjust pH]
- Lysis solution 1 (50 mM glucose, 25 mM Tris–HCl pH 8.0, 10 mM EDTA)
- Lysis solution 2 [0.2 M NaOH, 1% (w/v) 1% (w/v) SDS]. [Prepare fresh from stock solutions of 10 M NaOH and 10–20% SDS]
- Lysis solution 3 [Mix 600 ml 5.0 M potassium acetate, 115 ml glacial acetic acid, 285 ml H_2O]
- TE buffer (10 mM Tris–HCl pH 8.0, 1 mM EDTA)
- CsCl
- Ethidium bromide solution (10 mg/ml). *Caution*: ethidium bromide is a suspected carcinogen
- Water-saturated *n*-butanol
- 3 M sodium acetate, pH 5.2 or pH 7.0.

Method

1. Transform (11) *E. coli* host cells with 1–100 ng of the desired plasmid vector (e.g. pDB20) and plate serial dilutions of the transformants on LB ampicillin plates.
2. Inoculate 500 ml TBG broth[a] containing 100 μg/ml ampicillin from a single colony and grow overnight with vigorous shaking at 37°C.
3. Harvest the bacteria by centrifugation at 4°C for 10 min at 7000–10 000 *g*.[b]
4. Resuspend the bacterial pellet completely in 10 ml lysis solution 1. The resuspension can best be done by passing the mixture repeatedly in and out of a 25 ml pipette using a pump-driven pipetting aid.
5. Incubate for 10 min at room temperature.
6. Add 20 ml lysis solution 2. Swirl to mix.
7. Incubate for 10 min at room temperature.
8. Add 15 ml lysis solution 3. Swirl to mix.
9. Centrifuge at 4°C for 10 min at 10 000 *g*.
10. Decant the supernatant (which contains the plasmid) into a fresh centrifuge bottle. Be careful to transfer only clear supernatant.
11. Add 30 ml isopropanol to the supernatant. Shake vigorously.
12. Incubate for at least 10 min at room temperature.
13. Centrifuge at 4°C for 10 min at 10 000 *g*. Decant and discard the supernatant. Invert the bottle briefly on a paper towel to remove excess liquid, but do not let the pellet of nucleic acids dry appreciably.
14. Resuspend the pellet to a final volume of 8 ml with TE buffer (this requires about 6 ml buffer). Resuspension may require 10–30 min incubation at room temperature. Then transfer the solution to a graduated disposable plastic 15 ml centrifuge tube with cap.
15. Add 8 g CsCl. It is usually necessary to incubate briefly at 37°C to dissolve the CsCl completely. Adjust the volume if necessary with TE buffer to 10 ml using the markings on the tube as a guide.

Protocol 1. *Continued*

16. Add 0.8 ml ethidium bromide solution.
17. Prepare a loading reservoir for the Quick-seal tube by removing the plunger from a 10 ml syringe to which an 18-gauge 1.5 in needle has been attached. Place the reservoir's needle into the open neck of the Quick-seal tube. The syringe barrel will wobble, but will not fall off; the extra clearance around the needle permits air to escape the Quick-seal tube during loading.
18. Carefully pour the plasmid/CsCl/ethidium bromide solution into the reservoir.
19. To fill the Quick-seal tube to the top of its shoulder, add exactly 1.9 ml of fresh 'blank' solution prepared in the same proportions as the plasmid: 8 ml TE buffer, 8.0 g CsCl, 0.8 ml ethidium bromide solution.
20. Seal the tube, balance against another sample tube or 'blank' tube, and centrifuge for 36–48 h at 18°C at 175 000 *g*.[c]
21. Carefully remove the tube from the rotor and gently clamp it above a large beaker. A large red band of plasmid will be readily apparent without UV illumination (for pDB20, a 500 ml culture will produce a band about 0.5–0.75 in wide). A fainter, much narrower band of *E. coli* chromosomal DNA will be present higher in the tube. With *E. coli* strain XL-1 Blue (Stratagene), there is often no apparent chromosomal band.
22. Pierce the top of the tube once or twice with an 18-gauge syringe needle, leaving the needle in place after the final piercing.
23. Using a 5 ml syringe fitted with an 18-gauge, 1.5 in needle, pierce the tube below the plasmid band, but above the pelleted RNA. Rotate the needle so that the needle opening is directed upwards and slowly draw the plasmid band into the syringe. Avoid the chromosomal band. Withdraw the syringe from the tube with care (*note: ethidium-containing liquid will flow from the hole left behind; for safety reasons, be careful to position the beaker to catch this effluent*). The total volume of plasmid collected should be about 2.5 ml.
24. Remove the syringe needle from the syringe and discharge the plasmid solution into a disposable 50 ml plastic centrifuge tube.
25. To extract ethidium bromide from the plasmid solution, add 10 ml water-saturated *n*-butanol to the plasmid solution, cap the tube tightly, and vortex. Allow the phases to separate, then withdraw and discard the upper, alcohol phase into the waste beaker. *Observe proper disposal precautions*. Repeat the *n*-butanol extraction until the aqueous phase is completely clear.[d]
26. Bring the plasmid solution to 7.5 ml with TE buffer, add 750 μl 3 M sodium acetate and transfer to a 30–50 ml centrifuge tube.

27. Add 16 ml absolute ethanol. Cap and shake.
28. Centrifuge at 4°C for 30 min at 10 000 *g* to recover the plasmid DNA.
29. Decant and discard the supernatant. Invert the tube briefly on a paper towel to remove excess liquid. Briefly air dry the pellet of plasmid DNA. Do not allow the pellet to dry too much since this makes resuspension difficult.
30. Resuspend in 400 μl TE buffer. Transfer this to a microcentrifuge tube.
31. Precipitate the plasmid DNA a second time by adding 40 μl 3 M sodium acetate and 1 ml ethanol. Vortex. Spin at top speed in a microcentrifuge 30 min at 4°C. Aspirate the supernatant, briefly air-dry the pellet, and resuspend in 1 ml TE buffer. Full resuspension at this concentration (usually >1 mg/ml) may require overnight incubation at 4°C.

[a] TBG broth, a modification of 'terrific broth' (12), yields greater than fivefold the amount of plasmid provided by an equal volume culture of standard Luria-Bertani (LB) broth. It is important, as a consequence, not to lower the initial ampicillin concentration below 100 μg/ml.

[b] The liquid measurements in the following steps can be done most efficiently by adding the appropriate solution to the markings on a graduated, disposable, 50 ml centrifuge tube (such as Falcon, No. 2070) so these tubes should be chosen for centrifugation purposes.

[c] A wide variety of ultracentrifuge rotors are available in three principal geometries: fixed angle, near-vertical, and vertical. Fixed angle rotors require the longest spins, vertical rotors the shortest, with near-vertical somewhere in between. Although the choice will depend principally on the availability of rotors in any laboratory, we find that vertical rotors produce the least clean preparations.

[d] With plasmids such as pDB20, which replicate well, a 500 ml plasmid preparation will yield milligram quantities of plasmid DNA. A commensurately large amount of ethidium bromide will be complexed in this DNA, and multiple rounds of *n*-butanol extraction will be required for complete extraction.

Protocol 2. Linearization of vector and removal of 'stuffer' segment

Equipment and reagents

- Centricon-100 filter unit (Amicon)
- CsCl-prepared plasmid vector (see *Protocol 1*)
- *Bst*XI (New England Biolabs)
- TE buffer (10 mM Tris–HCl pH 8.0, 1 mM EDTA)
- Phenol:chloroform:isoamyl alcohol (25:24:1 by vol.)
- Chloroform:isoamyl alcohol (24:1 v/v)
- 3 M sodium acetate pH 5.2 or pH 7.0.

Method

1. Digest 5 μg of CsCl-prepared plasmid vector (e.g. pDB20) overnight at 55°C with 10 units *Bst*XI in a total volume of 100 μl. Use fresh enzyme and fresh enzyme buffer.
2. Run 10 μl of the reaction mixture on an agarose gel (see Chapter 6, *Protocol 6*) to confirm complete linearization of the vector.

Protocol 2. *Continued*

3. Extract once with phenol:chloroform:isoamyl alcohol and once with chloroform:isoamyl alcohol.
4. Add 10 μl 3 M sodium acetate (0.3 M final concentration) and 250 μl ethanol (2.5 vol.). Vortex. Spin at top speed 30 min at 4°C in a microcentrifuge to precipitate the DNA.
5. Resuspend the pellet in 1 ml TE buffer.
6. Transfer the DNA to a Centricon-100 filter unit.
7. Add 1 ml TE buffer to the original microcentrifuge tube, vortex, centrifuge briefly, and transfer this wash to the Centricon unit.
8. Centrifuge the Centricon-100 unit for 45–60 min at 1000 *g*.
9. Discard the filtrate.[a] Add 2 ml TE buffer to the retained DNA and mix gently.
10. Repeat steps 8–9 for a total of four rounds of ultrafiltration. After the final centrifugation, do not add TE buffer; apply the retentate cup, invert the filter unit, and centrifuge 5 min at 1000 *g* to collect the DNA.
11. Determine the concentration of linearized vector by spectrophotometry at 260 nm.

[a] The 22 bp 'stuffer' in the pDB20 series vectors passes through the Centricon-100 filter while the linearized vector is retained. Do *not* use a Centricon-30 unit as it may retain an unacceptably high proportion of the 'stuffer' fragments.

For cDNA synthesis, we have had excellent success with the cDNA synthesis kit from Life Technologies (BRL, No. 8090SA). As with other single-tube formats (and in stark contrast to the kit from Invitrogen), the yield of cDNA is maximized by eliminating precipitations and extractions between the first- and second-strand cDNA synthesis. In addition, the BRL reaction uses a modified M-MLV (Moloney murine leukemia virus) reverse transcriptase ('Superscript') that possesses increased thermal stability and lacks RNase-H activity, both of which should contribute to a higher percentage of full-length cDNA.

We do not present a protocol for cDNA synthesis, since we routinely follow the instructions of the manufacturer without alteration. However, we offer the following recommendations:

- Purify the mRNA using two rounds of oligo-dT selection (13) to minimize contamination by ribosomal RNA and transfer RNA.
- Use 2–5 μg mRNA for each synthesis, adjusting the enzyme concentrations for first-strand cDNA synthesis as described in the manufacturer's instructions.
- Perform the first-strand cDNA synthesis at 42–44°C.

- Do not include a trace radiolabel to monitor cDNA synthesis. Following the phenol extraction of the double-stranded cDNA, it is usually easy to visualize the cDNA by gel electrophoresis of 5 μl of the aqueous phase (160 μl total volume) with ethidium bromide in a 1% agarose gel. The key is to pour a gel of no more than 10 ml on a glass slide.
- There should be a sufficient quantity of double-stranded cDNA such that carrier (tRNA, glycogen, etc.) need not be added to produce a visible pellet following phenol extraction and precipitation.

Protocol 3 describes the ligation of double-stranded adaptors to the cDNA and the subsequent removal of excess adaptors. As with *Protocol 2*, the details are specific to the *Bst*XI strategy. No matter what cloning strategy one chooses, however, we strongly recommend the use of Sephacryl spun columns for linker removal (see *Protocol 3*). Spun column chromatography with Sephacryl in a single step removes unligated adaptors, adaptor dimers, and small cDNAs, at the same time resuspending the larger cDNAs directly in fresh ligation buffer for the subsequent ligation to vector. It is extremely efficient.

Protocol 3. Ligation of synthetic adaptors to cDNA

Equipment and reagents

- *Bst*XI adaptors (Invitrogen, No. N408-19, or alternatively No. N418-18)[a]
- T4 DNA ligase
- 5 × ligation buffer[b] [0.33 M Tris–HCl pH 7.6, 5 mM spermidine, 50 mM $MgCl_2$, 75 mM dithiothreitol (DTT), 1.0 mg/ml bovine serum albumin (BSA) (DNase-free; acetylated is fine), 5 mM ATP (Pharmacia)]
- Sephacryl S-300 cDNA spun columns (Pharmacia, No. 27-509901; do not confuse with the Pharmacia S-300 miniprep spun columns)[c]
- Sephacryl S-400 size select columns (Pharmacia, No. 27-510501; optional)[c]
- Sephacryl S-500 spun columns (Life Technologies/BRL cDNA synthesis kit No. 8090SA; optional)[c]
- TE buffer (10 mM Tris–HCl pH 8.0, 1 mM EDTA)
- Phenol:chloroform:isoamyl alcohol (25:24:1 by vol.)
- Falcon tubes (No. 2059).

Method

1. Add the following to a microcentrifuge tube:
   - 15 μl cDNA in water (half of the cDNA synthesized from 2–5 μg mRNA)
   - 6 μl of 5 × ligation buffer
   - 3 μl (3 μg) *Bst*XI adaptors[d]
   - 4 μl H_2O
   - 2 μl T4 DNA ligase (units will vary but will certainly be in excess).
2. Incubate at 16°C overnight.

Protocol 3. *Continued*

3. At the end of the overnight incubation, prepare Sephacryl spun columns exactly as described by the manufacturer.[c] Important considerations are:
   - Save the caps upon opening each column, since top and bottom caps are needed with each wash of the column.
   - Use Falcon No. 2059 tubes to hold the Pharmacia columns during gravity feed and centrifugation.
   - Do not allow the top of the gel to dry, otherwise fissures may form during the preparative centrifugation.
   - *Do not use a fixed angle rotor*, since this disturbs the geometry of the column.
   - If centrifuged correctly, the bed of Sephacryl will compact and separate at its top from the walls of the column but there should be no fissures. If there are cracks in the Sephacryl bed, perform an additional resuspension and centrifugation.
4. Add 70 μl TE buffer to the 30 μl ligation mixture.
5. Extract the ligation reaction with 100 μl phenol:chloroform:isoamyl alcohol. Discard the lower (organic) phase.
6. Apply the aqueous phase in dropwise fashion directly on to the centre of the compacted bed of Sephacryl. Do not allow the liquid to track around the bed.
7. Centrifuge *exactly* as during column preparation.
8. Discard the Sephacryl column and transfer the cDNA (80–100 μl) from the Falcon No. 2059 tube to a microcentrifuge tube.
9. Proceed directly with ligation of the cDNA to the vector (see *Protocol 4*), or store at −70°C to prevent degradation of the ATP in the buffer.

[a] The *Bst*XI adaptors originally offered by Invitrogen (No. N408-19) and used in the pDB20 series libraries (see *Table 1*) do not reconstitute the *Bst*XI restriction sites of the vector. The alternative adaptors (No. N418-18) can be substituted directly in the procedure and do regenerate *Bst*XI restriction sites on each side of the insert. The modified adaptors also contain internal *Eco*RI sites, providing an additional restriction site for insert excision.

[b] In theory, using the 5 × ligation buffer packaged with the BRL cDNA synthesis kit should present no problem.

[c] The BRL cDNA synthesis kit includes Sephacryl S-500 spun columns, which have a nominal exclusion limit of greater than 500 bp. Pharmacia offers spun columns containing either Sephacryl S-400 or S-300, which have nominal exclusion limits of 400 and 300 bp, respectively. For the original pDB20 series libraries (see *Table 1*) we used S-300 columns, which impose the least stringent size selection of the three. Further analysis of the cDNA inserts in the HeLa library (14) demonstrates that the inserts, on average, are fairly short. Given the difficulty of generating a sufficient number of yeast transformants to produce an adequate sampling of the cDNA library (see Sections 2.3 and 2.4), we now recommend more stringent size selection using the S-400 or S-500 columns. The S-500 columns are packed with a different geometry, though, so we cannot from our own experience vouch for their efficacy.

[d] The adaptors (and alternative adaptors) are partially double-stranded and fully phosphorylated; no prior kinasing step is required.

Protocols 4 and *5* describe ligation of the cDNA into the plasmid vector and the subsequent amplification and purification of the library. It is absolutely crucial in these protocols that the library be introduced into *E. coli* by electroporation (11). *Use no other procedure.* Now that electroporation cuvettes are available with a 0.1 cm gap (BioRad), a 0.15 cm gap (Life Technologies/BRL), and a 0.2 cm gap (Hoeffer; note that this is a square wave device), field strengths can be achieved that routinely allow transformation efficiencies of 10^{10} transformants/μg. These transformation efficiencies are only possible, however, with clean DNA: *Ligated DNA must be extracted with phenol: chloroform:isoamylalcohol* and the organic phase removed prior to transformation (see *Protocols 4* and *5*).

Using the protocols presented here we have produced libraries with 2–3 × 10^7 primary transformants with up to 100% inserts.

Protocol 4. cDNA ligation

Equipment and reagents

- Linearized vector (from *Protocol 2*)
- cDNA (from *Protocol 3*)[a]
- 5 × ligation buffer stock (see *Protocol 3*)
- T4 DNA ligase
- TE buffer (10 mM Tris–HCl pH 8.0, 1 mM EDTA)
- Phenol:chloroform:isoamyl alcohol (25:24:1 by vol.)
- Sephadex G-50 spun columns[b]
- Electroporation equipment[c]
- *E. coli* host cells for transformation (see *Protocol 4*, footnote a)
- *Not*I
- 3 M sodium acetate pH 5.2 or pH 7.0.

Method

1. Set up three test ligations using linearized vector from *Protocol 2* and cDNA from *Protocol 3*, as shown below:

	A	B	C
vector (100 ng) + 1 × ligation buffer (μl)	13	10	4
cDNA[a] (μl; approximates % by vol.)	1	4	10
T4 DNA ligase (μl; units will vary)	1	1	1

2. Incubate at 16°C overnight.
3. Add 5 μl TE buffer. Extract with 20 μl phenol:chloroform:isoamyl alcohol. Discard the lower organic phase.
4. Apply the aqueous phase to a Sephadex G-50 spun column[b] equilibrated with TE buffer and centrifuge this. Recover the eluted material (~20 μl).
5. Transform *E. coli* with 5 μl of each reaction by electroporation.[c] Plate serial dilutions to permit accurate titration. In a parallel experiment, transform supercoiled vector to assess the absolute transformation efficiency.
6. Choose 10 isolated colonies from each transformation and determine the

Protocol 4. *Continued*

number and size of cDNA inserts in each by plasmid miniprep (15) and *Not*I digestion.

7. Based on the titre and percentage of vector molecules containing inserts, choose the optimal insert:vector ratio (A, B, or C). Using the remaining linkered cDNA (~85% by vol.), set up *multiple parallel ligations* at that optimal ratio, each with 100 ng vector and the identical volumes as in step 1. Incubate at 16°C overnight.
8. Pool the ligations. Extract with an equal volume of phenol:chloroform: isoamyl alcohol.
9. Add 3 M sodium acetate to 0.3 M and 2.5 vol. ethanol. Allow to precipitate at −20°C for 60 min. Centrifuge at top speed in a microcentrifuge for 30 min at 4°C.
10. Resuspend the pellet in 20 μl H_2O.

[a] Spun column chromatography performed as in *Protocol 3* will deliver cDNA to this protocol in 1 × T4 DNA ligase buffer.

[b] See (16) or (17) for preparation of G-50 spun columns. Sephadex G-50 columns can also be purchased commercially. It is important in either case to verify, in advance, that the specific column geometry will function correctly with as little as 20 μl. If not, add 80 μl TE buffer to the aqueous phase prior to centrifugation. Since the goal is to remove trace organic solvents, one may precipitate instead (see step 8); take care, however, to precipitate quantitatively without the addition of carrier, and then resuspend in 20 μl TE buffer.

[c] Protocols provided by the manufacturer of each electroporation device will provide the appropriate variation of the original procedure (11). Be certain to request the most current version of the protocol; newer cuvettes permit higher field strengths and higher efficiencies.

Protocol 5. Transformation and amplification of the library

Equipment and reagents

- Ligated cDNA (see *Protocol 4*)
- LB ampicillin plates, 150 mm diameter (see *Protocol 1*)
- *E. coli* host cells for transformation[a]
- Electroporation apparatus (see *Protocol 4*, footnote *c*)
- SOC medium [2% (w/v) tryptone, 0.5% (w/v) yeast extract, 10 mM NaCl, 2.5 mM KCl, 10 mM $MgCl_2$, 10 mM $MgSO_4$, 20 mM glucose]
- 250 ml or 500 ml centrifuge bottles
- Reagents for bacterial lysis and plasmid purification (see *Protocol 1*).

Method

1. Assume that the transformation efficiency will equal that measured in the trial ligations (see *Protocol 4*) and calculate the number of transformants expected from the large-scale transformation. Divide by 1 × 10^6 to determine the number of LB ampicillin plates required.

2. Electroporate freshly-prepared *E. coli*[a] with the ligated cDNA (see *Protocol 4*). This may be done in a single cuvette with a single aliquot of bacteria or in multiple parallel transformations.
3. After growth for 1 h at 37°C, add a volume of SOC medium equal to 200 μl × the calculated number of plates. Mix thoroughly.
4. Spread 200 μl of transformed *E. coli* per LB ampicillin plate. Be certain to perform serial dilutions on one aliquot to determine the actual titre since the plates will contain nearly confluent lawn.
5. Incubate overnight at 37°C.
6. Add 10–20 ml water [the highest quality available, such as Milli-Q (Millipore Corp.)] to each plate. Scrape the colonies off the surface. Transfer and pool the suspensions in 250 ml or 500 ml centrifuge bottles. This procedure need not be performed under sterile conditions. Avoid the transfer of agar.
7. Centrifuge at 10 000–15 000 *g* at 4°C for 15 min. Discard the supernatant.
8. Lyse the bacteria[b] and prepare plasmid DNA as described in *Protocol 1*. Adjust the volumes according to the size of the bacterial pellet.
9. Purify the plasmid by one round of CsCl gradient centrifugation[c] as described in *Protocol 1*.
10. Store the purified plasmid frozen at −70°C in aliquots.

[a] There are many strains that can be used to propagate cDNA libraries; for the pDB20-series libraries we used XL-1 Blue (Stratagene), a general purpose *recA* cloning strain that transforms well by electroporation.

[b] It is not necessary to store any of the library in *E. coli*. Transformation by electroporation is so efficient that additional amplification, if desired, can be accomplished by electroporating 1 μg of the initial library and repeating this protocol.

[c] Be careful not to overload the CsCl gradient: we obtained 12–15 mg plasmid from 2.5–3.0 × 10^7 colonies in our initial libraries.

2.3 Yeast transformation

The efficient introduction of the cDNA library into yeast is a prerequisite to identifying the desired clone. It becomes especially important when attempting to clone transcription factors whose cDNA may be in low abundance in the library. Since strain variability precludes a universal approach, we present two protocols for transforming yeast and discuss a third. Strains respond idiosyncratically to each of these methods, and there is no way to predict strain behaviour *a priori*. Our first choice is based on a recent modification (R. D. Gietz and R. H. Schiestl, personal communication) of a lithium acetate protocol (18). Its advantages are that it gives high efficiency transformation for many strains, it is relatively easy, and it generates increasing numbers of transformants for increasing input of DNA (an important consideration for obtaining a large population of transformants). *Protocol 6* describes the preparation of high molecular weight carrier DNA for the transformation and *Protocol 7* describes the lithium acetate transformation procedure itself. Our second choice is electroporation (see *Protocol 8*) (19). This procedure can provide the highest efficiency of transformation for many strains (measured as transformants/μg plasmid), and it is the easiest of the techniques. Unfortunately, however, electroporation saturates at 10–100 ng DNA input, necessitating a large number of parallel reactions to generate a sufficient population of transformants. A third possible approach is spheroplast transformation. This method (20) can generate large numbers of transformants, but it has several major drawbacks. The principal problem is the need to embed the yeast in top agar following transformation. If the yeast need to be replated (see Section 2.4), recovery of these embedded colonies is cumbersome, inefficient, and it alters the representation of clones in the library. Even when the yeast do not require replating, variable growth of colonies in and upon the top agar precludes the analysis of transformants based on colony morphology.

Protocol 6. Preparation of high molecular weight single-stranded carrier DNA [a,b]

Reagents

- DNA from salmon testes (Sigma-D1626 type III sodium)
- TE buffer (10 mM Tris–HCl pH 8.0, 1 mM EDTA)
- Phenol buffered with TE
- Chloroform
- 3 M sodium acetate.

Method

1. Add TE buffer to the DNA to give a final DNA concentration of 10 mg/ml. Stir overnight at 4°C to dissolve. The solution of DNA will be extremely viscous. Withdraw a small amount (~10 μl) to serve as a reference for steps 2–3.
2. Sonicate using a large probe until the viscosity appears to decrease slightly. Electrophorese 1 μg sonicated DNA on a 0.8% agarose gel with markers of known size to determine the size distribution of sonicated fragments (see Chapter 6, *Protocol 6*). In a parallel lane, electrophorese 1 μg unsonicated DNA to evaluate the effects of sonication. The larger the fragments, the better the transformation efficiency will be, but the more viscous and unwieldy the solution. The optimum distribution of fragment sizes appears to be one in which fragments range from 2–15 kb, with mean fragment size of about 7 kb.
3. Repeat step 2 as necessary to achieve the appropriate size distribution.
4. Extract once with buffered phenol, once with phenol:chloroform, and once with chloroform.

5. Add 3 M sodium acetate to 0.3 M and 2.5 vol. ethanol to precipitate the DNA. With DNA of this size and at this concentration, the DNA will form flocculent masses immediately upon addition of ethanol. Centrifuge at 4°C for 30 min at 10 000 *g*.
6. Resuspend the pellet with TE buffer to 10 mg/ml final concentration. Resuspension may require stirring overnight at 4°C.
7. Transfer the DNA to a Pyrex flask. Heat in a microwave oven to boiling to denature the DNA. Continue to boil for 2–3 min.
8. Chill the flask rapidly in ice water. Aliquot and freeze the carrier DNA in sterile tubes at −20°C.

[a] Modified from (18) and reproduced by permission of Springer-Verlag.
[b] The addition of denatured high molecular weight carrier DNA is critical to achieving high transformation efficiency using the lithium acetate protocol (18). Both parameters are crucial: the DNA must be single-stranded, and the larger the fragments, the better the transformation efficiency. Since the preparation of carrier DNA is the most crucial step in achieving high transformation efficiency, and since each individual transformation requires 200 μg, prepare a large amount (1–2 g) and freeze aliquots at −20°C.

Protocol 7. Transformation of yeast using lithium acetate[a]

Reagents

- 10 × lithium acetate stock (1.0 M lithium acetate pH 7.5; sterilize by ultrafiltration)
- 10 × TE buffer stock (0.1 M Tris–HCl pH 7.5, 10 mM EDTA; sterilize by ultrafiltration)
- 50% (w/v) PEG 4000 (*do not use PEG 8000*; sterilize by ultrafiltration using a pump)
- YPAD medium [1% (w/v) yeast extract, 2% (w/v) peptone, 2% (w/v) glucose, 40 mg/l adenine]
- Carrier DNA prepared as described in *Protocol 6*
- Selection plates [0.67% (w/v) yeast nitrogen base without amino acids (Difco No. 0919-15; contains ammonium), 2% (w/v) glucose, 2% (w/v) agar (use Difco), auxotrophic supplements as dictated by strain genotype and desired selection]
- Sterile water [the highest quality available, such as Milli-Q (Millipore Corp.); sterilize by autoclaving].

Method

1. Inoculate the cells from an overnight culture into 300 ml YPAD medium.
2. Grow with vigorous shaking at 30°C to 5–10 × 10^6 cells/ml (A_{600} ~0.15–0.3, depending on the strain). For two- to threefold higher efficiency, dilute at this point to 2 × 10^6 cells/ml in fresh YPAD medium and grow for another two generations (3–5 h).
3. Harvest the cells by centrifugation at 4000–6000 *g* for 5 min at room temperature.
4. Resuspend in 10 ml highest quality sterile water.
5. Transfer to a smaller centrifuge tube and pellet the cells by centrifugation at 4000–6000 *g* for 5 min at room temperature.

Protocol 7. *Continued*

6. Resuspend[b] in 1.5 ml freshly prepared 1 × TE/LiAc (prepare from sterile water, sterile 10 × TE buffer stock, and sterile 10 × lithium acetate stock).
7. For each transformation, mix 200 μg carrier DNA and up to 5 μg transforming DNA in a sterile 1.5 ml microcentrifuge tube. Maximal transformation efficiency is achieved by repeating the denaturation cycle (boiling and chilling) of carrier DNA immediately prior to use.
8. Add 200 μl yeast suspension to each microcentrifuge tube.[c]
9. Add 1.2 ml 40% PEG solution freshly prepared as follows: 8 vol. 50% PEG 4000, 1 vol. of 10 × TE buffer stock, 1 vol. of 10 × lithium acetate stock.
10. Incubate for 30 min at 30°C with agitation.
11. Heat shock for exactly 15 min at 42°C.
12. Centrifuge at room temperature for 5 sec in a microcentrifuge.
13. Resuspend in 200 μl–1 ml of 1 × TE buffer and spread up to 200 μl on to selection plates.
14. Incubate at 30°C until transformants appear.

[a] Modified from (18) and reproduced by permission of Springer-Verlag.
[b] For strains that clump, sonicate the cells for 3 min in an ultrasonic (Branson 2200) water bath.
[c] If the volume of DNA exceeds 20 μl, add appropriate volumes of 10 × lithium acetate buffer stock and 10 × TE buffer stock to prevent further (>10%) dilution of the lithium acetate/TE concentrations.

Protocol 8. Transformation of yeast by electroporation[a]

Equipment and reagents

- BioRad Gene Pulser with Pulse Controller
- BioRad 0.2 cm gap disposable electroporation cuvettes
- YPD broth [1% (w/v) yeast extract, 2% (w/v) peptone, 2% (w/v) glucose]
- Sterile water [the highest quality available, such as Milli-Q (Millipore Corp.); sterilize by autoclaving]
- 1.0 M sorbitol (prepare using the highest quality water; sterilize by ultrafiltration or autoclaving)
- 10 × lithium acetate stock (see *Protocol 7*)
- 10 × TE buffer stock (see *Protocol 7*)
- 1.0 M DTT (prepare with highest quality water and sterilize by ultrafiltration. Store at −20°C)
- Selection plates (see *Protocol 7*; supplement additionally with sorbitol to 1.0 M final concentration).

Method

1. Inoculate 500 ml YPD broth with yeast from an overnight culture.[b]

2. Grow with vigorous shaking at 30°C to an A_{600} of 1.3–1.5 (~1 × 10^8 cells/ml).
3. Harvest the culture by centrifugation[c] and resuspend vigorously in 80 ml sterile H_2O.
4. Add 10 ml of 10 × TE buffer stock. Swirl to mix.
5. Add 10 ml of 10 × lithium acetate stock. Swirl to mix.
6. Shake gently for 45 min at 30°C.
7. Add 2.5 ml 1.0 M DTT.
8. Shake gently for 15 min at 30°C.
9. Dilute the yeast suspension to 500 ml with H_2O.
10. Concentrate and wash the cells with several centrifugations,[c] resuspending[d] the successive pellets as follows:

1st pellet	250 ml H_2O
2nd pellet	20–30 ml 1.0 M sorbitol
3rd pellet	0.5 ml 1.0 M sorbitol.

The final volume of resuspended yeast should be about 1.0–1.5 ml.

11. Mix 40 μl of the concentrated yeast with ≤100 ng transforming DNA[e] in ≤5 μl. The DNA should be in a low ionic strength buffer such as TE buffer or in water. The incubation time can be varied to convenience.
12. Transfer to a cold electroporation cuvette.
13. Pulse at 1.5 kV, 25 μF, 200 Ω; τ will vary from 4.2–4.9 msec.
14. Add 1 ml ice-cold 1.0 M sorbitol to the cuvette and recover the yeast, with gentle mixing, using a sterile 9-inch Pasteur pipette.
15. Spread aliquots of the yeast suspension directly on selection plates containing 1.0 M sorbitol.
16. Incubate at 30°C until colonies appear. Colonies take longer to appear than in *Protocol 7*.

[a] Modified from (19), copyright Academic Press, and reproduced with permission. The original protocol omits steps 4–8 for rapidity and simplicity, but provides 5–10-fold lower efficiency.
[b] The overnight culture need not be fresh; yeast may be used from a culture kept several weeks at 4°C. Alternatively, yeast can be inoculated directly from a colony, but growth to the appropriate density will require commensurately longer incubation.
[c] We routinely centrifuge for 10–15 min at 5000–10 000 *g* in Sorvall GSA and SS34 rotors. The rotor, the exact speed, and the duration of centrifuge spins are not critical; the principal consideration is to centrifuge sufficiently hard to pellet all of the yeast and to wash the yeast completely. Centrifugation should be carried out at 4°C.
[d] The resuspension should be vigorous enough to dissociate each pellet completely. Solutions should be ice-cold.
[e] Do not include carrier DNA in this procedure, as it substantially reduces transformation efficiency.

2.4 Screens and selections

There are two alternative approaches to identifying cDNAs encoding heterologous transcription factors.

(a) Screen yeast transformants for a phenotypic change. This requires that one distinguish within the entire population of yeast transformants those colonies that exhibit the desired phenotype.
(b) Apply a selection to the yeast transformants that permits growth of only those cells that express the desired heterologous protein.

In either case, we recommend selecting first for yeast transformants (e.g. in the case of pDB20, for uracil prototrophy) and then applying to the resulting colonies the screen or selection for the desired clone.

2.4.1 Screens

In screening expression libraries for transcription factors, one most commonly looks for increased expression of a reporter gene also present in the yeast strain. The reporter construct can be integrated stably into the yeast chromosome or can be present on a plasmid maintained in the strain by auxotrophic selection. In cases in which the reporter gene is on a plasmid, we recommend introducing the plasmid first and propagating the strain under selection to produce a uniform population. It may be prudent, though, to grow these yeast for 2–3 generations in rich (non-selective) media immediately prior to transforming with the library to generate a maximal number of library transformants.

The most common reporter gene used in yeast is *lacZ*, coding for β-galactosidase, and the most common *lacZ* reporter vectors are pLG670-Z (21) and pSLF178K (22). These vectors contain a minimal promoter from the yeast *CYC1* gene, which is insufficient to drive detectable expression, and cloning sites for the introduction of upstream elements. Both are high copy episomal plasmids containing the *URA3* selectable marker. Note that most of the expression vectors and expression libraries (see *Tables 1* and *2*) are also marked with *URA3*, precluding simultaneous use of these reporters with those libraries. Other auxotrophic markers could be substituted for *URA3* on the reporters, whereas the vectors for construction of cDNA libraries require *URA3* for utilization of the *Bst*XI cloning scheme. Other reporters, such as CAT (23) and β-glucuronidase (24), offer no significant advantage over β-galactosidase. *Protocol 9* describes a solid media β-galactosidase assay for screening transformants.

Protocol 9. Solid media β-galactosidase assay

Reagents

- 10 × X-gal buffer [In a fume hood, add 136 g KH_2PO_4, 20 g $(NH_4)_2SO_4$, 42 g KOH to 1 litre of water. Adjust the pH with HCl or H_2SO_4 to pH 7.0. Sterilize by ultrafiltration]
- Selection plates (see *Protocol 6* or *Protocol 8*)

- X-gal stock (20 mg/ml in dimethylformamide; does not require sterilization; store in the dark at −20°C)
- Nitrocellulose membrane filters.[a]
- X-gal indicator plates [To 1 litre of selection plate agar (see *Protocol 6*) at about 55°C, add 100 ml of 10 × X-gal buffer and 5 ml X-gal stock]

Method

1. Prepare the selection plates and allow them to dry for several days at room temperature.
2. Transform the yeast using *Protocol 7* or *Protocol 8* and spread them uniformly on nitrocellulose filters placed on the selection plates.[b]
3. Incubate at 30°C until colonies appear.[c]
4. Maintaining sterility, transfer the nitrocellulose filters to X-gal indicator plates.
5. Incubate at 30°C until blue colour develops.

[a] We use Millipore type HATF or Schleicher & Schuell BA85. Nylon filters have too rough a surface to permit fine discrimination in colony morphology. For rapid evaluation of individual plasmids, the filters can be used from the package without further sterilization. For screening of libraries, we recommend autoclaving the filters: remove the intercalated waxed paper (Millipore) and sandwich each filter between pieces of chromatography paper (e.g. Whatman 3MM). Autoclave in aluminium foil.

[b] The indicator plates used subsequently (step 4) are toxic and so would not permit initial growth of yeast transformants, necessitating the filter transfer. In addition, the white background provided by the filter enhances colour discrimination.

[c] Large well-spaced colonies will provide the greatest discrimination of β-galactosidase levels for evaluation of individual clones. For library screens, plate 30 000 to 50 000 transformants per plate and transfer the filters when the colonies are pinpoint in size, which should take no more than 2–3 days. Colonies may require an additional day or two to reach this size, however, if transformed using *Protocol 7*.

2.4.2 Selections

Selections offer a significant advantage over screens for identifying cDNAs encoding transcription factors. In selections, only those yeast cells meeting the selection criteria grow under the restrictive conditions. One can plate as many as 5×10^5 transformants per plate and easily identify a single positive colony in a selection, whereas one can plate at most 5×10^4 colonies on a filter and still hope to identify individual blue colonies in a screen. This is a profound difference when one is attempting to obtain a rare clone and can, for example, mean the difference between analysing 10 or 100 plates of transformants.

We used selections to identify the *S. pombe* HAP2 homologue (2, 25) and HAP3 homologue (J. Fikes, unpublished data), the HeLa HAP2 homologue (2), and the *S. pombe* TFIID gene (1). The HAP2/3/4 complex in *S. cerevisiae* is required for expression of respiratory genes; we selected for complementation of *hap* null mutations by requiring growth on lactate, a nonfermentable carbon source. To identify TFIID, we selected for a Lys^+ phenotype in a strain carrying a point mutation in TFIID and a Tyδ insertion at the *Lys2* locus (26).

Given that every selection will differ, we can offer only a few general guidelines.

- For temperature-sensitive phenotypes, the entire plate of primary transformants can simply be shifted to the restrictive temperature 1 to 2 days after transformation; use of the spheroplast protocol may necessitate a slightly longer outgrowth at the permissive temperature.
- In selections that require replating the primary transformants on new media, the fundamental choice is between scraping, washing, and replating colonies, or alternatively, plating the transformation initially on nitrocellulose filters and then moving the filter (see *Protocol 9*). Scraping and replating has the advantage of eliminating the carry-over of nutrients from the original plate, but increases the risk of overgrowth by contaminants. Filter lifts have the advantages of simplicity and of localizing fungal contaminants, but have the disadvantage of carrying nutrients from the nonselective to the selective media.

2.4.3 Critical issues

Yeast phenotypes

The paramount consideration in attempting to clone a particular gene by function is the availability of a discernible yeast phenotype for screen or selection. Reversion of the yeast mutation, and/or a 'leaky' phenotype, raises the background against which true positives must be identified.

Poor complementation

The heterologous gene product may incompletely complement the mutant phenotype. For example, the *S. pombe* HAP3 homologue complements a *hap3* null mutation poorly and requires more than two weeks' growth on selective media for observable (but not unambiguous) growth above background.

Non-complementation

The heterologous transcription factor may not complement the yeast mutation at all. We obtained the *S. pombe* TFIID gene based on function in *S. cerevisiae*, but failed in a parallel selection to obtain the human TFIID; the human TFIID has now been shown not to function in *S. cerevisiae*. A more general concern is that only acidic-type activation domains from heterologous transcription factors have been shown to activate transcription in yeast. Libraries fusing the yeast *GAL4* activation domain to the cDNAs (see Section 3.2) may circumvent the problem of non-complementation on the basis of incompatible activation domain.

Multimeric complexes

In the case of transcription factors that contribute to multimeric complexes, such as the HAP complex, complementation demands that the protein–protein interactions be sufficiently conserved to allow assembly of a functional hybrid complex.

2.5 Verification of clones

2.5.1 Phenotype confirmation

The first step in analysis of potential clones is to verify the initial phenotype. The first step in any verification is to restreak for isolates. For selections, restreaking on selective media provides pure isolates and confirms that the colonies exhibit the desired growth characteristics. For screens, restreak blue colonies on plates lacking X-gal to obtain isolates, then replica-plate or patch on to X-gal-containing plates. For screens, it is prudent in addition to test β-galactosidase activity of blue isolates with a liquid assay (see *Protocol 10*). The standard assay is given in the main body of this protocol with alternatives in the footnote.

Protocol 10. Liquid β-galactosidase assay

Reagents

- Minimal medium [0.67% (w/v) yeast nitrogen base without amino acids (Difco No. 0919-15; contains ammonium), 2% (w/v) glucose, auxotrophic supplements as dictated by strain genotype and desired selection]
- ONPG stock (4 mg/ml in H_2O; filter and store at 4°C)
- 0.1% SDS
- Z buffer (60 mM Na_2HPO_4, 40 mM NaH_2PO_4, 10 mM KCl, 1 mM $MgSO_4$, 30 mM 2-mercaptoethanol) [Prepare without 2-mercaptoethanol. Adjust to pH 7.0 with 50% (w/v) NaOH, and store at 4°C. Add 2-mercaptoethanol immediately before use]
- 1.0 M Na_2CO_3 stock
- Chloroform.

Method

1. Inoculate 3–5 ml minimal medium from a single yeast transformant or from an overnight culture grown under selective conditions and shake at 30°C to an A_{600} of 0.8–1.2.
2. Read and record the exact A_{600} value.
3. Transfer 1 ml to a clean glass culture tube and centrifuge at maximum speed in a table-top clinical centrifuge for 5 min at room temperature.
4. Aspirate the supernatant and resuspend the pellet in 1 ml Z buffer.
5. Add 1 to 2 drops of 0.1% SDS and 1–2 drops of chloroform. Vortex vigorously.

Protocol 10. *Continued*

6. Incubate at 30°C for 5 min. Warm the ONPG stock briefly to 30°C.
7. Add 200 μl ONPG stock. Record the time. Continue the reaction until the solution is moderately yellow.
8. Stop the reaction with 0.5 ml 1.0 M Na_2CO_3 (the colour will darken slightly). Record the time.
9. Centrifuge out cell debris and read the A_{420} of the supernatant.
10. Calculate Miller units of β-galactosidase activity as follows, with the volume of cells expressed in ml and time in min:

$$\text{Units} = 1000 \times \frac{A_{420}}{\text{vol. of cells assayed} \times \text{time of reaction} \times A_{600}}$$

[a] Two modifications of the standard technique can be applied, either singly, or in combination. (a) Do not read the A_{600} in step 2. Pellet and resuspend 2 ml culture (rather than 1 ml) in step 3. Read the A_{600} of 1 ml versus Z buffer as standard. Continue the basic protocol as written with the remaining 1 ml. These changes (B. Turcotte, personal communication) eliminate sample-to-sample variability in the efficiency with which the cells are recovered after centrifugation. (b) Normalize all values of β-galactosidase activity to protein concentration.

2.5.2 Confirming plasmid linkage

A consistent positive phenotype may be due to a mutational event on the yeast chromosome, an alteration of the reporter gene (in screens), or the expression of a heterologous protein. To confirm that the phenotype is dependent on the presence of the heterologous cDNA, it is necessary to recover the library plasmid (27) from the yeast isolate and show that retransformation of the parental strain by this plasmid confers the desired phenotype.

In our experience, yeast transformants often harbour multiple library plasmids, in addition to any reporter plasmids, and so it is necessary to retransform each of these in order to identify which, if any, confers the phenotype. Miniprep DNA is sufficient for retransformation of yeast. If there are a large number of potential positives (as, for example, in strains with a high background of mutational reversion or phenotypic 'leakage'), it may be prohibitively difficult to sort through the plasmids in this fashion. If the library vector is marked with *URA3*, as are most of the vectors presented in *Tables 1* and *2*, an alternative approach is to evict library plasmids by growth of the transformant on media containing 5-fluoroorotic acid (FOA) (28). Loss of phenotype provides strong evidence that a heterologous cDNA in fact confers the desired properties. Such loss of phenotype must still be followed by recovery of plasmids from the original isolate (not passaged over FOA) and retransformation of the parental strain.

3. Analysis of heterologous transcription factors in yeast

3.1 Advantages

Many of the strategies used to analyse mammalian transcription factors, as detailed elsewhere in this book, have also been applied to studies of yeast transcription factors. All of these approaches are available for studies of mammalian transcription factors expressed heterologously in yeast. We emphasize here, however, characteristics unique to yeast that make it particularly advantageous to study transcription factors from other species in yeast. These include the following.

Genetic analysis

The greatest advantage is the opportunity to exploit genetic strategies in yeast. For example, one can apply mutational analysis to yeast strains expressing heterologous transcription factors by treating the strain with ethylmethane sulphonate (29). This generates a large number of random mutations in the gene encoding the heterologous transcription factor and in the yeast chromosome. Either type of mutation can be identified by screening or selecting for phenotypic changes using approaches described in Sections 2.4.2 and 2.4.3. Mutations in the heterologous gene identify critical residues in the transcription factor; mutations in the yeast genome can identify interacting factors. Searching for suppressors of either type of mutation continues the analysis by identifying linked second-site mutations or unlinked extragenic suppressors.

Yeast molecular genetic tools

(a) Yeast genes identified by genetic analysis can be cloned rapidly by functional complementation. In many cases, one of the large number of existing yeast genomic libraries may be used. Alternatively, complementation with expression cDNA libraries (see Section 2) may directly provide the mammalian homologue.

(b) Homologous recombination in yeast is highly efficient, and there are trivial auxotrophic selections that permit one easily to target chromosomal locations. A major consequence is that one can generate chromosomal null mutations in almost any cloned yeast gene. One can also exploit these properties to integrate heterologous genes or reporters stably into the yeast genome. This ensures uniform copy number and expression levels within the population. We have, for example, integrated the human homologue of the yeast *HAP2* gene into a strain from which the yeast gene has been deleted (D. Becker, unpublished data). This strain allows us to study the function of the expressed human protein in the trimeric CCAAT-binding complex independent of other mammalian components.

(c) Expression levels of heterologous genes can be finely modulated by taking advantage of the wide variety of constitutive and inducible promoters (see Section 2.1), and the availability of high and low copy origins of replication for plasmid maintenance.

In vitro *transcription*

The ability to generate extracts from mutant yeast strains offers an unparalleled versatility to this *in vitro* system.

Economy

The growth, maintenance, and manipulation of yeast are inexpensive. In addition, the tools described above result in an economy of effort. For example, analysis of randomly mutagenized heterologous genes is facilitated by the ease with which one can select among millions or screen among hundreds of thousands of yeast transformants (see Section 2.4).

3.2 Gene fusion strategies in yeast

Use of gene fusion adds another dimension to the approaches discussed above. Here we collate information relevant to the design and application of gene fusion in yeast.

The yeast transcription factor Gal4 can provide a well characterized DNA binding domain. Residues 1–147 suffice to confer sequence-specific DNA binding upon fused heterologous protein domains (30). There is a wide variety of *gal4*$^-$ strains that are suitable recipients for such fusions. One can monitor a reporter containing Gal4-binding sites or assess the ability of the strain to grow on galactose, an extremely sensitive assay and a potentially powerful genetic selection.

The bacterial repressor *LexA* has also been used as a source of sequence-specific DNA binding activity in yeast (31). Unlike Gal4, *LexA* does not participate in any yeast regulatory network, which may in some circumstances be advantageous. Note, however, that residues 1–87, used for many studies (25), lacks the *LexA* dimerization domain found in the larger 1–202 fusions (32).

Two effective activation domains for transcription studies in yeast are provided by Gal4 residues 149–196 and 768–881 (33). A third, particularly potent activating domain for yeast is provided by residues 413–490 of the herpes virus VP16 protein (34). A wide variety of well characterized mutational derivatives of these acidic activation domains can be used to provide varying levels of activation function (35, 36).

Acknowledgements

J.D.F. was supported in this work by a postdoctoral fellowship from the American Cancer Society. D.M.B. was supported by a Special Fellowship from the Leukemia Society of America.

References

1. Fikes, J. D., Becker, D. M., Winston, F., and Guarente, L. (1990). *Nature,* **346,** 291–4.
2. Becker, D. M., Fikes, J. D., and Guarente, L. (1991). *Proc. Natl Acad. Sci. USA,* **88,** 1968–72.
3. Colicelli, J., Birchmeier, C., Michaeli, T., O'Neill, K., Riggs, M., and Wigler, M. (1989). *Proc. Natl Acad. Sci. USA,* **86,** 3599–603.
4. Schild, D., Brake, A. J., Kiefer, M. C., Young, D., and Barr, P. J. (1990). *Proc. Natl Acad. Sci. USA,* **87,** 2916–20.
5. Elledge, S. J., Mulligan, J. T., Ramer, S. W., Spottswood, M., and Davis, R. W. (1991). *Proc. Natl Acad. Sci. USA,* **88,** 1731–5.
6. Colicelli, J., Nicolette, C., Birchmeier, C., Rodgers, L., Riggs, M., and Wigler, M. (1991). *Proc. Natl Acad. Sci. USA,* **88,** 2913–17.
7. Minet, M. and Lacroute, F. (1990). *Curr. Genet.,* **18,** 287–91.
8. Lahue, E. E., Smith, A. V., and Orr-Weaver, T. L. (1991). *Genes Dev.,* **5,** 2166–75.
9. Leopold, P. and O'Farrell, P. H. (1991). *Cell,* **66,** 1207–16.
10. Aruffo, A. and Seed, B. (1987). *Proc. Natl Acad. Sci. USA,* **84,** 8573–7.
11. Dower, W. J., Miller, J. F., and Ragsdale, C. W. (1988). *Nucleic Acids Res.,* **16,** 6127–45.
12. Tartof, K. D. and Hobbs, C. A. (1989). *Focus,* **9,** 12.
13. Aviv, H. and Leder, P. (1972). *Proc. Natl Acad. Sci. USA,* **69,** 1408–12.
14. Lew, D. J., Dulic, V., and Reed, S. I. (1991). *Cell,* **66,** 1197–206.
15. Del Sal, G., Manfioletti, G., and Schneider, C. (1989). *BioTechniques,* **7,** 514–19.
16. Sambrook, J., Fritsch, E. F., and Maniatis, T. (1989). *Molecular cloning: a laboratory manual,* 2nd edn. Cold Spring Harbor Laboratory, New York.
17. Ausubel, F. M., Brent, R., Kingston, R. E., Moore, D. D., Seidman, J. G., Smith, J. A., and Struhl, K. (ed.) (1989). *Current protocols in molecular biology.* Greene and Wiley-Interscience, New York.
18. Schiestl, R. H. and Gietz, R. D. (1989). *Curr. Genet.,* **16,** 339–46.
19. Becker, D. M. and Guarente, L. (1991). In *Methods in enzymology,* Vol. 194 (ed. C. Guthrie and G. R. Fink), pp. 182–7. Academic Press, San Diego.
20. Sherman, F., Fink, G. R., and Hicks, J. B. (ed.) (1986). *Methods in yeast genetics.* Cold Spring Harbor Laboratory, New York.
21. Guarente, L. and Ptashne, M. (1981). *Proc. Natl Acad. Sci. USA,* **78,** 2199–203.
22. Forsburg, S. L. and Guarente, L. (1988). *Mol. Cell. Biol.,* **8,** 647–54.
23. Mannhaupt, G., Pilz, U., and Feldmann, H. (1988). *Gene,* **67,** 287–94.
24. Schmitz, U. K., Lonsdale, D. M., and Jefferson, R. A. (1990). *Curr. Genet.,* **17,** 261–4.
25. Olesen, J. T. and Guarente, L. P. (1990). *Genes Dev.,* **4,** 1714–29.
26. Eisenmann, D. M., Dollard, C., and Winston, F. (1989). *Cell,* **58,** 1183–91.
27. Hoffman, C. S. and Winston, F. (1987). *Gene,* **57,** 267–72.
28. Sikorski, R. S. and Boeke, J. D. (1991). In *Methods in enzymology,* Vol. 194 (ed. C. Guthrie and G. R. Fink), pp. 302–18. Academic Press, San Diego.
29. Lawrence, C. W. (1991). In *Methods in enzymology,* Vol. 104 (ed. C. Guthrie and G. R. Fink), pp. 273–81. Academic Press, San Diego.

30. Carey, M., Kakidani, H., Leatherwood, J., Mostashari, F., and Ptashne, M. (1989). *J. Mol. Biol.,* **209,** 423–32.
31. Brent, R. and Ptashne, M. (1985). *Cell,* **43,** 729–36.
32. Ruden, D. M., Ma, J., Li, Y., Wood, K., and Ptashne, M. (1991). *Nature,* **350,** 250–2.
33. Ma, J. and Ptashne, M. (1987). *Cell,* **48,** 847–53.
34. Berger, S. L., Cress, W. D., Cress, A., Triezenberg, S. J., and Guarente, L. (1990). *Cell,* **61,** 1199–208.
35. Gill, G. and Ptashne, M. (1987). *Cell,* **51,** 121–6.
36. Cress, W. D. and Triezenberg, S. J. (1991). *Science,* **251,** 87–90.

A1

Transcription controls: *cis*-elements and *trans*-factors

JOSEPH LOCKER

Gene expression is regulated by the interaction of specific proteins (*trans*-factors) with control signals in the DNA (*cis*-elements). This summary is organized around the *cis*-elements, integral parts of the genes they regulate, and presents mRNA (RNA polymerase II) controls from vertebrate species. The summary has been enlarged from an earlier one (1) to include the *trans*-factors as well as much new information. A recent summary (2) that focuses primarily on transcription factors is complementary to this listing.

The presentation of *cis*-elements is designed to facilitate their primary identification. The elements have been cross-searched to identify homologies, and additional versions of elements are included when they provide significantly different searches of DNA sequences. However, the computer identification of *cis*-elements in sequence data must be customized for each element, based on the length and degeneracy of the particular sequence motif. Moreover, microcomputer searching programs often fail to detect homology to variant motifs where bases are inserted or deleted. Also, known *trans*-factors may bind to more than one distinct motif or to related motifs that deviate significantly from a consensus. Thus, failure to identify a *cis*-element does not rule out a particular control system. On the other hand, detection of homology to a control element does not establish the presence of the control, only a model on which experiments can be based. Nevertheless, an ever increasing amount of information can be obtained directly from DNA sequences.

The control of tissue-specific gene expression remains a perplexing question. Many *cis*-elements were formerly thought to be tissue-specific. Such elements actually regulate gene expression in many tissues, often through distinct factors. Analysis should therefore never be limited to putative tissue-specific controls. Negative control elements also remain problematic. An element may be positive in one tissue and negative in a second, binding different factors in each. In addition, overlapping or closely spaced controls may be competitive. The safest approach is to assume that any element can have a positive or negative effect depending on its context in the gene and tissue.

Element	Sequence[1,2]	Factor[3]	Type[4]	Comments[2]
A. Promoter-linked controls				
1. TATA-Box	**TATAWAW**	TFIID	N	Location near −30 required, sets transcription initiation site; not present in some housekeeping genes (3)
2. Initiator (Inr)	**CTCANTCT**	TFIID	N	Alternate TFIID binding site, often from −3 to +5, in genes that lack a TATA box (4)
3. CAAT-Box/CP1	**YNNNNNNRRCCAATCANYK**	CP1	N	Distinct factor binding sites, each with its own high specificity (5); CBF another factor that binds similar sites (6)
4. CAAT-Box/CP2	**YAGYNNNRRCCAATCNNNR**	CP2	N	
5. CAAT-Box/NF1	**TTGGCNNNNNGCCAA**	NF1/TGCCA Proteins	N	Multiple factors binding sites with similar consensus, but with individual differences; also **TTGGCNNNNNNGCCA** (5, 7, 8)
6. CAP-Box	**CTTYTG**	Pol II Complex	N	Weak consensus sometimes found just downstream of initiation (capping) site; may be helpful for predicting transcription start site (9)
a. CAP-Box	**YCATTCR**	Pol II Complex	N	Cap site consensus of histone genes (10)
b. CAP-Box	**CANYYY**	Pol II Complex	N	Another weak Cap consensus (11)
7. GC-Box	**KRGGCGKRRY**	SP1	ZnF(Kru)	Found in many promoters and enhancers; involved in linking distant control elements; also **KGGGCGGRRY** (12)
8. CACCC-Box	**GCCACACCC**	U	U	Found in SV40, Ig, and Apo-AII genes; similar to β-globin consensus, **RRRNCCHCACCCTG** (13)

9. POB binding site	*AGAGAAGAGTGACAG*	POB	U	Stimulatory factor binds at −3 to −15 in proopiomelanocortin gene; factor found in divergent cell types (14); unusual location for a promoter binding site
B. General transcription controls—major systems				
1. Octamer	**ATTTGCAT**	NF3, OTF1, NFA1, NFA2, Oct1-6	HTH(POU)	Binds numerous distinct factors in different tissues: e.g. NF3 (=NFAI) ubiquitous, NFA2 B-cell specific; common consensus with Pit1 binding is **LLLLTNCAT** (15, 16, 17)
2. Enhancer core (AP3 site, acute phase response element)	**GGGRHTYYCC**	EBP1, TEFI, NFκB, H2TF1, KBF1, AGIEBP1, MBP1, PRDIIBF1, αACRYBP1	HTH, ZnF	Multiple factors bind consensus sequence originally described in SV40 core enhancer; some generally expressed, some cell specific (18, 19, 20); AP3 consensus, **GGAAAGTCC**; alternate core consensus, **GTGGWWWG**
a. Sph-Box	**AAGYATGCA**	TEFI	HTH	Two sites in SV40 enhancer that also bind TEFI; do not fit core consensus (21)
3. AP1 site (TRE, TPARE, GCN4, PEA1)	**TGAGTCAG**	Fos, FosB, Fra1, Fra2, Jun, JunB, JunD	bZIP	Ubiquitous factor which responds to TPA and kinase C (22, 23)
4. AP2 site (class I Enh, KBFI enhancer)	**GSSWGSCC**	AP2	N	TPA, kinase C and cAMP inducible; frequently next to other enhancer elements (24, 25, 26)
5. AP4 site (Polyoma PvuII Box)	**CAGCTGTGG**	AP4, GTIIB	HTH/bZIP	Overlaps with AP5 in SV40 enhancer; also **CAGCTG** (27, 28, 29)
6. AP5 site	**CTGTGGAATG**	AP5, EF.E, GTIIC	U	Sites in SV40 and polyoma enhancers (27)

Appendix 1 *continued*

Element	Sequence[1,2]	Factor[3]	Type[4]	Comments[2]
7. Delta site (κE3)	**CNGCCATC**	δ, YY1, NFE1	ZnF(Kru)	Site at immediate downstream (+1 to +20) of ribosomal proteins and other genes (some without TATA); also upstream site (30, 31)
8. Serum response element (CARG Box, SRE, distal serum element, DSE)	**CCWWWWWWGG**	p67/SRF, RSRF	HTH	Found in many genes; RSRF a family of proteins that bind related consensus, **CTAWWWWTAG** (32)
9. ATF site	**RTGACGTMR**	ATF1, ATF2, Adeno E4TF3, E2A.E	bZIP	A family of cellular transcription factors, often cAMP inducible, some E1A inducible; CREB in same family, binds related consensus (33, 34)
10. PEA2 (polyoma enhancer site A2)	*AACTGACCGCA*	PEA2, PEB2, PEB3, PEBP4	U	Stimulatory factor site overlaps PEA1=Ap1; repressing factor PEBP4 has larger footprint (*AACTGACCGCAGCTGGCCGT*) (35, 36)
11. PEA3 (polyoma enhancer site A3)	**CMGGAAGT**	Ets1, Ets2, PeA3, URTF	Ets	Found as compound with AP1 site in polyoma enhancer and several other genes; Ets protooncogenes often act in combination with AP1 (37, 38, 39); may be related to PU Box, **AGGAAG**
a. RPGα site	**MACTTCCGG**	RPGα	U	Binding competed by PEA3 binding at these sites (40)
b. SV40 Pu-Box	*CTGAAAGAGGAA*	PEA3?	U	Distinct purine-rich binding site near −300 in SV40 early promoter (41)
c. Elk1 site	*CAGGA*	Elk1, p62, TCF	Ets	Site immediately upstream (3 bases) of SRF binding site (CArG Box); binding SRF dependent (42)

12. μE3	**AGGTCATGTGGCAAC**	TFE3	HTH/bZIP	Ubiquitous factor; homology to κE3 (*GTCCCATGTGGTTAC*), MLTF, AND c-Myc sites (43, 44)
13. Adenovirus major late transcription factor (MLTF) site	**GGCCACGTGACC**	MLTF, UEF, USF	HTH/bZIP	Sites in adeno, α- and γ-fibrinogen, and mouse MT-I (43, 45)
C. General transcription controls—regulators of cell proliferation				
1. c-Myc binding site	**CACGTG**	Myc/Max	HTH/bZIP	Binds Myc-Max dimer; both have related binding sites; palindromic sequence, may be more effective within a larger G-C rich palindromes; also found in MLTF and μE3 (43, 46)
2. E2F site	**TTTCGCGC**	E2F	N	Cellular factor that binds adeno E2A and E1A enhancers, and cellular genes; activity found in infected HeLa and in undifferentiated F9 cells; forms complexes with Rb and cyclin A (27, 47, 48, 49)
3. TCE (TGF-β control element)	*GCGTGGGGA*	U	U	Inhibitory site in at −79 in c-Myc gene; binding stimulated by TGF-β and Rb protein; closely related to TIE that binds fos (50); common consensus with TIE is **GMGTKGGKGA**
4. TGF-β inhibitory element (TIE)	**GNNTTGGTGA**	Fos	bZIP	Found upstream of genes repressed by TGF-β; binding competed by AP1 binding site oligonucleotides, despite lack of homology to AP1 consensus (51)
5. p53 binding site	**TGCCT**	p53	N	Binding site contains multiple repeats of **TGCCT**; required number and spacing uncertain (52)

Appendix 1 *continued*

Element	Sequence[1,2]	Factor[3]	Type[4]	Comments[2]
6. Myc-CF1 site (common factor 1)	*AGAAAATGGT*	CF1	U	Conserved sequence at −261 in murine c-Myc; binds a widely distributed protein; may be associated with repression and myc-PRF binding (53)
7. c-Myc P2 promoter site	*TCGCGTGAGTATAAAAGCGCGGTT*	MBP1	N	Binds c-Myc P2 promoter site just 5′ of TATA box; may be a negative regulator; sequence from footprint data (54)
D. General transcription controls—miscellaneous activators				
1. Heat shock element (HSE)	**NGAAN**	HSF1, HSF2	bZIP	Heat shock element appears as multiple copies of **NGAAN** with variable spacing, often as inverted repeats; e.g. −105 to −91 region of HSP70 gene; binding is heat inducible (55)
2. GSG element	**GCGGGGGCG**	NGFIA, NGFIC, Krox20, Krox24, Wilm's tumor gene product	ZnF	Family of transcription activators related to development (56)
3. Class I gene enhancer	*ACATTCAAAATAACTTTGAG*	U	U	Novel enhancer in a porcine class I gene; neighbouring inverted form is *ACATGTATTTTTAATAACCTTTTCAG* (57)

4. Interferon-stimulated response elements and virus response elements (ISRE, VRE, PRDI)	**YAGTTTCWYTTTYCC**	PRDBF1, ISGF3, IRF1	ZnF	Related but distinct response elements that can be described by common consensus; PRDI is central region, **TTTCWYTTT**; ISREs in genes stimulated by Ifn α and β; VREs in IFN genes; repressed by PRDIBFI (58, 59, 60, 61, 62); HSV IE repeat (**GCGGAA**) appears to bind same factor(s) (63)
5. CMV enhancer	*CCCCATTGACGTCAATGGG*	U	U	Strong enhancer expressed in many cells; probably composite binding sites (64)
6. Ultraviolet response element (URE)	*TGACAACA*	U	U	Element in polyoma promoter that binds multiple UV-induced cellular proteins (65); resembles several other binding sites
7. 3′-Enhancer	**GCTTTTCACAGCCCTTGTGGATGC**	U	U	Conserved sequence, probably composite binding site, found 3′ of globin genes and 5′ of AFP gene; homology to Core, Silencer 3, erythroid NFE1 (66)
8. ETS2 S/S2 site	*GCGCGCCGCGTG*	S/S2	U	−5 to −16 unique stimulatory binding site in widely expressed ETS2 gene; factors in diverse cell types (67)
9. ETS2 H2 site	*GAGACTGACGA*	H2a, H2b	U	Unique regulator of ETS2 gene; H2a, expressed in many cells, inhibits; H2b, expressed in few cell types, stimulates (67)
10. U2/U6 proximal element	**YACCGYRACTTTGAAWGT**	U	U	Region required for Pol.II transcription of U2 gene, also found in U6 gene (68)
11. fos BLE-1 (basal level enhancer)	*GCGCCACC*	U	U	GC-rich element found at −90 in human c-fos gene; resembles CACCC Box (69)

Appendix 1 *continued*

Element	Sequence[1,2]	Factor[3]	Type[4]	Comments[2]
12. fos BLE-2	*AAGCCTGGGGCGTA*	U	U	GC-rich element of mouse c-fos gene at −147, similar to BLE-1 (69)
13. E2aE-Cβ site (adeno)	**TGGGAATT**	E2aECβ, E2A, E4F2	U	Common binding in E2 and E4 promoters (27)
14. E4TF1 site (adeno, CMV)	**AGGAAGTGAAA**	E4TF1	U	Distinct E4 promoter binding site, also in CMV (27)
15. EF.C site	**GTTGCNNGGCAAC**	EF.C	U	Binding site in polyoma, hepatitis B (27)
16. NRF1 site	**YGCGCAYGCGCR**	NRF1	U	Transcriptional activator system common to cytoplasmic and mitochondrial genes of cellular respiration (70)
E. General transcription controls—silencers				
1. Polyoma fPyF9 repressor	**GCATTCCATTGTT**	U	U	3 repeats within a genomic sequence inserted into enhancer of fPyF9 polyoma mutant; represses SV40 constructs in embryonal cells; strong homology to silencer 3 (71); subsequence *CATTCCA* is target for E1A-induced repression of c-neu (72)
2. HIV negative enhancer	*CATTTCATCACATGG*	U	U	Binding site in HIV LTR at −173; binding of negative regulator competes stimulatory factor with overlapping site (73)
3. Silencer 1	**ANCCTCTCY**	U	U	Chicken lysozyme gene element; located −1.0 to −0.2 kb; may be cell specific; repression is position and distance independent; homology to AP2 and silencer 2 (74)

4. Silencer 2	**ANTCTCCTCC**	U	U	Second lysozyme silencer element; some homology with GRE, AP2 (74)
5. Silencer 3	*AACAATGGCTATGCAGTAAAA*	U	U	Third lysozyme silence element; some homology with 3′ enhancer, AP2, AP3 (74)
6. N-ras NRE	*TTTTATGTTAATGG*	U	U	Negative element at −176 in N-ras gene (75)
7. Class I gene silencer	*CCAAAATTATCTGAAAAGGTTATTAAAA*	U	U	Novel silencer in a porcine class I gene (57)
8. Sterol response element (SRE)	**STGSSGYG**	CNBP	ZnF	Negative response element in cholesterol-related genes; repeated in promoter region; represses in presence of sterols (76, 77)
9. Fos intragenic regulatory element (FIRE)	*TCCCCGGCCGGGGA*	U	U	Negative response element contained within fos first exon (78)
F. Hormone-response controls				
1. cAMP response element (CRE)	**TGACGTCA**	CREB	bZIP	Inverted repeat consensus binds at least two forms of CREB; element may have basal activity without hormone induction; CREB in ATF protein family (34, 79, 80)
2. Estrogen response element (ERE)	**AGGTCANNNTGACCT**	Estrogen Receptor	ZnF	Inverted repeat consensus, though often imperfect in functional EREs (81, 82)
3. Glucocorticoid/ mineralocorticoid/ progesterone/androgen response element (GRE/MRE/PRE/ARE)	**GGTACANNNTGTTCT**	GR/PR/MR/AR	ZnF	All four receptors bind same consensus, though there are selective affinities for individual elements; current information does not allow discrimination of response elements by sequence alone (82)
a. Glucocorticoid repressor (GRE-)	**ATYACNNNNTGATCW**	GR	ZnF	Consensus of GREs that mediate repression rather than stimulation of transcription (82)

Appendix 1 *continued*

Element	Sequence[1,2]	Factor[3]	Type[4]	Comments[2]
4. Thyroid hormone/retinoic acid response element (TSE, TRE, RARE)	**AGGTCATGACCT**	TR, ErbA, RARα, RARβ, RARγ, RXRα, RXRβ	ZnF	Inverted repeat of 6-base half element (**RGKTCA** or ideal **AGGTCA**) without spacing; mediates TR and RAR stimulation (83, 84)
5. Retinoic acid response element (RARE)	**RGKTCAYYNRNRGKTCA**	RARα, RARβ, RARγ, RXRβ	ZnF	Direct repeat of 6-base half element (**RGKTCA** or ideal **AGGTCA**), separated by 5 bases (**YYNRN**) mediates selective RAR stimulation (83, 84)
a. Cytoplasmic retinol binding protein response element (CRBPE, RXRE)	**AGGTCANAGGTCA**	RXRα, RXRβ	ZnF	Coregulators of RARs bind to direct 6-base half element (**AGGTCA**) separated by 1 base (84)
6. Thyroid hormone response element (TSE, TRE)	**RGGTSANNNNAGGWCA**	TR, ErbA	ZnF	Direct repeat of 6-base half element (**RGKTSA** or ideal **AGGTCA**), separated by 4 bases, mediates selective TR stimulation (83, 84)
7. Vitamin D response element (VDRE)	**RGGTSANNRRGGNCA**	Vitamin D Receptor	ZnF	Direct repeat of 6-base half element (**RGGTSA** and **RGGNCA**) separated by 3 bases (**NNR**) (83, 84)
8. Insulin response element (IRE)	**CCCGCCTC**	IREABP	HMG	Directs insulin-inducible gene expression; consensus of two footprints in liver GAPDH gene; larger response element, *AACTTTCCCGCCTCTCAGCCGAAG*, is proposed (85, 86)
9. Thyroid hormone inhibitory element (TIE)	*AGGGTATAAAAAGGGC*	U	U	Negative element encompassing TATA box of human and rat growth hormone genes (87)

G. Tissue-specific controls—liver

Site	Sequence	Factors	Motif	Comments
1. HNF1 site	**GGTTAATNATTAAC**	HNF1, HNF1α, HNF1β, LFB1, PE, HP1, APF, ABF, HS, AFP	HTH(POU)	Major transcription control system for genes of differentiated hepatocytes; HNF1β, expressed in variety of tissues, acts as negative regulator (88, 89)
2. ATBF1 site	*TGATTAATAATTACA*	ATBF1, AFP1	ZnF/HOM	Unusually large factor, binds to site in human AFP enhancer; also binds weakly to HNF1 sites (90, 91)
3. C/EBP site	**TKNNGYAAK**	C/EBP, C/EBPα, C/EBPβ, CEBPδ, EBP20, NFIL6, LAP, LIP, IL6DBP, AGP/DBP, CRP1, CRP2, CRP3	bZIP	Binding sites for important family of proteins, most liver (and adipocyte) enriched; also binds to retrovirus LTRs and other viral enhancers; consensus (92) recognizes most example sequences but also HNF1 and other sites; alternate consensus for strong binding sites, **ATTGCGCAAT**, does not identify most examples (93, 94)
a. C/EBP site examples				C/EBP binding sites (95, 96, 97, 98, 99, 100); large footprints suggest composite sites with additional factors, e.g. Alb D includes at least C/EBB+DBP; set is more useful than consensus for searching gene sequences (101)
TTR-2 (transthyretin)	*GTTCAAACATGTCCTAATACTCTGT*			
TTR-3	*AGTAGTTTTCCATCTTACTCAACATC*			
α1AT-A	*TTCGTCAGGTGGGCACATAACCTACTCT*			
α1AT-C	*TAACTGCTTTGCTTAAGACTCCATTGATTTAGG*			
SV40 Core-C	*GTAGGGTGTGGAAAGTCCCA*			
Albumin D Element	*TATGATTTTGTAATGGGGTA*			

Appendix 1 *continued*

Element	Sequence[1,2]	Factor[3]	Type[4]	Comments[2]
HBV Box-α	*CAAGGTCTTACATAAGAGGACTCTT*			
HBV Box-β	*CCTACTTCAAAGACTGT*			
Gene 248	*CAAAGTTGAGAAATTTCTATT*			−149 in adipose-specific 248 gene (98)
SCD1	*AGGGGGCTGAGGAAATACTGAACA*			−80 in adipose specific stearoyl-CoA desaturase (SCD1) gene (98)
4. Albumin D-site	*GATTTTATAATA*	DBP	bZIP	Major family of liver-enriched factors; binds to albumin promoter D-site (−98 to −109) overlapping with C/EBP site; may interact with C/EBP; binds to VBP site and vice versa; DBP/VBP consensus **GWTTWYATAAWM** (102, 103, 104)
a. VBP binding site	*GTTTACATAAAC*	VBP	bZIP	Protein related to DBP; binds at albumin D-site and vitellogenin promoter (−94 to −82) (104)
5. LTFE Site (HBV E site enhancer, AFP Box 2)	**TGTTTGCT**	LTFE, eHTF	U	Liver-stage specific factor; in HBV and related viruses, AFP, α1AT, albumin promoters and enhancers (105, 106)
a. HBLF site	*AGTAAACAGTA*	HBLF	U	HBV binding site, upstream of E-site, binds a distinct factor; sequence related to LTFE consensus (107)
6. GA binding protein site	**CGGAARCGGAAR**	GABP, GABPα, GABPβ, VP16	Ets, Notch	Liver-enriched protein, binds repeats (often imperfect) of **CGGAAR** motif; HSV VP16 also binds prototypic site in HSV IE gene (108)
7. HNF3 site	**TATTGAYYYWG**	HNF3, HNF3α, HNF3β, HNF3γ	HOM(FKH)	Liver-enriched protein family also expressed in lung and intestine (109, 110)

8. HNF4 site	**KGCWARGKYCAY**	HNF4	ZnF	Liver enriched factor, resembles thyroid receptor family, but no HNF4 ligand has been demonstrated; regulator of HNF1 (111, 112)
9. HNF5 site	**RCAAAYA**	HNF5	U	In TAT gene, binding overlaps with GRE, and may require prior GRE binding to open chromatin (113)
10. LFA1 site	**TGRACYTGGCCC**	LFA1, HNF2, tfLF1, tfLF2	U	Bipartite binding site (**TGRACY** . . . **TGGCCC**) found in numerous hepatic genes; spacing between parts is 0–4 bp; additional consensus sequences **TGRACYYTGGCCC**, **TGRACYYWTGGCCC**, **TGRACTYWWTGGCCC**, **TGRACYYWWGTGGCCC** will identify most variants (114)
11. ANF site	**CTTKAWCTSG**	ANF	U	Two negative regulatory elements in albumin gene enhancer; factor present in many tissues (115)
12. AFP Box	**CTTTGAGCAA**	U	U	Highly conserved 10 base element in three mouse, three rat, and one human AFP enhancer (101, 116)
13. Xenobiotic response element (XRE)	**TCAGCG**	XREBF, Dioxin receptor	U	6 nucleotide core of XRE1 and XRE2 binding sites, that activate cytochrome P-450IA1 gene (117)
14. Metal response element (MRE)	**CTNTCCRCNCGGCCC**	MTF1	U	Consensus of multiple regulatory elements in metallothionein genes; binding stimulated by Cd^{2+} or Zn^{2+}; alternate consensus **TYTGCGCCCGGCCC** (118, 119)

Appendix 1 *continued*

Element	Sequence[1,2]	Factor[3]	Type[4]	Comments[2]
15. Interleukin-1 response element	*AATGTTGGAA*	NFAB	U	Unique binding site in distal regulatory element or cytokine response element of AGP gene; same sequence in IL6 response element of CRP gene (120); acute phase response mediated by this plus NFκB binding acute phase response element (20)
16. α_1I3 gene element I	*TCCTTTACCAACACTGT*	U	U	Binding site at −196 in a_1I3 gene (121)
H. Tissue-specific controls—lymphocytes				
1. Immunoglobulin gene E boxes				Multiple strong enhancer motifs from immunoglobulin genes; may represent general cellular rather than lymphocyte-specific genes, i.e. μE3 = κE3; it remains unclear how many distinct motifs and specificities are represented (122)
a. κE2 = μE4	**CAGGTGKY**	E12, E47, ITF1, ITF2	HTH	Sequence motif closely related to MLTF, PEA2, pancreatic enhancer, μE3, μE2, and insulin enhancer
b. μE1	*AAGATGGC*	U	U	Related to insulin enhancer
c. μE2	*CAGCTGGC*	U	U	Related to PEA2, κE1, insulin enhancer
d. μE4	*CACCTGGG*	U	U	Unique motif
e. κE1	*CATCTGGC*	U	U	Related to μE3, μE2, and insulin enhancer
2. B-cell E6 enhancer	*CCGAAACTGAAAAGG*	U	U	B-cell-specific enhancer upstream of μE4 and octamer in human heavy chain gene (123)

3. LEF1 site	*CCTTTGAA*	LEF1, TCF1	HMG	Pre-B and pre-T cell factor involved in TCRα gene regulation; TCF may be a different HMG factor that binds a similar consensus, as does testis-specific SRY; alternate consensus is **GTTTGTT**, suggesting **GTTTGWW** is common HMG consensus (124, 125)
4. NFAT site	**ARGARATTCCA**	NFAT	U	T-cell restricted factor that binds to IL2 gene at −285; consensus derived form this and conserved sites in related genes (126, 127, 128)
5. Class II gene X boxes	**CTAGCAACWGANG**	U	U	Upstream regulatory sites in Class II genes; consensus suggested by similarity of elements is useful for analysis, but the individual boxes appear to bind different factors (129)
a. IAβX	*CCAGAGACAGACG*	U	U	
b. IAαX	*CTGGCAACTGTGA*	U	U	
c. IEαX	*CTAGCAACAGATG*	U	U	
d. IEβX	*CTAGCAACTGATG*	U	U	
e. IAβX′	*CTAGCAACAGAAG*	U	U	
6. GM-CSF promoter site	**MATTAWTCATTTCCT**	U	U	GM-CSF gene expressed in mitogen activated but not quiescent T-cells; activation requires this site which contains repeats of **CATTW**; consensus with similar site in IL5 genes (130)
7. Myc-PRF	*CGTAC*<u>*AGAAAGGGAAAGGA*</u>*CTAG CGC*	U	U	−270 element in murine c-Myc associated with plasmacyte-specific repression; underlined segment common to Myc-CF1 (53)

Appendix 1 *continued*

Element	Sequence[1,2]	Factor[3]	Type[4]	Comments[2]
I. Tissue-specific controls—erythrocytes				
1. Erythroid-specific promoter element	**WGATAR**	GATA1, NFEI, NFE1a, NFEIb, NFE1c, Eryf1, GF1	ZnF	In promoters and enhancers of erythroid-specific genes (131)
2. Erythroid-specific NFE4 site	**RAGAGGRGG**	NFE4	U	In chicken β globin enhancer and promoter (132)
3. Erythroid pyr factor site	*CCTTCCTTCC*	pyr factor	U	Binds specific (usually symmetrical) sites in pyrimidine-rich control regions of γ, δ, and β globin genes (133)
4. G string	*GGGGGGGGGGGGGGGG*	BGP1	U	Unique 16 bp G motif in chicken β globin that binds an erythrocyte-specific protein (134)
5. Myb recognition element	**CCGTTA**	Myb	HTH	Nuclear oncoprotein transcription activator, limited to differentiated hematopoietic cells; activates mim1, c-myc, cdc2, po1α, and HIV LTR (135, 136)
J. Tissue-specific controls—miscellaneous				
1. Muscle-specific gene E box (MyoD family sites)	**CANNTG**	MyoD, myogenin, Myf5, MRF4	HTH	Family of HTH proteins that regulate myogenesis; consensus binding site defines a core sequence that is found in several other enhancer motifs (137, 138, 139)
2. Myocyte enhancer	**YTAWAAATAR**	MEF2	U	Consensus of conserved sites in muscle gene regulatory regions; motif resembles ATBF, Class I enhancer (140)

3. Exocrine pancreatic (XP) enhancer	**GWCACCTGTSCTTTTCCCTG**	XPF1	U	Consensus of enhancer for several pancreatic genes; binding site defines a bipartite core **CACCTGNNNNTTCCC** which resembles binding sites of several HTH proteins (141, 142)
4. Insulin gene enhancer (IEB)	**GCCATCTGSC**	IEF1	U	2 copies upstream of rat insulin genes; β-cell specific (143, 144)
5. Thyroid-specific enhancer	**GNNCACTCAAG**	TTF1, T/EBP	NK2	Found in promoters of 8 thyroid specific genes (145, 146)
6. Anterior pituitary enhancer	**AWWTATNCAT**	Pit1	HTH(POU)	Multiple upstream sites in prolactin and growth hormone genes; closely related to octamer (POU) consensus (147, 148)
7. Mammary gene MGF site	**ANTTCTTGGNA**	MGF	U	Conserved site of positive regulation in casein genes from several species (149)
8. Testis-determining protein (SRF) site	**AACAAAG**	SRF	HMG	Closely related to lymphocyte LEF1 site, which binds another HMG factor; common consensus is **GTTTGWW** (86, 125)
9. Keratinocyte enhancer	**AARCAAA**	U	U	Conserved promoter motif found in all genes for bovine, human, and murine cytokeratins; also found in C (constitutive) enhancer of HPV18 (150, 151)
10. Keratinocyte KRF1 site	*GCATAACTATATCCACTCCC*	KRF1	U	Footprint of constitutive keratinocyte factor, binds to a site in HPV18 C enhancer; partially overlaps octamer site and probably competes for binding (151)
K. Viral elements				
1. HSV IE promoters	**TAATGARAT**	VP16	N	Binds VP16, HSV early gene activator; binds in complex with Oct-1; VP16 also binds GABP sites (108)

Appendix 1 *continued*

Element	Sequence[1,2]	Factor[3]	Type[4]	Comments[2]
2. HSV late promoter	*GGGTATAAATTCCGG*	α4, α27	U	Modified TATA box required for HSV late gene expression (152)
3. Papillomavirus E2 enhancer	**ACCNNNNNNGGT**	E2	N	Binding sites for viral E2 gene product; has positive or negative effect depending on position in promoter (153, 154)
4. Papillomavirus PVF site	**AGGCACATAT**	PVF	U	Conserved binding sites in viral LCR enhancers (155)
L. Miscellaneous				
1. Topoisomerase II cleavage site	**RNYNNCNNGYNGKTNYNY**	TopoII	N	May modify enhancer interactions, especially in transient assays; linked to nuclear matrix association regions (MAR) (156)
2. Methylated DNA binding protein (MDBP) site	**RTYRYYAYRGYRAY**	MDBP	U	Ubiquitous factor; binds element when Y = methylcytosine or thymine; found in several enhancers and c-Myc regulatory first intron (157)
3. Translation initiation site	**GCCRCCATGG**	Ribosome	N	Site of ribosome binding to mRNA (158); motif is useful for defining transcription start sites in relation to 5′-untranslated regions
4. Splice junction donor site	**MAGGTRAGT**	RNA splicing factors	N	Splicing consensus (159); motif is useful for defining transcription start sites in relation to long 5′-untranslated regions that contain introns

5. Splice junction acceptor site	**YYYYYYYNYAGG**	RNA splicing factors	N	Consensus motif actually **YYYYYYYYYYYNYAGG** (159), but shorter consensus listed here more useful for searching

Notes

Consensus sequences are shown in bold type; specific example sequences are in italic.

IUB Code: R = A, G; M = A, C; W = A, T; Y = T, C; K = T, G; S = G, C; B = T, G, C; V = A, G, C; H = A, T, C; D = A, T, G.

Listings include both synonyms of individual factors and distinct factors (both positive and negative) that bind the same motif.

Protein-binding domains: N, protein defined, but not categorized; U, protein undefined; HTH, helix turn helix; ZnF, zinc fingers; LZnH, loop-zinc-helix; bZIP, leucine coiled zipper; βrib—β ribbon (160); POU, common HTH domain in octamer and Pit1; HOM, homology to *Drosophila* homeodomain (17); Kru, related to Krüppel ZnF protein (161); Ets, DNA-binding domain common to Ets gene family (39); Notch, 33 aa motif found in *Drosophila* notch and numerous other regulatory proteins (162); Fkh, *Drosophila* fork head homeodomain (110); HMG, homology to the chromosomal high mobility group 1 (HMG1) protein (124); NK2, homology to *Drosophila* NK2 homeodomain (145). A slash separates distinct domains in the same protein; a comma separates domains of different factors that bind the same motif; parentheses separate special features of a domain.

References

1. Locker, J. and Buzard, G. (1990). *DNA Sequence*, **1,** 3.
2. Faisst, S. and Meyer, S. (1992). *Nucleic Acids Res.*, **20,** 3.
3. Breathnach, R. and Chambon, P. (1981). *Annu. Rev. Biochem.*, **50,** 349.
4. Smale, S., Schmidt, M. C., Berk, A. J., and Baltimore, D. (1990). *Proc. Natl Acad. Sci. USA*, **87,** 4509.
5. Chodosh, L. A., Baldwin, A. S., Carthew, R. W., and Sharp, P. A. (1988). *Cell*, **53,** 11.
6. Maity, S. N., Vuorio, T., and DeCrombrugghe, B. (1990). *Proc. Natl Acad. Sci. USA*, **87,** 5378.
7. Santoro, C., Mermod, N., Andrews, P., and Tjian, R. (1988). *Nature*, **334,** 218.
8. Rupp, R. A. W., Kruse, U., Multhaup, G., Gobel, U., Beyreuther, K., and Sippel, A. E. (1990). *Nucleic Acids Res.*, **18,** 2607.
9. Baralle, F. E. and Brownlee, G. G. (1978). *Nature*, **274,** 84.
10. Hentschel, C. C. and Birnstiel, M. L. (1981). *Cell*, **25,** 301.
11. Bucher, P. and Trifonov, E. N. (1986). *Nucleic Acids Res.*, **14,** 10009.
12. Briggs, M. R., Kadonaga, J. T., Bell, S. P., and Tjian, R. (1986). *Science*, **234,** 47.
13. Shelley, S. C. and Baralle, F. E. (1987). *Nucleic Acids Res.*, **15,** 3801.
14. Riegel, A. T., Remenick, J., Wolford, R. G., Berard, D. S., and Hager, G. L. (1990). *Nucleic Acids Res.*, **18,** 4513.
15. Pruijn, J. M., van Driel, W., van Miltenberg, R. T., and van der Vliet, P. C. (1987). *EMBO J.*, **6,** 3771.
16. O'Neill, E. A., Fletcher, C., Burrow, C. R., Heintz, N., Roeder, R. G., and Kelly, T. J. (1988). *Science*, **219,** 1210.
17. Ruvkun, G. and Finney, M. (1991). *Cell*, **64,** 475.
18. Clark, L., Pollock, R. M., and Hay, R. T. (1988). *Genes Dev.*, **2,** 991.
19. Lenardo, M. and Baltimore, D. (1989). *Cell*, **58,** 227.
20. Ron, R., Brasier, A., and Habener, J. F. (1991). *Mol. Cell. Biol.*, **11,** 2887.
21. Xiao, J. H., Davidson, I., Matthes, H., Garnier, J., and Chambon, P. (1991). *Cell*, **65,** 551.
22. Angel, P., Imagawa, M., Chiu, R., Stein, B., Imbra, R. J., Rahmsdorf, H. J., Jonat, C., Herrlich, P., and Karin, M. (1987). *Cell*, **49,** 729.
23. Yen, R., Wisdom, R. M., Tratner, I., and Verma, I. M. (1991). *Proc. Natl Acad. Sci. USA*, **88,** 5077.
24. Mitchell, P. J., Wang, C., and Tjian, R. (1987). *Cell*, **50,** 847.
25. Imagawa, M., Chiu, R., and Karin, M. (1987). *Cell*, **51,** 251.
26. Leask, A., Byrne, C., and Fuchs, E. (1991). *Proc. Natl Acad. Sci. USA*, **88,** 7948.
27. Jones, N. C., Rigby, P. W. J., and Ziff, E. B. (1988). *Genes Dev.*, **2,** 267.
28. Mermod, N., Williams, T. J., and Tjian, R. (1988). *Nature*, **332,** 557.
29. Hu, Y., Luscher, B., Admon, A., Mermod, N., and Tjian, R. (1990). *Genes Dev.*, **4,** 1741.
30. Hariharan, N., Kelley, D., and Perry, R. (1991). *Proc. Natl Acad. Sci. USA*, **88,** 9799.
31. Park, K. and Atchison, M. (1991). *Proc. Natl. Acad. Sci. USA*, **88,** 9804.
32. Pollock, R. and Treisman, R. (1991). *Genes Dev.*, **5,** 2327.

33. Hai, T., Liu, F., Coukos, W., and Green, M. (1989). *Genes Dev.*, **3,** 2083.
34. Liu, F. and Green, M. (1990). *Cell*, **61,** 1217.
35. Piette, J. and Yaniv, M. (1987). *EMBO J.*, **6,** 1331.
36. Furukawa, K., Yamaguchi, Y., Ogawa, E., Shigesada, K., Satake, M., and Ito, Y. (1990). *Cell Growth Differentiation*, **1,** 135.
37. Wasylyk, B., Wasylyk, C., Flores, P., Begue, A., Leprince, D., and Stehelin, D. (1990). *Nature*, **346,** 191.
38. Rorth, P., Nerlov, C., Blasi, F., and Johnsen, M. (1990). *Nucleic Acids Res.*, **18,** 5009.
39. Fisher, R., Mavrothalassitis, G., Kondoh, A., and Papas, T. (1991). *Oncogene*, **6,** 2249.
40. Lennard, A. C. and Fried, M. (1991). *Mol. Cell. Biol.*, **11,** 1281.
41. Pettersson, M. and Schaffner, W. (1987). *Genes Dev.*, **1,** 962.
42. Hipskind, R., Rao, V., Mueller, C., Reddy, E., and Nordheim, A. (1991). *Nature*, **354,** 531.
43. Kerkhoff, E., Bister, K., and Klempnauer, K. (1991). *Proc. Natl Acad. Sci. USA*, **88,** 4323.
44. Beckmann, H., Su, L., and Kadesch, T. (1989). *Genes Dev.*, **4,** 167.
45. Chodosh, L. A., Carthew, R. W., Morgan, J. G., Crabtree, G. R., and Sharp, P. A. (1987). *Science*, **238,** 684.
46. Prendergast, G. C. and Ziff, E. B. (1991). *Science*, **251,** 186.
47. Hiebert, S., Chellappan, S., Horowitz, J., and Nevins, J. (1992). *Genes Dev.*, **6,** 177.
48. Devoto, S., Mudryl, M., Pines, J., Hunter, T., and Nevins, J. (1992). *Cell*, **68,** 167.
49. Shirodkar, S., Ewen, M., DeCaprio, J., Morgan, J., Livingston, D., and Chittenden, T. (1992). *Cell*, **68,** 157.
50. Pietenpol, J., Munger, K., Howley, P., Stein, R., and Moses, H. (1991). *Proc. Natl Acad. Sci. USA*, **88,** 10227.
51. Kerr, L. D., Miller, D. B., and Matrisian, L. M. (1990). *Cell*, **61**, 267.
52. Kern, S. E., Kinzler, K. W., Bruskin, A., Jarosz, D., Friedman, P., Prives, C., and Vogelstein, B. (1991). *Science*, **252,** 1708.
53. Kakkis, E., Riggs, K. J., Gillespie, W., and Calame, K. (1989). *Nature*, **339,** 718.
54. Ray, R. and Miller, D. M. (1991). *Mol. Cell. Biol.*, **11,** 2154.
55. Sarge, K. D., Zimarino, V., Holm, K., Wu, C., and Morimoto, R. L. (1991). *Genes Dev.*, **5,** 1902.
56. Crosby, S. D., Puetz, J. J., Simburger, K. S., Fahrner, T. J., and Milbrandt, J. (1991). *Mol. Cell. Biol.*, **11,** 3835.
57. Weissman, J. D. and Singer, D. S. (1991). *Mol. Cell. Biol.*, **11,** 4217.
58. Levy, D. E., Kessler, D. S., Pine, R., Reich, N., and Darnell, J. E. (1988). *Genes Dev.*, **2,** 383.
59. Whittemore, L. and Maniatis, T. (1990). *Proc. Natl Acad. Sci. USA*, **87,** 7799.
60. Kessler, D. E., Levy, D. E., and Darnell, J. E. (1988). *Proc. Natl Acad. Sci. USA*, **85,** 8521.
61. Kessler, D. S., Veals, S. A., Fu, X. Y., and Levy, D. E. (1990). *Genes Dev.*, **4,** 1753.
62. Keller, A. D. and Maniatis, T. (1991). *Genes Dev.*, **5,** 868.
63. LaMarco, K. L. and McKnight, S. L. (1989). *Genes Dev.*, **3,** 1372.

64. Dorsch-Häsler, K., Keil, G. M., Weber, F., Jasin, M., Schaffner, W., and Koszinowski, U. H. (1985). *Proc. Natl Acad. Sci. USA*, **82,** 8325.
65. Ronai, Z. A. and Weinstein, I. B. (1990). *Cancer Res.*, **50,** 5374.
66. Bodine, D. M. and Ley, T. J. (1987). *EMBO J.*, **6,** 2997.
67. Mavrothalassitis, G. J. and Papas, T. S. (1991). *Cell Growth Differentiation*, **2,** 215.
68. Lobo, S. M., Ifill, S., and Hernandez, N. (1990). *Nucleic Acids Res.*, **18,** 2891.
69. Verma, I. M. and Sassone-Corsi, P. (1987). *Cell*, **51,** 513.
70. Evans, M. E. and Scarpulla, R. C. (1990). *Genes Dev.*, **4,** 1023.
71. Ariizumi, K., Takahashi, H., Nakamura, M., and Ariga, H. (1989). *Mol. Cell. Biol.*, **9,** 4032.
72. Yu, D., Suen, T. C., Yan, D. H., Chang, L. S., and Hung, M. C. (1990). *Proc. Natl Acad. Sci. USA*, **87,** 4499.
73. Garcia, J. A., Wu, F. K., Mitsuyasu, R., and Gaynor, R. B. (1987). *EMBO J.*, **6,** 3761.
74. Baniahmad, A., Muller, M., Steiner, C., and Renkawitz, R. (1987). *EMBO J.*, **6,** 2297.
75. Paciucci, R. and Pellicer, A. (1991). *Mol. Cell. Biol.*, **11,** 1334.
76. Osborne, T. F., Gil, G., Goldstein, J. L., and Brown, M. S. (1988). *J. Biol. Chem.*, **263,** 3380.
77. Rajavashisth, T., Taylor, A., Andalibi, A., Svenson, K., and Lusis, A. (1989). *Science*, **245,** 640.
78. Lamb, N., Fernandez, A., Tourkine, N., Jeanteur, P., and Blanchard, J. (1990). *Cell*, **61,** 485.
79. Roesler, W. J., Vandenbark, G. R., and Hanson, R. W. (1988). *J. Biol. Chem.*, **263,** 9063.
80. Berkowitz, L. A. and Gilman, M. Z. (1990). *Proc. Natl Acad. Sci. USA*, **87,** 5258.
81. Martinez, E., Givel, F., and Wahli, W. (1987). *EMBO J.*, **6,** 3719.
82. Beato, M. (1989). *Cell*, **56,** 335.
83. DeLuca, L. (1991). *FASEB J.*, **5,** 2924.
84. Yu, V., Delsert, C., Andersen, B., Holloway, J. M., Devary, O. V., Näär, A., Kim, S., Boutin, J., Glass, C., and Rosenfeld, M. (1991). *Cell*, **67,** 1251.
85. Nasrin, N., Ercolani, L., Denaro, M., Kong, X. F., Kang, I., and Alexander, M. (1990). *Proc. Natl Acad. Sci. USA*, **87,** 5273.
86. Nasrin, N., Buggs, C., Kong, X. F., Carnazza, J., Goebl, M., and Alexander-Bridges, M. (1991). *Nature*, **354,** 317.
87. Wight, P. A., Crew, M. D., and Spindler, S. R. (1988). *Mol. Endocrinol*, **2,** 536.
88. Mendel, D. and Crabtree, G. (1991). *J. Biol. Chem.*, **266,** 677.
89. Mendel, D., Hansen, L., Graves, M., Conley, P., and Crabtree, G. (1991). *Genes Dev.*, **5,** 1042.
90. Sawadaishi, K., Morinaga, T., and Tamaoki, T. (1988). *Mol. Cell. Biol.*, **8,** 5179.
91. Morinaga, T., Yasuda, H., Hashimoto, T., Higashio, K., and Tamaoki, T. (1991). *Mol. Cell. Biol.*, **11,** 6041.
92. Ryden, T. and Beemon, K. (1988). *Mol. Cell. Biol.*, **9,** 1155.
93. Agre, P., Johnson, P., and S. M. (1989). *Science*, **246,** 922.

94. Williams, S., Cantwell, C., and Johnson, P. (1991). *Genes Dev.*, **5**, 1553.
95. Grayson, D. R., Costa, R. H., Xanthopoulos, K. G., and Darnell, J. E. (1988). *Science*, **239**, 786.
96. Costa, R. H., Lai, E., Grayson, D. R., and Darnell, J. E. (1987). *Mol. Cell. Biol.*, **8**, 81.
97. Costa, R. H., Grayson, D. R., Xanthopoulos, K. G., and Darnell, J. E. (1988). *Proc. Natl Acad. Sci. USA*, **85**, 3840.
98. Christy, R. J., Yang, V. W., Ntambi, J. M., Geiman, D. E., Landschulz, W. H., Friedman, A. D., Nakabeppu, Y., Kelly, T. J., and Lane, M. D. (1989). *Genes Dev.*, **3**, 1323.
99. Grayson, D. R., Costa, R. H., Xanthopoulos, K. G., and Darnell, J. E. (1988). *Mol. Cell. Biol.*, **8**, 1055.
100. Yuh, C. and Ting, L. (1991). *Mol. Cell. Biol.*, **11**, 5044.
101. Buzard, G. and Locker, J. (1990). *DNA Sequence*, **1**, 33.
102. Mueller, C., Maire, P., and Schibler, U. (1990). *Cell*, **61**, 279.
103. Wuarin, J. and Schibler, U. (1990). *Cell*, **63**, 1257.
104. Iyer, S., Davis, D., Seal, S., and Burch, V. (1991). *Mol. Cell. Biol.*, **11**, 4863.
105. Shaul, Y. and Ben-Levy, R. (1987). *EMBO J.*, **6**, 1913.
106. Liu, J., DiPersio, C., and Zaret, K. (1991). *Mol. Cell. Biol.*, **11**, 773.
107. Trujillo, M., Letovsky, J., MaGuire, H., Lopez-Cabrera, M., and Siddiqui, A. (1991). *Proc. Natl Acad. Sci. USA*, **88**, 3797.
108. LaMarco, K., Thompson, C., Byers, B., E., W., and McKnight, S. (1991). *Science*, **253**, 789.
109. Costa, R., Grayson, D., and Darnell, J. (1989). *Mol. Cell. Biol.*, **9**, 1415.
110. Lai, E., Prezioso, V., Tao, W., Chen, W., and Darnell, J. (1991). *Genes Dev.*, **5**, 416.
111. Sladek, F., Zhong, W., Lai, E., and Darnell, J. (1990). *Genes Dev.*, **4**, 2353.
112. Tian, J. and Schibler, U. (1991). *Genes Dev.*, **5**, 2225.
113. Rigaud, G., Roux, J., Pictet, R., and Grange, T. (1991). *Cell*, **67**, 977.
114. Ramji, D., Tadros, M., Hardon, E., and Cortese, R. (1990). *Nucleic Acids Res.*, **19**, 1139.
115. Herbst, R., Boczko, E., Darnell, J., and Babiss, L. (1990). *Mol. Cell. Biol.*, **10**, 3896.
116. Godbout, R., Ingram, R. S., and Tilghman, S. M. (1988). *Mol. Cell. Biol.*, **8**, 1169.
117. Hapgood, J., Cuthill, S., Soderkvist, P., Wilhelmsson, E., Pongrantz, I., Tukey, R., Johnson, E., Gustafsson, J., and Poellinger, L. (1991). *Mol. Cell. Biol.*, **11**, 4314.
118. Karin, M., Haslinger, A., Heguy, A., Dietlin, T., and Cooke, T. (1987). *Mol. Cell. Biol.*, **7**, 606.
119. Mueller, P., Salser, S., and Wold, B. (1988). *Genes Dev.*, **2**, 412.
120. Won, K. and Baumann, H. (1991). *Mol. Cell. Biol.*, **11**, 3001.
121. Abraham, L., Bradshaw, A., Shiels, B., Northemann, W., G., H., and Fey, G. (1990). *Mol. Cell. Biol.*, **10**, 3483.
122. Staudt, L. (1991). *Annu. Rev. Immunol.*, **9**, 373.
123. Wang, J., Oketani, M., and Watanabe, T. (1991). *Mol. Cell. Biol.*, **11**, 75.
124. Giese, K., Amsterdam, A., and Grosschedl, R. (1991). *Genes Dev.*, **5**, 2567.
125. Harley, V. R., Jackson, D. I., Hextall, P. J., Hawkins, J. R., Berkovitz, G. D.,

Sockanathan, S., Lovell-Badge, R., and Goodfellow, P. N. (1992). *Science*, **255,** 453.
126. Granelli-Piperno, A. and McHugh, P. (1991). *Proc. Natl Acad. Sci. USA*, **88,** 11431.
127. Fraser, J. D., Irving, B. A., Crabtree, G. R., and Weiss, A. (1991). *Science*, **251,** 313.
128. Flanagan, W. M., Corthesy, B., Bram, R. J., and Crabtree, G. R. (1991). *Nature*, **352,** 803.
129. Celada, A. and Maki, R. (1989). *Mol. Cell. Biol.*, **9,** 5219.
130. Nimer, S., Fraser, J., Richards, J., Lynch, M., and Gasson, M. (1990). *Mol. Cell. Biol.*, **10,** 6084.
131. Martin, D. and Orkin, S. (1990). *Genes Dev.*, **4,** 1886.
132. Gallarda, J., Foley, K., Yang, Z., and Engel, J. (1989). *Genes Dev.*, **3,** 1845.
133. O'Neill, D., Bornschlegel, K., Flamm, M., Castle, M., and Bank, A. (1991). *Proc. Natl Acad. Sci. USA*, **88,** 8953.
134. Lewis, C., Clark, S., Felsenfeld, G., and Gould, H. (1988). *Genes Dev.*, **2,** 863.
135. Luscher, B., Christenson, E., Litchfield, D., Krebs, E., and Eisenman, R. (1990). *Nature*, **344,** 517.
136. Gabrielson, O. S., Sentenac, A., and Fromageot, P. (1991). *Science*, **253**, 1140.
137. Sartorelli, V., Webster, K. A., and Kedes, L. (1990). *Genes Dev.*, **4,** 1811.
138. Lin, H., Yutzey, K. E., and Konieczny, S. F. (1991). *Mol. Cell. Biol.*, **11,** 267.
139. Weintraub, H., Davis, R., Tapscott, S., Thayer, M., Krause, M., Benezra, R., Blackwell, T. K., Turner, D., Rupp, R., Hollenberg, S., Zhuang, Y., and Lassar, A. (1991). *Science*, **251,** 761.
140. Cserjesi, P. and Olson, E. N. (1991). *Mol. Cell. Biol.*, **11,** 4854.
141. Boulet, A. M., Erwin, C. R., and Rutter, W. J. (1986). *Proc. Natl Acad. Sci. USA*, **83,** 3599.
142. Weinrich, S. L., Meister, A., and Rutter, W. J. (1991). *Mol. Cell. Biol.*, **11,** 4985.
143. Karlsson, O., Walker, M. D., Rutter, W. J., and Edlund, T. (1989). *Mol. Cell. Biol.*, **9,** 823.
144. Whelan, J., Poon, D., Weil, A., and Stein, R. (1989). *Mol. Cell. Biol.*, **9,** 3253.
145. Guazzi, S., Price, M., De Felice, M., Damante, G., Mattei, M. G., and Di Lauro, R. (1990). *EMBO J.*, **9,** 3631.
146. Mizuno, K., Gonzalez, F. J., and Kimura, S. (1991). *Mol. Cell. Biol.*, **11,** 4927.
147. Rosenfeld, M. G., Glass, C. K., Adler, S., Crenshaw, E. B., He, X., Lira, S. A., Elsholtz, H. P., Mangalam, H. J., Holloway, J. M., Nelson, C., Albert, V. R., and Ingraham, H. A. (1988). *Cold Spring Harbor Symp. Quant. Biol.*, **53,** 545.
148. Mangalam, H. J., Albert, V. R., Ingraham, H. A., Kapiloff, M., Wilson, L., Nelson, C., Elsholtz, H., and Rosenfeld, M. G. (1989). *Genes Dev.*, **3,** 946.
149. Schmitt-Ney, M., Doppler, W., Ball, R. K., and Groner, B. (1991). *Mol. Cell. Biol.*, **11,** 3745.
150. Blessing, M., Zentgraf, H., and Jorcano, J. L. (1987). *EMBO J.*, **6,** 567.
151. Mack, D. H. and Laimins, L. (1991). *Proc. Natl Acad. Sci. USA*, **88,** 9102.
152. Homa, F. L., Glorioso, J. C., and Levine, M. (1988). *Genes Dev.*, **2,** 40.
153. Hirochika, H., Hirochika, R., Broker, T. R., and Chow, L. T. (1988). *Genes Dev.*, **2,** 54.

154. Dostatni, N., Lambert, P., Sousa, R., Ham, J., Howley, P., and Yaniv, M. (1991). *Genes Dev.*, **5,** 1657.
155. Chong, T., Chan, W., and Bernard, H. (1990). *Nucleic Acids Res.*, **18,** 465.
156. Spitzner, J. R. and Muller, M. T. (1988). *Nucleic Acids Res.*, **16,** 5533.
157. Zhang, X., Supakar, P. C., Wu, K., Ehrlich, K. C., and Ehrlich, M. (1990). *Cancer Res.*, **50,** 6865.
158. Kozak, M. (1987). *Nucleic Acids Res.*, **15,** 8125.
159. Mount, S. M. (1982). *Nucleic Acids Res.*, **10,** 459.
160. Harrison, S. C. (1991). *Nature*, **353,** 715.
161. Bray, P., Lichter, P., Thiesen, H., Ward, D., and Dawid, I. (1991). *Proc. Natl Acad. Sci. USA*, **88,** 9563.
162. Thompson, C., Brown, T., and McKnight, S. (1991). *Science*, **253,** 762.

A2

Fractionation of nucleic acids by gel electrophoresis

B. DAVID HAMES and STEPHEN J. HIGGINS

1. DNA

Electrophoresis of DNA fragments is most frequently carried out in agarose or polyacrylamide gels although agarose-polyacrylamide composite gels have also been used. A linear DNA fragment will migrate more slowly as the gel concentration is increased due to the sieving effect of the gel. The relationship between electrophoretic mobility and gel concentration is such that choice of appropriate gel concentrations allows the resolution of a wide size range of DNA molecules.

1.1 Agarose gels

The electrophoretic mobility of linear double-stranded DNA fragments in agarose gels is proportional to their size. Thus a plot of electrophoretic mobility versus $\log_{10}$ molecular mass produces a straight line but this relationship becomes markedly non-linear as the molecular mass of the DNA increases. In practice, agarose gels are able to fractionate double-stranded DNA fragments up to about 800 kb in size (1). However, the fractionation of very large DNAs requires such low percentage gels (<0.3%) that handling them becomes difficult and so pulsed field gel electrophoresis (PGFE; ref. 2) is now often the method of choice for separating DNA molecules more than about 50 kb in size.

A guide to the most appropriate agarose gel concentration to choose for fractionating double-stranded DNA by electrophoresis is given in *Table 1*.

The relationship between molecular mass of DNA and electrophoretic mobility can be used to estimate the molecular mass of sample DNAs. Standard DNA fragments of known molecular mass are co-electrophoresed in a parallel lane with the sample DNA. Typical DNA markers are a *Hin*dIII digest of phage lambda DNA (giving fragments 23130, 9416, 6557, 4361, 2322, 2027, 564, and 125 bp in size) and a *Hae*III digest of phage ϕX174 DNA (giving fragments 1353, 1078, 872, 605, 310, 281, 271, 234, 194, 118, and 72 bp). Other suitable DNA size markers which can easily be prepared in

Table 1. Electrophoretic resolution of double-stranded DNA fragments in agarose gels

Agarose gel concentration (% w/v)	Size fractionation range for linear double-stranded DNA (kb)
0.3	5–60
0.6	1–20
0.9	0.5–7.0
1.2	0.4–6.0
1.5	0.2–3.0
2.0	0.1–2.0

the laboratory are cited in a number of references (e.g. ref. 3) but many of these are also available commercially. After electrophoresis, the migration positions of the standard fragments are measured and used to construct a standard curve of electrophoretic mobility versus $\log_{10}$ molecular mass. The distance of migration of the sample DNA fragment is then used to read off from the standard curve its estimated size. Although this method of size estimation is sufficient for many routine applications, a more accurate approach to determining DNA sizes by agarose gel electrophoresis has been proposed by Southern (4, 5) who showed that, for a double-stranded DNA molecule of length l and electrophoretic mobility m, a plot of l versus $1/(m-m_0)$ is linear over a wide range of molecular masses when the gel is run at low voltage gradients (m_0 is an empirical correction factor). Schaffer and Sederoff (6) found accuracies better than 1% in estimating molecular masses using Southern's procedure.

1.2 Polyacrylamide gels

Polyacrylamide gels have a higher sieving power than agarose gels and so are useful for fractionating small DNA molecules from 6 bp (ref. 7) to about 1000 bp. Most double-stranded DNA molecules migrate through non-denaturing polyacrylamide gels in such a way that the distance migrated is inversely proportional to the $\log_{10}$ of their size. As with agarose gel electrophoresis, a plot of electrophoretic mobility versus $\log_{10}$ molecular mass produces a straight line graph. However, the electrophoretic mobility of these DNA molecules also depends on their base composition and sequence to some extent. Thus, two DNAs of the same size may have mobilities that differ by up to about 10%, probably due to secondary structure differences between the two molecules. Because of this effect, it is impossible to determine the size of double-stranded DNA molecules accurately in non-denaturing polyacrylamide gels and so denaturing gels (usually containing urea) must be used for accurate size determination.

In denaturing polyacrylamide gels, single-stranded DNA migrates essentially

Table 2. Resolution of DNA fragments in polyacrylamide gels

Percentage acrylamide[a] (% w/v)	Size range for effective separation		Dye mobility[b]	
	Double-stranded DNA (bp)	Single-stranded DNA (nt)	Xylene cyanol	Bromophenol Blue
3.5	100–1000	800–2000	450	100
5.0	75–500	100–1000	250	65
8.0	50–400	50–400	150	45
12.0	35–250		70	20
15.0	20–150		60	15
20.0	5–100		45	12

[a] Acrylamide:bisacrylamide at 29:1.
[b] Approximate sizes (bp) of double-stranded DNA with which the dye comigrates in TBE buffer. For TBE composition, see Chapter 2, *Protocol 6*.

independently of base composition and sequence; a plot of electrophoretic mobility versus $\log_{10}$ molecular mass gives a straight line.

A guide to the most appropriate polyacrylamide gel concentration for fractionating DNA by electrophoresis is given in *Table 2*.

2. RNA

Electrophoresis of RNA molecules can be carried out in agarose or polyacrylamide gels although agarose-polyacrylamide composite gels have also been used for the fractionation of large RNA molecules or RNA-protein complexes such as ribosomes or spliceosomes (8). Electrophoresis in agarose gels is particularly popular because, after electrophoresis, one can blot the separated RNAs on to a nitrocellulose or nylon membrane and then examine individual RNAs by hybridization with suitable radiolabelled DNA probes. This Northern blotting methodology has become a fundamental practical tool of molecular biology.

As with DNA, the electrophoresis of RNA can be carried out either under non-denaturing or denaturing conditions. Under non-denaturing conditions, RNA molecules usually exhibit substantial secondary structure. Since the unfolded form of an RNA molecule typically migrates slower than more compact molecules during gel electrophoresis, estimates of RNA molecular mass will depend on the degree of secondary structure, which in turn is determined by the precise conditions of electrophoresis (temperature, ionic strength, etc.). Whilst this phenomenon can sometimes be utilized to enhance the electrophoretic resolution between two or more RNA species, accurate determination of RNA molecular mass requires the use of denaturing gels. Urea, formamide, methyl mercuric hydroxide, and glyoxal/dimethyl sulphoxide (DMSO) have all been used as denaturants. Note, however, that prolonged exposure of RNA to urea at high temperatures (e.g. 60°C) can cause strand breakage.

2.1 Agarose gels

As a guide, use denaturing 1.4% agarose gels for RNAs up to 1000 nt long and 1.0% agarose gels for longer RNAs. RNAs of known size (such as 28S rRNA, 18S rRNA, rabbit β-globin mRNA; 6333, 23666, and 710 nt respectively) must be co-electrophoresed with the sample RNA(s). A plot of electrophoretic mobility versus $\log_{10}$ molecular mass gives a straight line with no anomalies caused by secondary structure and with no effect of base composition. One clear exception to this, however, is the electrophoresis of non-linear RNA molecules such as lariat splicing intermediates or other RNA molecules with covalent branch points. DNAs of known size have been used as standard for RNA electrophoresis but one should be aware that, although this may give an approximate guide to the size of the sample RNA(s), it may not be accurate; for example, RNA migrates faster than DNA of equivalent size in agarose gels containing formaldehyde (9).

2.2 Polyacrylamide gels

Non-denaturing and denaturing polyacrylamide gel electrophoresis of RNA has been fully described elsewhere (e.g. ref. 10). Typically the gel concentration used is in the range 2.0–10.0% with the choice of concentration depending on the sizes of the RNAs to be separated. Thus, for example, 2.2–2.4% gels resolve RNAs in the 17S–45S RNA range (about 2000–13 000 nt range) whereas 10.0% gels are needed to resolve smaller RNA molecules such as tRNA efficiently.

References

1. Fangman, W. L. (1978). *Nucleic Acids Res.*, **5**, 653.
2. Anand, R. and Southern, E. M. (1990). In *Gel electrophoresis of nucleic acids: a practical approach*, 2nd edn (ed. D. Rickwood and B. D. Hames), p. 101. Oxford University Press.
3. Minter, S. J, Sealey, P. G. and Arrand, J. E. (1985). In *Nucleic acid hybridisation: a practical approach* (ed. B. D. Hames and S. J. Higgins), p. 211. Oxford University Press.
4. Southern, E. M. (1979). *Anal. Biochem.*, **100**, 319.
5. Sealey, P. G. and Southern, E. M. (1990). In *Gel electrophoresis of nucleic acids: a practical approach*, 2nd edn (ed. Rickwood, D. and Hames, B. D.), p. 51. Oxford University Press.
6. Schaffer, H. E. and Sederoff, R. R. (1981). *Anal. Biochem.*, **115**, 113.
7. Jovin, T. M. (1971). In *Methods in enzymology*, Vol. 21 (ed. L. Grossman and K. Moldave), p. 179. Academic Press, London and New York.

8. Dahlberg, A. E. and Grabowski, P. J. (1990). In *Gel electrophoresis of nucleic acids: a practical approach*, 2nd edn, (ed. D. Rickwood and B. D. Hames), p. 275. Oxford University Press.
9. Wicks, R. J. (1986). *Int. J. Biochem.*, **18,** 277.
10. Grierson, D. (1990). In *Gel electrophoresis of nucleic acids: a practical approach*, 2nd edn (ed. D. Rickwood and B. D. Hames), p. 1. Oxford University Press.

A3

Addresses of suppliers

Alcatel CIT, Division technologie, VIDE, 98 Avenue de Brogny, BP 69 74009, Annecy Cedex, France.

Aldrich Chemical Company Ltd, The Old Brickyard, New Road, Gillingham SP5 4BR, UK.

Aldrich Chemical Company Inc., 940 West St. Pool Avenue, Milwaukee, Wisconsin 53233, USA.

Amersham International PLC, Lincoln Place, Green End, Aylesbury HP20 2TP, UK.

Amersham Corporation, 2636 Saint Clearbrook Drive, Arlington Heights, Illinois 60005, USA.

Amicon Ltd, Upper Mill, Stonehouse, Gloucester GL10 2BJ, UK.

Amicon Division, WR Grace & Co., 72 Cherryhill Drive, Beverley, MA 01915-1065, USA.

Azlon, Silobent Industrial Park 205–1, Kelsey Lane, Tampa, Florida 33619, USA.

BDH Chemicals Ltd, Shaw Road, Speke, Liverpool L24 9LA, UK.

Beckman Instruments UK Ltd, Progress Road, Sands Industrial Estate, High Wycombe HP12 4JL, UK.

Beckman Instruments Inc., 2500 Harbor Boulevard, PO Box 3100, Fullerton, CA 92634, USA.

Berthold Instruments (UK) Ltd, 35 High Street, Sandridge, St Albans AL4 9DD, UK.

Bethesda Research Laboratories (BRL); see Gibco BRL

Bio-101 Inc., La Jolla, CA, USA and c/o Stratech Scientific Ltd, 61–63 Dudley Street, Luton LU2 0HP, UK.

BioRad Laboratories Ltd, Maylands Avenue, Hemel Hempstead HP2 7TD, UK.

BioRad Laboratories, Division Headquarters, 3300 Regatta Boulevard, Richmond, CA 94804, USA.

Boehringer Mannheim UK (Diagnostics/Biochemicals) Ltd, Bell Lane, Lewes BNY 1LG, UK.

Boehringer Mannheim Biochemicals, PO Box 50414, Indianapolis, IN 46250, USA.

Branson Sonic Power, Eagle Road, Danbury, Connecticut 06813, USA and c/o Lucas Dawes Ultrasonics Ltd, Concord Road, Western Avenue, London W3 0SD, UK.

Cambridge Bioscience Ltd, 25 Signet Court, Stourbridge Common Business Centre, Swans Road, Cambridge CB5 8LA, UK.

Campden Instruments, 185 Campden Hill Road, London W8 1TH, UK.

Carl Zeiss, D-7082, Oberkochen, Germany.

Carl Zeiss (Oberkochen) Ltd, PO Box 78, Woodfield Road, Welwyn Garden City ALY 1LU, UK.

Cetus, see Perkin Elmer Cetus

David Kopf, Tujunga, CA, USA.

Difco Laboratories Ltd, Central Avenue, East Molesley KT8 0SE, UK.

Difco Laboratories, PO Box 331058, Detroit, Michigan 48232-7058, USA.

DuPont Co., Biotechnology Systems Division, PO Box 80024, Wilmington, Delaware 19880-0024, USA.

DuPont UK Ltd, Wedgwood Way, Stevenage SG1 4QN, UK.

Ealing Electro-optics, Greycaine Road, Watford WD2 4PW, UK.

Eppendorf, 2000 Hamburg 65—Postfach 650670, Germany.

Eppendforf, c/o Merck Ltd in UK.

Falcon, c/o Becton Dickinson Labware, 2 Bridgewater Lane, Lincoln Park, NJ 07035, USA and c/o Becton Dickinson UK Ltd, Between Towns Road, Oxford OX4 3LY, UK.

Fisher Scientific Co., 50 Fadem Road, Springfield, New Jersey 07081, USA and c/o Arnold R. Horwell Ltd, 73 Maygrove Road, West Hampstead, London NW6 2BP, UK.

FMC BioProducts, 5 Maple Street, Rockland, ME 0481, USA.

FMC BioProducts Europe, Risingevej 1, DK-2665 Vallensbaek Strand, Denmark.

Gilson (c/o Anachem Ltd), 20 Charles Street, Luton LU2 0KB, UK.

Gilson Medical Electronics, 3000 West Beltline Highway, Middleton, Wisconsin 53562, USA.

Gibco BRL (c/o Life Technologies Ltd), Trident House, Renfrew Road, Paisley PA3 4EF, UK.

Arnold R. Horwell Ltd, 73 Maygrove Road, West Hampstead, London NW6 2BP, UK.

ICN Flow, Eagle House, Peregrine Business Park, Gomm Road, High Wycombe HP13 7DL, UK.

ICN Biomedical Inc., 3300 Highland Avenue, Costa Mesa, CA 92626, USA.

Interpet Ltd, Vincent Avenue, Dorking, Surrey RH4 3YX, UK.

Invitrogen Corporation, 3985 Sorrento Valley Blvd, Suite B, San Diego, CA 912121, USA.

Invitrogen (c/o British Biotechnology Ltd, 4–10 The Quadrant, Barton Lane, Abingdon, Oxon OX14 3YS, UK.)

Leitz Instruments Ltd (see Wild Leitz).

Life Sciences Laboratories Ltd, Sedgwick Road, Luton LU4 9DT, UK.

Life Technologies Inc., 8400 Helgerman Court, Gaithersburg, MD 20877, USA.

Life Technologies Ltd, Trident House, Renfrew Road, Paisley PA3 4EF, UK.

Merck Ltd, Merck House, Poole, Dorset BH15 1TD, UK.

Millipore (UK) Ltd, The Boulevard, Blackmoor Lane, Watford, Herts WD1 8YN, UK.

Millipore Corporation, Bedford, Massachusetts 01730, USA.

Molecular Dynamics Ltd, 4 Chaucer Business Park, Kemsing, Sevenoaks, Kent TN15 6PL, UK.

Molecular Dynamics Inc., 880 East Arques Avenue, Sunnyvale, CA 94086, USA.

Narishige Scientific Instrument Laboratory, 9.28 Kasuya, 4 Chome Setagayaku, Tokyo, Japan.

New England Biolabs (NBL), 32 Tozer Road, Beverley, MA 01915-5510, USA and c/o CP Labs Ltd, PO Box 22, Bishop Stortford, Herts CM23 3DH, UK.

New England Nuclear (NEN), Dupont de Nemours & Co., Wilmington, Delaware, USA and c/o DuPont UK Ltd, Wedgwood Way, Stevenage SG1 4QN, UK.

Nikon (UK) Ltd, Instrument Division, Haybrook, Halesfield 9, Telford, Shropshire TF7 4EW, UK.

Nikon Inc., Instrument Group, Walt Whitman Road, PO Box 9050, Melville, NY 11747-9050, USA.

Nunc Inc., 2000 North Aurora Road, Naperville, Illinois 60563-1796, USA and c/o Life Technologies Ltd, Trident House, Renfrew Road, Paisley PA3 4EF, UK.

The Perkin-Elmer Corporation, 761 Main Avenue, Norwalk, CT 0689-0251, USA.

Perkin-Elmer Ltd, Maxwell Road, Beaconsfield HP9 1QA, UK.

Pharmacia Biosystems Ltd (Biotechnology Division), Davy Avenue, Knowlhill, Milton Keynes MK5 8PH, UK.

Pharmacia LKB Biotechnology Inc., 800 Centennial Avenue, PO Box 1327, Piscataway, NJ 08855-1327, USA.

PL Biochemicals; contact Pharmacia

Promega Biotech Ltd, Delta House, Chilworth Research Centre, Southampton SO1 7NS, UK.

Promega, 2800 Woods Hollow Road, Madison, WI 53711-5399, USA.

Raymond Lamb Ltd, 6 Sunbeam Road, London NW10 6JL, UK.

Savant Instrument Inc., 110-113 Bi-County Blvd, Farmingdale, NY 11735, USA and c/o Life Sciences Ltd, Dunstable LU6 1BD, UK.

Schleicher and Schuell, Postfach 4, D-3354, Dassell, Germany.

Schleicher and Schuell (c/o Anderman & Co. Ltd), 145 London Road, Kingston-upon-Thames, Surrey KT2 6NH, UK.

Schleicher and Schuell, Keene, New Hampshire, USA.
Schott Glaswerke, Werk Wiesbaden, Postfach 13 03 67, D-6200, Wiesbaden 13, Germany.
Sigma Chemical Co. Ltd, Fancy Road, Poole, Dorset, BH17 7NH, UK.
Sigma Inc., PO Box 14508, St Louis, MO 63178, USA.
Sorvall; c/o Dupont (see above).
Sterilin (Bibby Sterilin Ltd), Tilling Drive, Stone ST15 0SA, UK.
Stratagene Ltd, Unit 140, Cambridge Innovation Centre, Milton Road, Cambridge CB4 4FG, UK.
Stratagene Inc., 11011 North Torrey Pines Road, La Jolla, CA 92037, USA.
Stratech Scientific Ltd, 61–63 Dudley Steet, Luton LU2 0HP, UK.
United States Biochemical (USB) Corporation, PO Box 22400, Cleveland, Ohio 44122, USA and c/o Cambridge Bioscience, 25 Signet Court, Stourbridge Common Business Centre, Swans Road, Cambridge CB5 8LA, UK.
Whatman Scientific Ltd, Whatman House, St Leonards Road, Maidstone, Kent ME16 0LS, UK.
Wild Leitz UK Ltd, Davey Avenue, Knowlhill, Milton Keynes, MK5 8LB, UK.
Worthington Biochemical Corporation, Halls Mill Road, Freehold, NJ 07728, USA and c/o Cambridge Bioscience Ltd (see above).

Index

Index